Population Fluctuations in Rodents

Population Fluctuations in Rodents

CHARLES J. KREBS

THE UNIVERSITY OF CHICAGO PRESS CHICAGO AND LONDON

Charles J. Krebs is professor emeritus of zoology at the University of British Columbia and Thinker in Residence at the University of Canberra.

The University of Chicago Press, Chicago 60637
The University of Chicago Press, Ltd., London

Printed in the United States of America

22 21 20 19 18 17 16 15 14 13   1 2 3 4 5

ISBN-13: 978-0-226-01035-9 (cloth)
ISBN-13: 978-0-226-01049-6 (e-book)

Library of Congress Cataloging-in-Publication Data
Krebs, Charles J., author.
Population fluctuations in rodents / Charles J. Krebs.
pages cm
Includes bibliographical references and index.
ISBN 978-0-226-01035-9 (cloth : alkaline paper) — ISBN 978-0-226-01049-6 (e-book)
1. Rodent populations. 2. Voles—Northern Hemisphere. 3. Lemmings—Northern Hemisphere. 4. Population biology. I. Title.
QL737.R6K94 2013
599.35—dc23   2012037021

⊗ This paper meets the requirements of ANSI/NISO Z39.48-1992 (Permanence of Paper).

TO THE LATE DENNIS CHITTY, MENTOR EXTRAORDINAIRE,
AND TO ALICE KENNEY FOR HER SKILLS, ASSISTANCE, AND INSPIRATION

Contents

Preface

How did rodent outbreaks in Germany help to end World War I? What caused the destructive outbreak of rodents in Oregon and California in the late 1950s and produced the large outbreak of rodents in Scotland in the spring of 2011? Population fluctuations or "outbreaks" of rodents constitute one of the classic problems of animal ecology, recognized by Charles Elton more than 80 years ago. Since that time, the number of publications on this topic must be among the largest within the subdisciplines of ecology. As these publications have proliferated, strong divisions have arisen between different schools of thought about what the problem is, what changes occur during population fluctuations, how general these demographic events are, and what evidence there is for one possible mechanism as opposed to another. The result is a rather confusing mass of literature that makes some critics say that the problem is not soluble, or that controversy has gone on so long that the problem is no longer of interest.

The purpose of this book is to draw together all of the literature on rodent population fluctuations with the aim of reaching a general consensus on problem definition, demographic attributes, and potential explanations. Another aim is to try to outline the missing links that must be found for a satisfactory ecological theory of rodent population fluctuations. This book is designed for upper-division undergraduate ecology majors, graduate students, and ecologists who wish to know the state of the art of rodent population studies. I have deliberately kept my focus narrow by concentrating on the voles and lemmings of the Northern Hemisphere, and I will use the literature on rats and mice as a source of enlightenment about specific questions that arise in the analysis of vole and lemming population fluctuations. In the same vein, I will not discuss bird populations that may have similar cyclic fluctuations. I emphasize that we should obtain a general theory of fluctuations in vertebrate populations, but that at present there are practical limits on how much material one can consolidate in a single book.

If I am critical of particular studies, I am not critical of particular scientists. I evaluate and critique ideas, not people, and the most important message of this book is that we have to state our ideas in a clear manner with clear predictions, and then conduct rigorous experiments to find out whether they are correct. The rodent literature is rife with papers that do not follow this approach, and I will not speak highly of such work.

There are many excellent studies of small rodents that I cannot review in this book while keeping it down to a reasonable size. I apologize to all ecologists who seem to be neglected in the studies cited here. All have been a part of the zeitgeist of small mammal ecology, and their contribution is hidden but still present.

These reflections on rodent population dynamics have evolved over 45 years of fieldwork in the temperate and polar regions. I make no claim to be infallible, and the greatest tribute a scientist can give to another is to test his or her ideas and find out if they are right or wrong. If we can agree on experimental protocols for deciding the controversial issues of rodent dynamics, we will have made much progress in understanding these fascinating animals.

I thank my colleagues and students for their help in gathering much of the data for this book. In particular I thank Rudy Boonstra, Mike Gaines, Xavier Lambin, Judy Myers, Bob Tamarin, Tony Sinclair, Tom Sullivan, Heikki Henttonen, Hannu Ylönen, Erik Framstad, Nils Stenseth, Grant Singleton, Lowell Getz, George Batzli, and the late Jamie Smith for their help and inspiration. Many colleagues shared their data with me; I am grateful to all of them, and to all the small mammal ecologists who have worked on the problem of rodent fluctuations over many years. Renaldo Migaldi of the University of Chicago Press provided excellent editorial help in the final preparation of this book, and I am most grateful to him for this assistance.

Charles J. Krebs
Vancouver, British Columbia
5 April 2012

CHAPTER ONE

Classifying Rodent Population Changes

Key Points:

- Rodent populations vary from relatively stable to highly cyclic.
- Populations of the same species may fluctuate strongly in one part of their geographic range and minimally in other areas.
- Critical methodological issues regarding size of the sampling area, the sampling interval and length of study, the methods of estimation, and the habitat mosaic of the study site affect the value of any particular study.
- The time scale of rodent dynamics is monthly or even weekly, and annual census estimates are of limited use for deciphering population processes.
- The key question for all rodent population changes, whether cyclic or relatively stable, is what factors determine the population growth rate.

Small rodents have been a focus of interest for humans throughout history because of the diseases they have carried and their impact as pests of agricultural crops. The rise and fall of rodent populations have become part of myth and legend, from the Pied Piper and the Black Death to lemming migrations and rodents falling from the sky during storms. Underlying each of these legends is some germ of truth, and science in general seeks to uncover these germs and explain how they came to be. Population fluctuations or "cycles" became noticed when these natural history observations were analyzed by early ecologists. When Robert Collett in

Norway and Charles Elton in Britain began to write about periodic eruptions of lemming numbers in Norway and vole numbers in Britain, few could imagine the complexity their insights would later uncover (Collett 1895; Elton 1924).

Population ecologists have carried out more research on small rodent dynamics than on the dynamics of most other ecological groups. This is partly because rodents are convenient for graduate theses and other short research projects, and partly because of the damage rodents inflict on forest regeneration (e.g., Huitu et al. 2009) and agriculture (e.g., Singleton et al. 2005). Reviews of rodent population changes have become more specialized in light of the increasing literature, and therefore I think it critical to step back and take a broad overview of the current state of progress and the uncertainties and controversies that continue in this area of research. Progress has been uneven in understanding rodent population changes, partly because of bandwagons that develop and partly because of geography. Rodents in many parts of the world are poorly studied while those in other areas, like the North Temperate Zone of Europe and North America, have provided the bulk of our information base.

In this chapter I will explore the various population changes that have been described for small rodents, and attempt to classify them into four categories of population patterns. There are two approaches one can take to any set of ecological patterns. First, one can assume that all patterns in the set are variants of a single general pattern, and then search for a single general explanation. In principle, this is what ecologists have done with population regulation. All populations are regulated, this approach asserts, and the regulation is produced by some type of density dependence in the broad sense. In principle, if the density dependence has some type of delay built in, the populations affected will fluctuate periodically, and may become "cyclic." But if the density dependence operates with little time delay, the population may show numerical stability or only seasonal fluctuations. In this simple yet powerful model, all types of population traces can be envisaged, and the ecologist's problem becomes one of how to find the density-dependent mechanism behind the population events (Krebs 2002). Does birth rate change? Does juvenile survival change? Do more individuals emigrate as density rises? These kinds of questions become the critical focus of ecological investigations.

An alternative approach is to view each of the uncovered patterns as distinct, and then to develop a separate theory for each recognized pattern. For example, stable populations can be viewed as one particular case of rodent population dynamics, and explanations for stability would become the focus of these studies. Alternatively, cyclic populations can be viewed as a second pattern, and a different explanation would be proposed to cover that case. Clearly, these two different approaches to population analysis may converge, but the philosophical differences that underlie them will have a major effect on our ability to develop general theory and general models. One particular example of the alternative approach was developed by Turchin et al. (2000) in an analysis of vole and lemming cycles in Fennoscandia (box 1.1).

BOX 1.1 **Patterns of vole and lemming population changes in Fennoscandia**

Turchin et al. (2000) pointed out that models of population fluctuations had the characteristic feature that if the population changes were driven by a herbivore-food interaction, the population peaks would characteristically be sharp. By contrast, if the population changes were driven by a predator-prey interaction, the peaks would be smooth and rounded. The models make these kinds of predicted patterns:

By analyzing time series of population indices of Norwegian lemmings and voles (*Microtus*) in Fennoscandia, Turchin et al. (2000) concluded that the spiked pattern described lemming cycles and the rounded pattern described vole cycles. Consequently, there must be different causal mechanisms for the population changes of voles and lemmings in this region.

I use this paper of Turchin et al. (2000) to illustrate one approach to analyzing population fluctuations in rodents. Note a few simple messages that come from this type of analysis:

- We need no data on predators or food plants to make these decisions about causal agents.
- We need not worry about the precision of the population indices or the frequency of sampling to obtain estimates of rodent numbers.
- We need not worry about birth rates, death rates, movement rates, or any of the details of these species' biology.

We will discuss these issues in the following chapters.

Background Issues

Four background issues must be reviewed briefly before we can begin to classify patterns of rodent population changes. Throughout this book we will talk about populations and population density, and discuss hypotheses about regulation.

What Is a Population?

We assume that we can define a population of small rodents unambiguously; this assumption is common in population ecology. In practice we define the population operationally by the area we choose to study: often a small trapping area of less than a few hectares. We implicitly assume that the population we are studying is closed—or, if it is open, that it has equal amounts of emigration and immigration across its boundaries (White et al. 1982). We also assume that the population we are studying is typical, and although statisticians caution us to use random sampling, virtually no one selects their study population site randomly. We assume that a site is representative of a broad class of local populations in the region with absolutely no idea whether or not this is correct.

Not all rodent populations are closed, and if we have an open population we will find it difficult to measure survival rates, because of movements in and out of the area. In principle we can solve the problem of open populations by adopting a larger study site, but one soon runs into practical problems of time and budget. Consequently, we pray that our population is closed or nearly so. Few ecologists seem to lose any sleep over this fundamental issue.

Most studies take little note of the landscape surrounding the study area; this is an issue that has recently come under scrutiny (Delattre et al. 2009). We tend to assume that the population under study is little affected by events in the surrounding landscape—an assumption that conservation ecologists never make (Ferraz, et al. 2007), and which we will explore later in chapter 6.

What Is Population Density?

Once we get over the first hurdle of defining a population, we run into a second problem of measuring its density. This involves two issues. First we need to measure population size accurately, and then we need to know exactly the area on which this population lives, so that we can measure abundance per unit area. This measurement is simple only with island populations, when we can sample an entire island.

There are now numerous statistical techniques that can be used to estimate absolute population size (Krebs 1999; Pollock et al. 1990), and there is an extensive literature on this particularly for small mammals (e.g., Parmenter et al. 2003; Efford 2004). There is no excuse for rodent ecologists not to use these methods. Nevertheless, many studies continue to be published without adequate statistical methods of estimation, and much of the population data we will use in this book can be described only as indices of abundance, with all the problems associated with the use of indices (Anderson 2003). If there is one practical recommendation for future work on small rodents, it is simply to use modern methods of population estimation.

Once one has good estimates of population size from Program MARK (White 2008), there is still the problem of estimating population density. The two alternative methods of density estimation from mark-recapture data are boundary strip estimation to determine effective grid size (e.g., Jett and Nichols 1987) and spatially explicit methods developed by Murray Efford (Efford 2004; Borchers and Efford 2008; Efford et al. 2009). Efford's suggestion was to fit a simple spatial model of animal trapping that estimated the probability of an animal being caught in a trap at a given distance from its home range center. It requires data on recaptures of individually marked animals to estimate the spatial model, but appears to be the best method now available for density estimation in live trapping studies of small rodents (Krebs et al. 2011).

Given that we have good population estimates and a good measure of the area used by the rodents, a third problem arises: What part of the population is being sampled? We do not usually count babies in the nest as part of the population; and juveniles must reach a minimum size before they can be trapped, so they are also underrepresented in population measures. The age structure and, in particular, the number of breeding adults should be used when possible, so that density estimates have some biological reality. We assume in most studies that we are discussing the adult segment of the population and most of the younger age classes are not included. Rarely is this stated explicitly.

What Is the Time Step of Sampling Population Attributes?

The biologically relevant time step over which to measure population attributes is on the order of a few days or weeks. But when we look for long-term data on population changes, we find that one data point per year, or two data points, represents much of these data for rodents. There are almost no data on winter populations for species that live in cold climates, due to the obvious difficulty of gathering such information. Nevertheless, these limitations on available data are often forgotten. The assumption that, for example, winter losses occur at a constant rate over the snow period in northern voles and lemmings may be convenient, but it must be recognized as a very large assumption.

The advantages of using small rodents as experimental animals in field studies of population dynamics are offset by the requirement to sample at a relevant time frequency. It should be a basic principle of ecological experimental design that the sampling frequency reflects the timing of biological events rather than human convenience. For small rodents this frequency is on the order of two to three weeks, which for many species is the gestation interval and slightly less than the generation time.

What Data are Needed to Test Hypotheses about Population Limitation?

When we proceed to test a particular hypothesis about rodent population changes, we adopt a uniformity-of-nature assumption: that whatever variable we can find to explain population changes will be the critical variable at other times and in other places. This assumption of repeatability is rarely tested, and is common to all of science but nevertheless worth remembering.

Whatever hypothesis or model of rodent population change anyone wishes to test, it is critical that they have a set of explicit predictions for the hypothesis. There are excellent sources for discussions of experimental design and hypothesis testing in science, and thus no need to belabor the subject here (Anderson et al. 2000; Eberhardt 2003). The history of studies of rodent population changes can be read as a haltingly slow progression of making more precise predictions from hypotheses about limitation and regulation. One of the goals of this book is to try to make these predictions more explicit, and to review which predictions seem to fit the available data and which are still untested.

The use of mathematical models to explore hypotheses of regulation in vertebrates has had the beneficial effect of making hidden assumptions more explicit (Berryman 2002; Turchin and Batzli 2001). This is a most desirable development, particularly because it can start a feedback cycle between empirical investigations and mathematical models, to the benefit of both.

Empirical Patterns of Population Change

It is useful to look at some good examples of long-term data on population abundance for several rodent species before we begin to discuss the possible ways of classifying patterns of change. I have picked out four data sets to illustrate the patterns we need to consider. All of these data sets will reappear in later chapters as we discuss mechanisms of population change.

Lemmings in Siberia and Norway

Russian ecologists have provided some of the longest time series of fluctuations of small rodents. The relative abundance of the collared lemming (*Dicrostonyx torquatus*) and Siberian lemming (*Lemmus sibiricus*) for 41 years is shown in figure 1.1 from Kokorev and Kuksov (2002). Fourteen peak years were described in 41 years of study, and almost all the fluctuations are periodic with a three-year interval. These lemming data may be considered typical of the classical three- to four-year vole and lemming cycle that Elton (1942) described many years ago. The pattern is clear but the quantitative dynamics are missing because these are indices of abundance, and we cannot interpret an increase from two to three in the index as being a numerical rise equivalent to a change from three to four.

The Norwegian lemming (*Lemmus lemmus*) has been monitored in the mountains of south central Norway for more than 30 years (Framstad et al. 1993, 1997, personal communication). Figure 1.2 shows that these lemming populations in the mountains of southern Norway have fluctuated periodically, with three- to four-year intervals during this period. Again, the lemming population estimates are an index, but in this case snap traps were used to provide a catch-per-100-trap-nights index. We do not know if

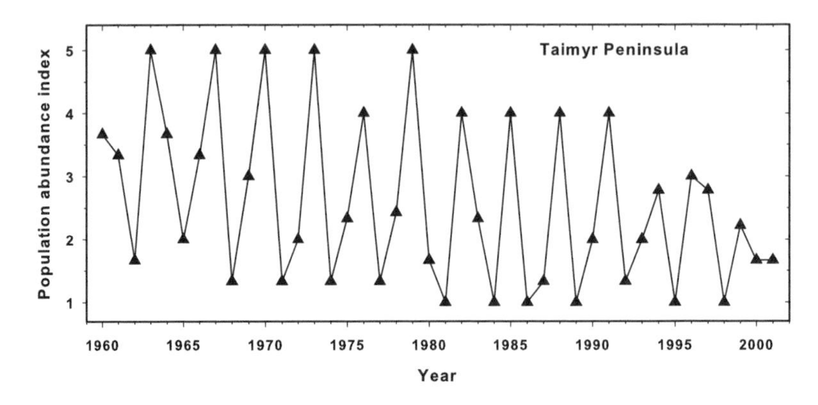

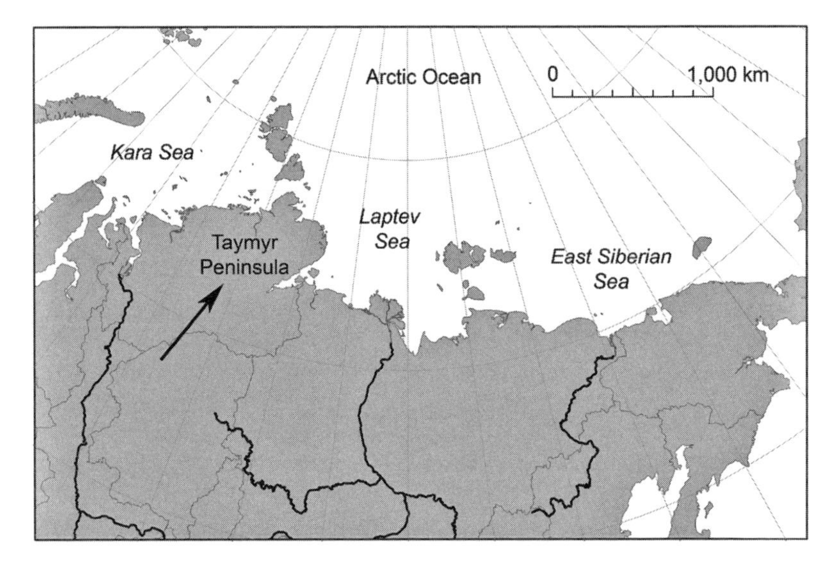

FIGURE 1.1 Abundance of brown and collared lemmings in the tundra regions of the Taimyr Peninsula, Siberia, from 1960 to 2001. Populations were indexed once each year on a scale of 1 (very low) to 5 (extremely common). A strong three- to four-year cycle is clear even from these relatively crude data. An abundance of data from Scandinavia and the old Soviet Union are available with qualitative estimates of lemming abundance. Only in more recent years have rigorous, quantitative data been taken. Data from Kokorev and Kuksov 2002.

snap-trap indices show a linear relationship to absolute densities, although ecologists studying small rodents often make this assumption. Under that assumption, lemming cycles at Finse vary considerably in amplitude, so that cyclic amplitude is not a constant. These lemming data illustrate one common pattern of population change: cyclic fluctuations with a three- to

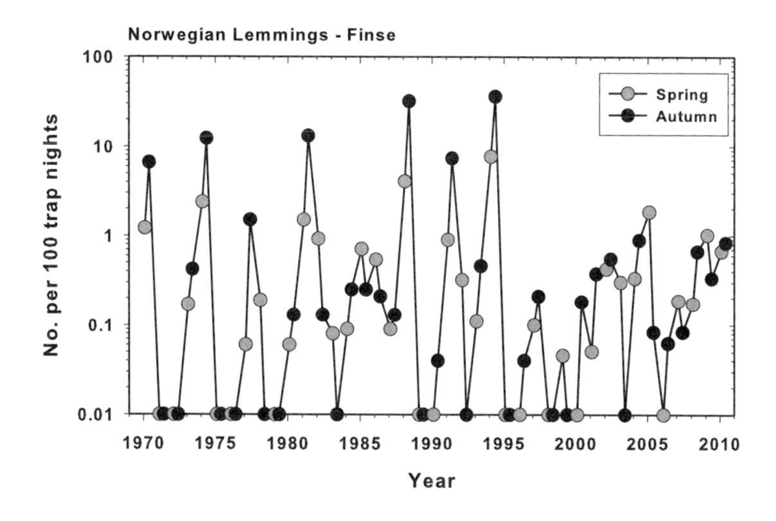

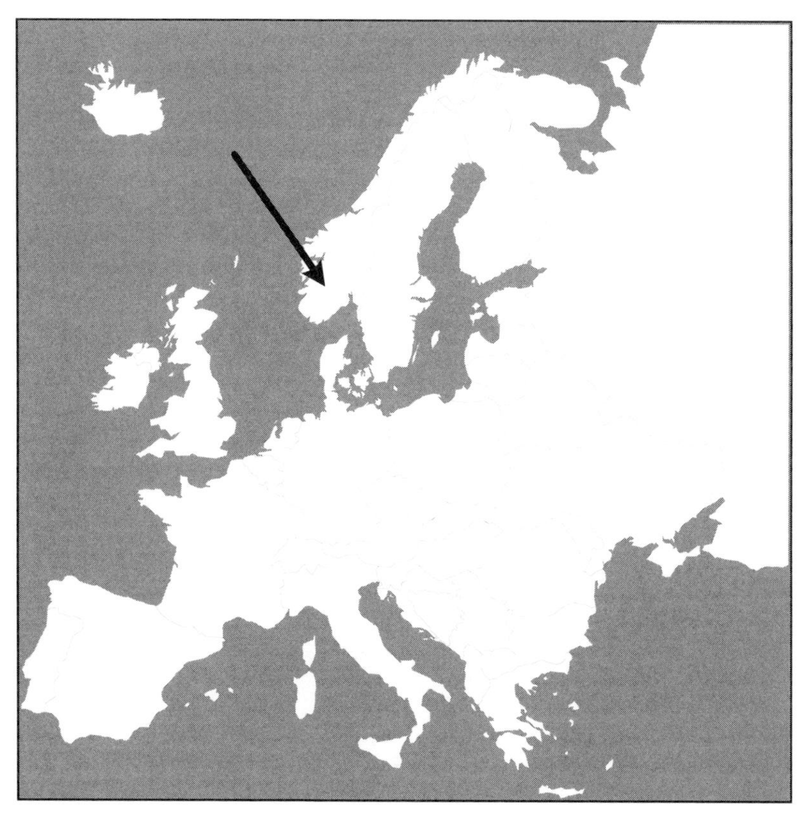

FIGURE 1.2 Abundance of Norwegian lemmings in the Finse region, Hardangervidda, Norway, from 1970 to 2010. Data were obtained by snap traps on fixed lines in the tundra areas of the Hardangervidda Plateau between Oslo and Bergen in southern Norway. Data courtesy of Eric Framstad.

four-year period. Not all lemming populations show this pattern (Reid et al. 1997), but many do.

Gray-Sided Voles on Hokkaido, Japan, and at Kilpisjärvi, Finland

Gray-sided voles (*Myodes* [*Clethrionomys*] *rufocanus*) are the most common voles in the coniferous forests of Hokkaido, Japan. Because they eat the seedlings of forest trees in plantations, the Forestry Agency on Hokkaido began in 1954 to conduct censuses on 1,000 areas all over the island. Until 1992 these censuses were carefully carried out, proving to be a gold mine of population data on this species for 31 years (Saitoh et al. 1998; Bjørnstad et al. 1999). Population abundance was measured by snap trapping, and these data are thus an index rather than an estimate of absolute density. Figure 1.3 gives two of the 225 time series that have been analyzed in detail by Takashi Saitoh and his colleagues (e.g., Saitoh et al. 1997, 1998). These two locations are 200 km apart, and both populations fluctuate but do not always reach a peak in the same year (e.g., peaks reached in 1981 and not again until 1983). Populations around Heian appear to fluctuate more strongly than those at Ebishima. Peak populations occur at three- to five-year intervals, and these peaks are not as easily defined as those observed in lemming populations. Amplitudes of fluctuation are highly variable, as we found for Norwegian lemmings.

BOX 1.2 **The fallacy of the s-index for cyclicity**

Some rodent populations fluctuate regularly, and others less regularly or not at all. Some fluctuate wildly and others only slightly. In an attempt to bring some quantitative order to these differences, Henttonen et al. (1985) proposed the *s*-index: the standard deviation of the log-transformed measures of rodent abundance:

$$s = \sqrt{\frac{\sum_{1}^{n}\left(\log N_i - \overline{\log N_i}\right)^2}{n-1}}$$

where s = index of variability,
N = population density or index,
n = number of years in time series,
and logs are to base 10.

BOX 1.2 **(Continued)**

Because many of the time series produce estimates of zero when rodents are at low numbers, it has become customary to add 0.1 to each population estimate so that logarithms can be taken.

Lewontin (1966) suggested *s* as a general measure of variation. Henttonen et al. (1985) analyzed *Myodes* (*Clethrionomys*) data from Finland and concluded that noncyclic populations had *s*-indices below 0.3 while all cyclic population had indices above 0.3. The *s*-index was widely adopted in the small rodent literature as an index of cyclicity, with the implicit assumption that two different explanations were required for cyclic and non-cyclic populations of the same species. The great advantage of the *s*-index was that it could be used on short time series (four years or more).

Unfortunately, the *s*-index measures only population variability, and not cyclicity, which requires standard time series (spectral) analysis (Bjørnstad et al. 1996). But time series analysis requires 25 years of data or more, so it cannot be applied to most existing small rodent data in the literature; hence the pressure to adopt the *s*-index as a measure of cyclic variation.

We can illustrate the problem with two simple hypothetical data sets:

A.) 10, 9, 11, 10, 9, 1, 1, 2, 1, 2 *s*-index = 0.46, therefore "cyclic."

B.) 10, 8, 6, 10, 8, 5, 10, 7, 5, 10 *s*-index = 0.12, therefore "non-yclic."

Clearly, the first time series moves from a high-density state to a low-density state over a period of 10 years, while the second time series is a nearly perfect cycle of low amplitude. The key point is that we cannot use the *s*-index as an indication of cyclicity. We are left with the serious problem of having to decide whether or not to classify a time series as periodic or "cyclic" if we have only a few years of data.

The gray-sided vole occurs across Asia and Fennoscandia, and by chance it has also been the subject of another long-term study in Finnish Lapland at Kilpisjärvi. Kalela (1957) began this work, and it has been continued by several Finnish ecologists (Henttonen 1986 and personal communication). Population abundance has been measured by kill trapping with the results shown in figure 1.4. Peak populations occurred at intervals of three to five years, and during the 1990s there was a decrease in the amplitude of fluctuations. A similar pattern was found by Hornfeldt et al. (2005) in northern Sweden, who suggested that cycles had been dampening out in northern Fennoscandia during the previous 20 years. This issue was explored further by Ims et al. (2008), and we will discuss it further in

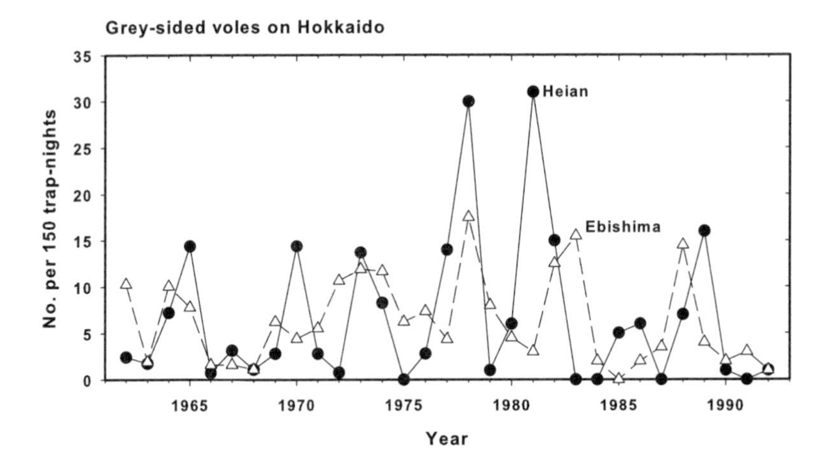

FIGURE 1.3 Autumn abundance of grey-sided voles (*Myodes [Clethrionomys] rufocanus*) on two areas of Hokkaido, Japan, from 1962 to 1993. Data were obtained by snap trapping two to six grids with 50 snap traps for three nights in each area. Fluctuations in the western coastal region (Ebishima) are not as strong as those in eastern, more interior areas (Heian). Data from Saitoh et al. 1998.

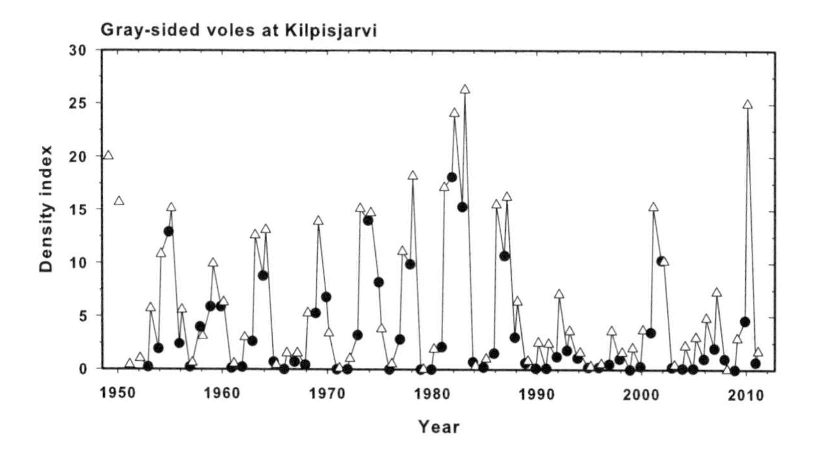

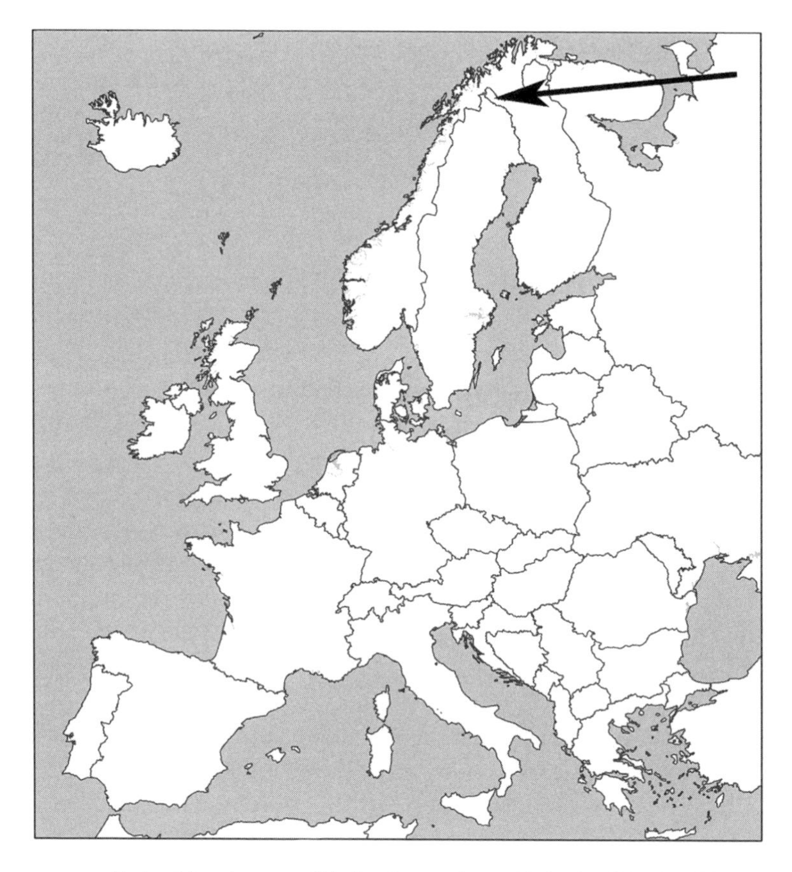

FIGURE 1.4 Spring (•) and autumn (Δ) abundance of grey-sided voles (*Myodes [Clethrionomys] rufocanus*) at Kilpisjärvi, Finland, from 1949 to 2010. This is one of the longest quantitative time series available on cyclic rodents. The density index is the number of individuals caught in snap traps per 100 trap nights. Data courtesy of H. Henttonen 2011.

chapter 6. High amplitude fluctuations returned in 2010, so the decrease in amplitude was only temporary (Brommer et al. 2010).

Meadow Voles in Central Illinois, USA

Both *Microtus pennsylvanicus* and *Microtus ochrogaster* occur in grasslands of the central United States, and Lowell Getz and his colleagues studied populations of these two microtines from 1972 to 1997 (Getz et al. 2001). Their study sites were on farmland in central Illinois, and three habitats were monitored: tallgrass prairie (the original habitat of this region), bluegrass (*Poa praetensis*, an introduced grass), and alfalfa (*Medicago sativa*). Each of these habitats occupied about four to five hectares in a farming area dominated by corn and soybean crops. Live mark-and-release trapping was carried out monthly throughout the 25 years of study, so that absolute density estimates were available for a detailed picture of seasonal and yearly dynamics. These data represent the most detailed long-term data on vole populations in North America.

Figure 1.5 shows the monthly estimates for the prairie vole (*Microtus ochrogaster*) from 1972 to 1997 for the alfalfa habitat. Densities were on average highest in the alfalfa habitat (50 voles/ha) over this whole time period, but the population fluctuation patterns were similar in bluegrass (average: 18/ha) and native tallgrass prairie (average: 7/ha). The pattern of fluctuations shown in figure 1.5 is much more complex because it includes seasonal fluctuations which can obscure the peak years that are so clear in, for example, the Norwegian lemming in figure 1.2. In Figure 1.6 this can be seen clearly from the population changes between 1981 and 1984. The prairie vole population increased to a peak in early winter 1981, and then dropped suddenly to a low level. A slight increase over the summer months of 1982 was followed by another peak in midwinter 1982–83, followed in turn by another rapid drop in numbers. For the next year this population maintained moderate numbers, and then decreased slowly to lower densities by mid-1984.

In terms of the conventional classification of population cycles (Krebs and Myers 1974) into four phases—increase, peak, decline, and low—this particular time trace is difficult to categorize. Two peaks might be recognized: one in late 1981 and another in midwinter 1982–83, only one year apart. The summer of 1982 could be called a low phase, and the two declines in January 1982 and April 1983 each occurred over one month or less. The moderate population of summer and early winter 1983 are like

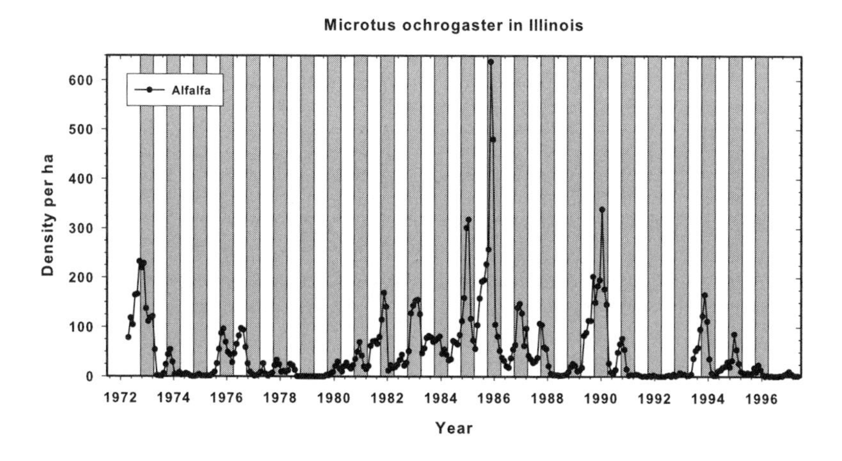

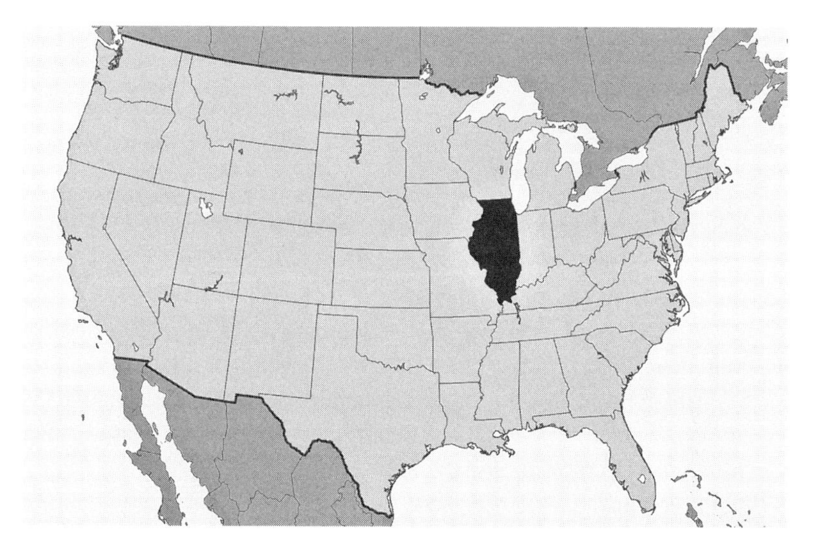

FIGURE 1.5 Monthly population estimates for the vole *Microtus ochrogaster* in an alfalfa field in central Illinois. Live trapping and mark-and-release methods were used for three days each month to estimate population size on two areas (1.0–1.4 ha) planted in alfalfa. Winter months are shaded. Data courtesy of Lowell Getz.

the low phase in showing no population increase, but this was happening at a moderate rather than low density. The bottom line is that some but not all rodent population time series show clear phases, and we need to consider other methods of description for time series like those shown in figure 1.6.

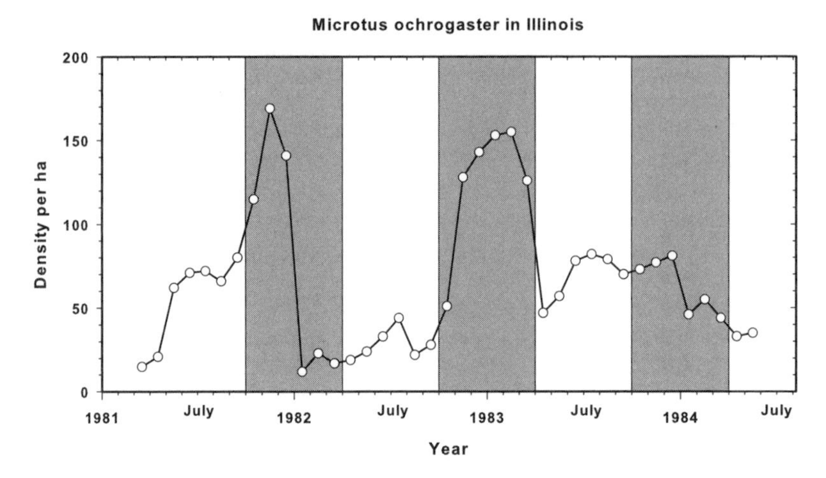

FIGURE 1.6 Details of population changes for the vole *Microtus ochrogaster* in an alfalfa field in central Illinois from March 1981 to June 1984. Data obtained by live trapping. Winter months are shaded. The breeding season of voles in this area generally lasts from March through November. (Data courtesy of Lowell Getz.)

Time Series Analysis of Population Changes

The great variety of population changes that have been described during the last 70 years have led to a variety of mathematical methods for describing them in a rigorous quantitative manner. Time series analysis has been one favorite method, and has been particularly developed for small mammals by Norwegian ecologists and their colleagues (Stenseth et al. 1996; Bjørnstad and Grenfell 2001; Erb et al. 2001).

The basic approach of time series analysis is to fit an autoregressive model to the logs of annual indices or estimates of population size. Given population estimates (N), we convert these values into natural logs (X) so that we can stabilize the variance. The autoregressive model in the general sense is given by

$$X_t = a_0 + a_1 X_{t-1} + a_2 X_{t-2} + a_3 X_{t-3} + \text{error}$$

where $X = \log_e$ (population size estimate) for each year,
t = time (year), and
a_0, a_1, a_2, a_3 = fitted constants.

Depending on the number of time lags used in the regression, these are typically second-order models (terms in X_{t-1} and X_{t-2}) or third order models (adding a term in X_{t-3}). Second-order models are most common in the literature (Stenseth 1999), which means that we attempt to explain this year's density statistically by referring to density observed in the last two years. We are not concerned here with the details of estimation from these autoregressive models; many statistical packages contain routines to fit such models to data. The key point for us is that the estimated constants for second-order autoregressive models (a_1 and a_2) measure direct density-dependence and delayed density-dependence (Royama 1992). The statistical estimates of direct and delayed density dependence can be displayed in a triangle diagram, as seen in figure 1.7. Under the half circle in this figure, multiannual population fluctuations will occur, and the periodicity of the cycles will increase as a_1, or direct density dependence, increases.

We can illustrate these calculations for several populations of *Myodes (Clethrionomys)* that have been studied in North America with at least

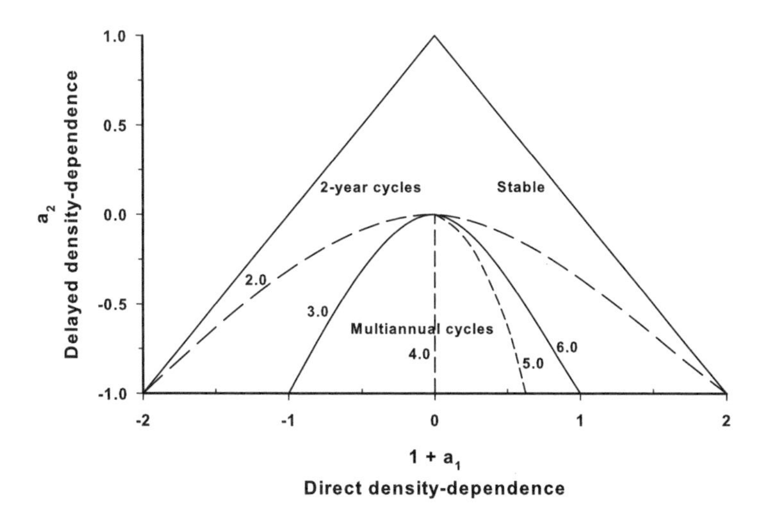

FIGURE 1.7 The dynamics of a statistical autoregressive model of order 2. The statistical measure of direct density dependence is given on the x-axis ($1+a_1$) and the measure of delayed density dependence is given on the y-axis (a_2). Below the semicircle, population cycles will occur; the contour lines indicate the cycle's periodicity in years, depending on the location of the time-series estimates in the parameter space. Royama (1992) discusses the mathematical details of this model.

14 years of consecutive estimates (Boonstra and Krebs 2012). We used estimates from late summer or autumn in preference to spring estimates, but the same results were found with both kinds of estimates when both were available. We used the software program Statistica 6.0 to analyze time series by calculating autocorrelation coefficients and by fitting autoregressive models of order 2 to the data. Following Bjørnstad et al (1995), we log-transformed population estimates and fitted them to the autoregressive model

$$R_t = a_0 + a_1 X_{t-1} + a_2 X_{t-2} + e_t$$

where $R_t = X_t - X_{t-1} = \log(N_1/N_2)$,
$X_t = \log(N_t)$,
N_t = abundance estimate for year t, and
t = time in years.

The coefficients a_1 and a_2 are measures of direct density dependence and delayed density dependence.

To determine whether there was a cyclic periodicity in the population indices, we estimated spectral densities in Statistica 6.0 after detrending the data, and smoothed the resulting periodogram with a five-point moving average based on Hamming weights. I appreciate that all but two of these time series have small sample sizes, and that it is desirable to have a sample of 25 data points to produce a robust autoregressive model. For the time series of shorter duration, I think the models are indicative rather than precise. The first indications of cyclic dynamics can be obtained from autocorrelation analyses. Table 1.1 gives the time lag one to two autocorrelation coefficients for six populations of *M. gapperi* and one population of *M. rutilus*. For a cyclic population, the ideal configuration would be to have a positive autocorrelation at a lag of one year and negative autocorrelations at lags of two to four years, suggesting a three- to four-year cycle in numbers.

There were only two time series for North American populations that had statistically significant autoregressive models of order 2—the Maine data for *M. gapperi* and the Kluane data for *M. rutilus* (table 1.2). Almost all others had significant order 1 regressions indicating some direct density dependence in the annual data. The positions of all the populations in phase space can be seen most easily on the periodicity diagram developed

TABLE 1.1 **Autoregressive model estimates for *Myodes* (*Clethrionomys*) populations from North America, following Bjørnstad et al (1995). Autumn population estimates were used unless otherwise noted. Statistically significant parameters appear in bold type. The parameter a_1 is a measure of direct density dependence, and a_2 a measure of delayed density dependence.**

Population	No. of years	a_1	S.E. (a_1)	P (a_1)	a_2	S.E. (a_2)	P(a_2)	R^2
Myodes* (*Clethrionomys*) *gapperi								
Algonquin Park, Ontario, Canada	34	**−0.42**	.18	.03	−0.03	.18	.87	0.22
Maine, USA	21	**−1.02**	.21	.001	**−0.92**	.42	.04	0.58
Northwest Territories, Canada	12	−0.62	.34	.10	−0.18	.38	.64	0.30
Pennsylvania, USA	13	**−0.69**	.25	.01	−0.07	.25	.77	0.36
Manitoba, Canada	14	**−0.72**	.27	.02	−0.32	.28	.27	0.49
Minnesota, USA	11	**−0.93**	.34	.03	−0.50	.64	.45	0.50
Myodes* (*Clethrionomys*) *rutilus								
Yukon, Canada	33	−0.11	.15	.50	**−0.59**	.16	.001	0.35

S.E. = standard error of the estimated parameter. P = probability that the estimated parameter differs significantly from zero. R2 = fraction of the variation in the data that is captured by the statistical autoregressive model.

by Royama (1992). Figure 1.8 shows that Maine, Yukon, and Minnesota populations fall in the zone of expected three- to four-year cycles, while the other populations align on a gradient from possibly cyclic around four to five years (Manitoba) to borderline populations that could fluctuate at time periods of more than four years (Northwest Territories and Pennsylvania) to expected stability (Ontario).

Spectral analyses of the population indices supported the suggestion that clear evidence of three- to four-year fluctuations could be found in only two time series (figure 1.9). The Algonquin time series in particular shows no indication of any periodicity in the range of two to five years. Both the Minnesota and the Manitoba time series show spectral peaks in the three- to five-year range, but they are based on a small number of observations.

It is clear from the analysis of the autoregressive model of Royama (1992) that in *Myodes (Clethrionomys)* populations in North America there is a gradient from cyclic to irregular fluctuations. The extremes of this pattern are shown by populations with more than 30 years of data, and while we can argue about time series of shorter duration, the Algonquin and Kluane data both come from periods long enough to enable us to recognize patterns of population change. The message shown in

TABLE 1.2 **Autocorrelations of *Myodes* (*Clethrionomys*) time series from North America, for autumn estimates up to lags of five years. Statistically significant autocorrelations appear in bold type.**

Population	Lag 1 r	S.E. (1)	Lag 2	S.E. (2)	Lag 3	S.E. (3)	Lag 4	S.E. (4)	Lag 5	S.E. (5)
Algonquin, Ontario, Canada	**0.39**	.17	0.15	.19	0.03	.19	−0.04	.19	−0.14	.19
Maine, USA	0.05	.20	**−0.50**	.19	−0.24	.19	0.25	.18	0.05	.18
Northwest Territories, Canada	0.07	.27	0.13	.27	−0.16	.27	−0.13	.28	−0.30	.28
Pennsylvania, USA	0.12	.22	−0.22	.22	0.07	.23	−0.14	.23	−0.30	.24
Manitoba, Canada	0.10	.25	−0.40	.25	−0.23	.29	0.13	.30	0.05	.30
Minnesota, USA	0.04	.28	−0.37	.28	−0.35	.31	0.30	.34	−0.01	.36
Yukon, Canada	−0.08	.17	**−0.53**	.17	0.13	.22	0.35	.22	−0.06	.24

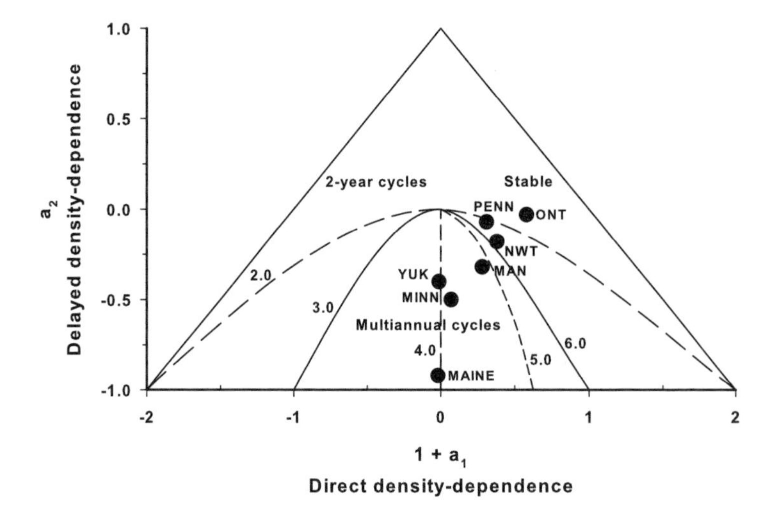

FIGURE 1.8 Coefficients of the second-order autoregressive model for North American *Myodes* (*Clethrionomys*) data. Below the semicircle, the dynamics are predicted by this model to be cyclic; the cyclic period increases from left to right, as in figure 1.7. Four contours of cycle length (three-, four-, five-, and six-year periods) are indicated: Maine, Minnesota (MINN), Manitoba (MAN), Northwest Territories (NWT), Pennsylvania (PENN), and Ontario (ONT) data for *C. gapperi*; and Yukon (YUK) data for *C. rutilus*. From Boonstra and Krebs 2012.

figure 1.9 is clear. Gradients of fluctuations do not demand different hypotheses for cyclic and noncyclic populations. It is more parsimonious to assume a uniformity of cause and effect, and to search for quantitative differences in the factors limiting the rates of population change in different environments.

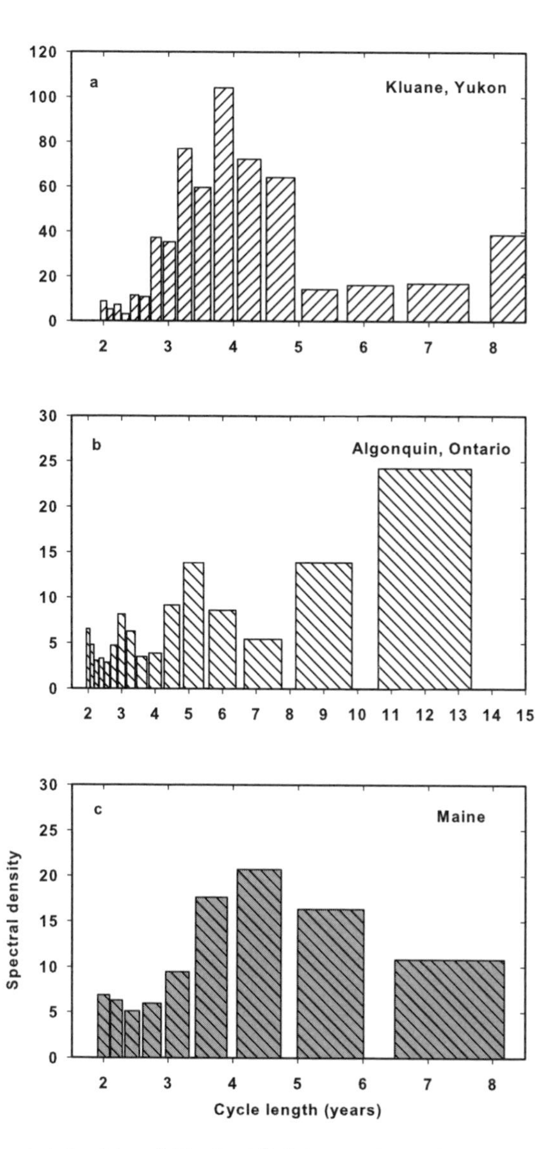

FIGURE 1.9 Spectral densities of *Myodes (Clethrionomys)* populations. Spectral densities were smoothed by a weighted moving average. (a) The Kluane population of *M. rutilus* shows a clear spectral peak in the three- to five-year range. (b) The Algonquin *M. gapperi* shows no clear spectral peak until 11 to 13 years, well beyond the range of three- to four-year cycles. (c) The Maine data for *M. gapperi* shows a spectral peak in the three- to five-year range, not unlike that of the Kluane population.

Conclusion

The history of cycles in small rodents has overemphasized the distinction between cyclic and noncyclic populations, with the implicit conclusion that two quite different sets of limiting factors must be involved. The original concept that propelled Charles Elton and Dennis Chitty was that cyclic populations were useful because they enabled one to plan field research in accordance with an experimental design that allowed replication in time, if not in space. You could ask what factors were operating during the phase of increase, and what changed in the decline phase of the cycle (Chitty 1960). This convenience was misinterpreted as a belief that cyclic populations were a special case, to be ignored if one wished to believe that they were restricted to polar habitats or to only a few species. It is more useful to investigate the general problem of what causes instability in small rodent populations, to suggest general explanations and alternative hypotheses, and to apply these ideas widely to small rodents or wider ensembles of species (Sinclair and Krebs 2002; Krebs 2009). In this spirit I will attempt in the next chapters to show what limits the population growth rate of field populations of small rodents, without the preconception that cyclic populations differ fundamentally from those that fluctuate more irregularly. The overall significance of research on rodent populations has been that it raises the general question of how one should reach an understanding of the population dynamics of any species—an issue that has been under debate for more than 60 years (Chitty 1996; Krebs 2002;, Stenseth 1999). Such knowledge will help us to understand and manage rodent damage problems.

CHAPTER TWO

Biogeography of Rodent Population Fluctuations

Key Points:

- Rodent population fluctuations have been described most clearly from temperate and polar ecosystems.
- The conventional wisdom is that rodent fluctuations show a latitudinal gradient with strongly variable fluctuations at high latitudes, and less variable fluctuations at temperate and tropical latitudes.
- There is evidence for this supposed latitudinal gradient in population variability in the *Myodes* species group in Eurasia and North America, but no evidence of it in the *Microtus* species group on either continent.
- The conventional belief is that rodent populations in productive tropical ecosystems are relatively stable in density, but data to support this belief are few.
- Peak densities of microtine rodents could map directly onto the productivity of the habitats they occupy, or onto the abundance and diversity of predators.

In this chapter I discuss the amplitude of population change in a variety of small rodent species. The main focus is on latitudinal variations in population variability and population density. Strong fluctuations in small rodent numbers have typically been associated with polar environments, and the belief in a north-south gradient in rodent dynamics in the Northern Hemisphere is firmly established in ecology textbooks. Odum (1971), for example, states that cycles "are most pronounced in the less complex ecosystems of northern regions" (page 193).

If we wish to test this commonly held belief, we run into problems if we try to separate population data on the basis of their being "cyclic" or "noncyclic." For many sets of population data, the time series are not long enough to decide in a rigorously mathematical manner whether or not the fluctuations shown are cyclic (cf. Bjørnstad et al. 1996, 1998). It is perhaps easier to retreat to the inclusion of all available time series data by measuring the amplitude of population fluctuations while ignoring the question of whether or not we might consider the time series cyclic. The implicit assumption is that strict mathematical cyclicity is not particularly significant ecologically.

Hypotheses of Rodent Biogeography

The conventional wisdom for polar-tropical gradients is encapsulated in the first hypothesis of rodent population biogeography:

Hypothesis 1. *The amplitude of population fluctuations is positively correlated with latitude.* This hypothesis can be tested either interspecifically by showing that over a variety of populations of different species it well describes the geographic pattern of fluctuations, or within a single species with a wide north-south distribution by showing that the more equatorial the population, the smaller the amplitude of fluctuations.

Of course, no one considers latitude per se the relevant ecological variable, and most ecologists believe that a pattern with latitude requires a search for a mechanistic explanation. As we saw in chapter 1, many rodent populations fluctuate strongly, and any variation in the amplitude of those fluctuations must have some ecological explanation, even if latitude is irrelevant. Two alternative hypotheses for these kinds of patterns emphasize a more mechanistic and ecosystem-oriented view of rodent population fluctuations.

Hypothesis 2. *The amplitude of population fluctuations is negatively correlated with ecosystem complexity.* Ecosystem complexity could be measured in a variety of ways to test this hypothesis, but the simplest measures would involve the diversity and abundance of predators, parasites, and diseases as well as the number of closely related species that compete for food or space. A corollary is that rodent populations in complex tropical ecosystems will show little or no fluctuation in density.

Hypothesis 3. *The amplitude of population fluctuations is a function of landscape connectivity and the size and configuration of suitable habitats within the larger landscape.* Landscape hypotheses are among the most difficult to test in ecology, but a simple approach would be to map the abundance of grassland in areas like central Illinois and Indiana and determine whether fragmentation of these habitats was affecting rodent fluctuations. This idea could be a subset of hypothesis 2, but hypotheses 2 and 3 could be distinguished by careful measurements of predator, parasite, and disease abundance in fragments. From a productivity viewpoint, a second set of interesting biogeographic questions involves the factors that control absolute density or the maximum standing crop of small rodents at the peak of a fluctuation. This issue can be cast as a top-down, bottom-up issue (Abrams 1993), in which case we have two additional alternative hypotheses to evaluate.

Hypothesis 4. *The maximum standing crop at the peak of a rodent population fluctuation is directly related to the productivity of the species' food resources.* This hypothesis states that the key to understanding the maximum abundance of rodent populations lies in plant productivity, a bottom-up effect. Note that this hypothesis does not require that the overexploitation of food resources is the factor causing the population fluctuations. It could be tested by experimentally increasing or reducing the food resources of a fluctuating population. One alternative to this hypothesis is the mirror-image, top-down hypothesis:

Hypothesis 5. *The maximum standing crop at the peak of a rodent population cycle is inversely related to the abundance of their predators and their numerical and functional responses to changing rodent density.* This hypothesis implies a limitation of maximum rodent density by predation, and can be tested directly by removing or reducing predator abundance at the population peak.

In this chapter we will explore the patterns shown by fluctuating rodent populations to determine whether there is clear support for one or more of these five hypotheses. Data are not available to test all the hypotheses conclusively, and much of the discussion should be considered as a preliminary assessment of possible patterns. Additional hypotheses could be added to the mix; these five are only a starting point in the analysis. Two different aspects of rodent fluctuations are involved: the variability of density over an extended time period, and the maximum density reached. Variability involves both the minimum and maximum densities, and in

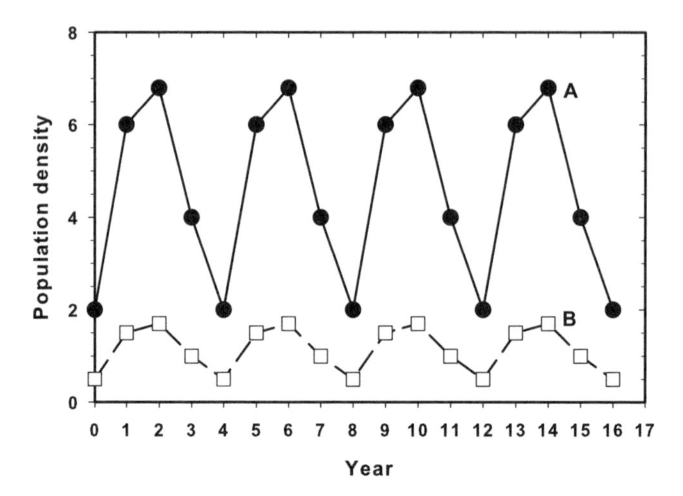

FIGURE 2.1 Hypothetical data to illustrate two components of population changes. Both time series have identical coefficients of variation (CV = 0.44) so that they are mirror-image populations in terms of variability. But the maximum density of population A is four times that of population B.

principle it is possible for two populations to have identical variability but quite different maximum densities. Figure 2.1 illustrates this distinction with hypothetical data.

Methodological Issues

The data available on rodent population changes are highly variable in their precision and potential accuracy. In many cases only an index of rodent density has been obtained, and some strong assumptions must be made to convert the indices to measurements of absolute density. In other cases only one or two data points per year are available, and one can infer amplitudes and absolute densities only by assuming that the sampling was done exactly at the high or low point in density. There is a particularly vexing problem with landscape issues that are mostly unresolved in the literature. The dynamics of a *Microtus* population being studied in a one-hectare grassland area surrounded by unsuitable agricultural fields might be assumed to be unrepresentative of a similar population that occupied hundreds of hectares of continuous grassland habitat. The most difficult data to evaluate come from populations in enclosures, and I do not use any of these enclosure population densities to evaluate potential carrying capacities for the species under study.

The amplitude or variability of any population can be measured by the *s*-index, the standard deviation of the log-transformed measure of rodent abundance (see chapter 1, page 10):

$$s = \sqrt{\frac{\sum_{1}^{n}\left(\log N_i - \overline{\log N_i}\right)^2}{n-1}}$$

where s = index of variability,
N = population density or index,
n = the number of years in a time series,
and logs are to base 10.

As we noted in chapter 1, this is not an index of cyclicity. We consider it here simply to quantify the amplitude of population changes, regardless of whether these fluctuations follow a cyclic pattern. An better measure of variability is the coefficient of variation of the measures of rodent abundance,

$$CV = \left(1 + \frac{1}{4n}\right)\left(\frac{s}{\bar{X}}\right)$$

where CV = the coefficient of variation with Haldane's correction (Krebs 1999, p. 565)
n = the number of data points,
s = the standard deviation of the n measures of abundance, and
$\bar{X}$ = the observed mean of the n measures of abundance.

The major problem with using the *s*-index is that rodent abundance data often contain zeros, which can have two meanings: that the population in that area is extinct, or that it is very small and cannot be sampled properly (McArdle and Gaston 1993). Most authors assume the second meaning for zeros and, to avoid the undefined log of zero, add 0.1 or 1.0 to all the estimates of abundance. But the value of the *s*-index changes with the size of the constant added to each measure of abundance, and this can make any comparisons of relative variability invalid (McArdle and Gaston

1992). In this chapter I shall use the coefficient of variation to measure population variability. The 95% confidence limits for the coefficient of variation can be computed by the protocols given by Verrill (2003).

Data Required for the Coefficient of Variation

The first question is: What data are needed to calculate the coefficient of variation? Is one density sample point per year sufficient? Are monthly samples better than annual samples? Should spring estimates of density be preferred over autumn estimates? We need to answer these questions to avoid biases in comparing data from different populations. To look into these problems we need a long-term data set with density estimates taken weekly or monthly all year round. Few of these data sets are available. The best data I have found is from Lowell Getz's group at the University of Illinois (http://www.life.uiuc.edu/getz/), which has provided monthly data spanning 25 years in three grassland habitats for *Microtus pennsylvanicus* and *M. ochrogaster*.

Figures 2.2 and 2.3 address the seasonality of data needed to calculate the coefficient of variation for a population. For *Microtus pennsylvanicus*, figure 2.2 shows that either April or September annual data provide essentially the same estimate of the coefficient of variation as the full monthly density data. Figure 2.3 shows the same results for *M. ochrogaster*. The only disadvantage of using annual data is illustrated by the wider confidence limits of the annual data. The Finse Norwegian lemming data shown in figure 1.2, which provides two indices per year, provides a coefficient of variation of 2.22 for spring densities (n = 29 years) and 2.39 for autumn densities (difference not significant, $p > 0.10$). The average coefficient of variation for all the Finse data is 2.93, which does not differ significantly from the spring and autumn values. Additional data on long-term monitoring would be useful to test further the conclusion that annual data are an adequate basis on which to estimate the coefficient of variation.

If this is the case, and if the frequency and season of population data do not affect the estimated coefficient of variation, the next issue is how long a sampling period is needed to obtain a relatively precise estimate of the coefficient of variation. If fluctuations are regular, the sampling period need extend only over one cycle, which would typically be three to four years in a rodent species. But if cycles are irregular with variable peak and low densities, then about ten to fifteen years would appear to be the

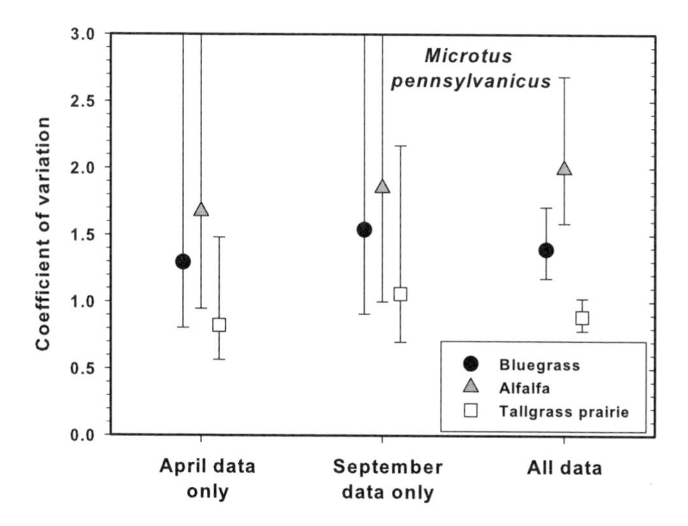

FIGURE 2.2 The coefficient of variation of population density for *Microtus pennsylvanicus* in central Illinois. Error bars are 95% confidence limits. The coefficient of variation is essentially the same whether only the April data are used ($n = 23$ years), only the September data ($n = 23$) are used, of all the data are used ($n = 277$ months). The average density over the entire study was 29.5 per hectare for tallgrass prairie, 14.1 for bluegrass, and 7.2 for alfalfa. Data courtesy of Lowell Getz.

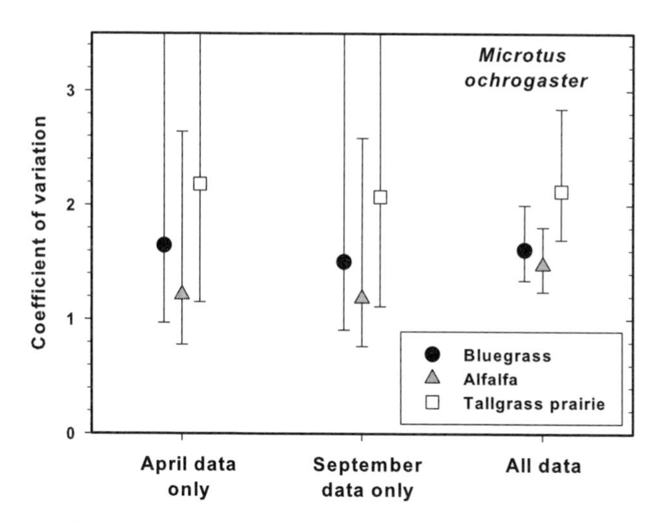

FIGURE 2.3 The coefficient of variation of population density for *Microtus ochrogaster* in central Illinois. Error bars are 95% confidence limits. The coefficient of variation is essentially the same whether only the April data are used ($n = 25$ years), only the September data ($n = 25$) are used, of all the data are used ($n = 302$ months). The average density over the entire study was 7.4 per hectare for tallgrass prairie, 18.1 for bluegrass, and 49.9 for alfalfa. Data courtesy of Lowell Getz.

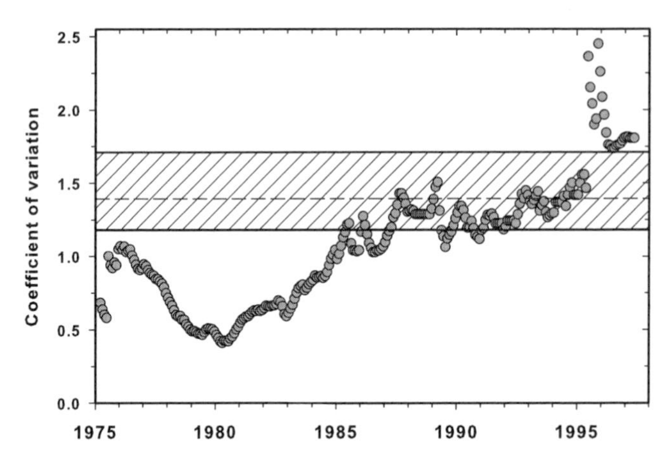

FIGURE 2.4 A three-year running average of the coefficient of variation of population density for *Microtus pennsylvanicus* in central Illinois. Error bars are 95% confidence limits around the overall average value of 1.39 (dashed line). The coefficient of variation drifts over time, indicating that a single value for this measure does not adequately provide a stable measure of population variability for this particular population. The average density over the entire study was 14.1 for bluegrass. Data are plotted over the last month of each three-year period. Data courtesy of Lowell Getz.

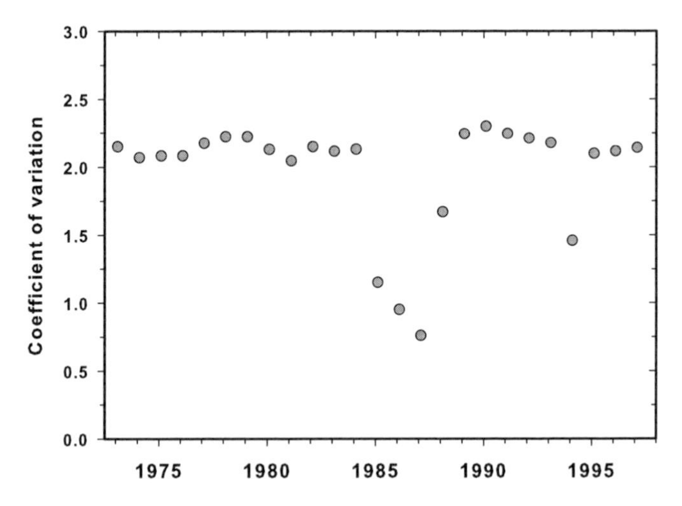

FIGURE 2.5 A three-year running average of the coefficient of variation of population density for the Norwegian lemming *Lemmus lemmus* at Finse in southern Norway. The coefficient of variation is nearly stationary over time, except during three years in the late 1980s, thus indicating that a single value for this measure does provide a reasonably stable measure of population variability. Data courtesy of Nils Stenseth.

length of time needed to estimate population variability. Figure 2.4 shows the coefficient of variation of the *Microtus pennsylvanicus* population in bluegrass habitat in central Illinois. The time series is far from stable, providing a complex trend with an irregularly rising coefficient of variation over time. The Finse Norwegian lemming data (figure 2.5), by contrast, stabilizes almost immediately. The bottom line is not definite, but in general I think that a sample size of about 10 years is needed to get a reliable estimate of the coefficient of variation.

A related methodological issue is whether coefficients of variation measured on indices (such as number per 100 trap nights) are equivalent to those measured with mark-recapture live trapping. At present we do not have the data available to answer this question.

Do Small Rodent Populations Fluctuate More at High Latitudes?

The Global Population Dynamics Database, held at Imperial College, London, has collated many sets of small rodent data. To test hypothesis 1, I selected data from this database in January 2010, with the criterion of reliable estimates of at least 10 years (http://www3.imperial.ac.uk/cpb/research/patternsandprocesses/gpdd). To these data I have added time series data obtained via personal communication from sources not yet published, and I am grateful to my colleagues for sharing these data.

Table 2.1 gives estimates of the coefficient of variation for 95 small rodent populations for which I have found adequate data. I have used the very crude measure of latitude to test hypothesis 1, that population fluctuations are stronger at high latitudes. Latitude is a crude measure because European latitudes are, in general, ecologically offset from those in North America. For example, latitude 68° N in Finland is about ecologically equivalent to latitude 61° N in western North America and latitude 55° N in eastern North America. But ignoring the lack of exact ecological equivalence between latitudes, we can look for a common pattern in these data.

Figure 2.6 shows for the four *Myodes* species in table 2.1 that populations at higher latitudes are more variable, as predicted in hypothesis 1. European populations have in general been studied at higher latitudes than have North American *Myodes*, but a single regression line is the best statistical descriptor of these data. The slope of the regression suggests as

TABLE 2.1 **Coefficients of variation of small rodent populations measured over a minimum of 10 years. The Global Population Dynamics Database at Imperial College dataset number is given for datasets obtained from this site in January 2010. Accessed at 3.imperial.ac.uk/cpb/research/patternsandprocesses/gpdd.**

Species	Location	Latitude	No. of years	Season	Mean	Standard. Deviation	CV	GPDB dataset number
Apodemus sylvaticus	Wytham Woods, UK	51° 42'	28	Summer	18.36	14.09	**0.77**	44
Apodemus sylvaticus	Wytham Woods, UK	51° 42'	28	Winter	39.68	17.11	**0.44**	45
Arvicola terrestris	Arboldswill, Switzerland	47° 24'	25	Annual	1587.21	1301.14	**0.83**	10012
Arvicola terrestris	Bulle, Switzerland	46° 37'	20	Annual	2987.70	4485.30	**1.52**	10009
Arvicola terrestris	Chateau d'Oex, Switzerland	46° 28'	42	Annual	20866.00	45262.00	**2.18**	10010
Arvicola terrestris	La Brevine, Switzerland	46° 59'	27	Annual	2021.79	1872.01	**0.93**	10008
Arvicola terrestris	Le Chenit, Switzerland	46° 35'	30	Annual	5006.93	5600.94	**1.13**	10006
Arvicola terrestris	Le Pont, Switzerland	46° 40'	41	Annual	821.47	1052.27	**1.29**	10007
Arvicola terrestris	Rougemont, Switzerland	46° 29'	49	Annual	5555.39	6724.59	**1.22**	10011
Arvicola terrestris	Ste Croix, Switzerland	46° 49'	54	Annual	6927.26	14552.10	**2.11**	10005
Arvicola terrestris	Stuttgart, Germany	48° 47'	27	Summer	2.91	1.69	**0.59**	9998
Arvicola terrestris	Titterten, Switzerland	47° 25'	25	Annual	2508.72	2544.14	**1.02**	10013
Myodes gapperi	Algonquin Park, Ontario	48° 30'	36	Spr/Au	2.04	2.09	**1.04**	
Myodes gapperi	Algonquin Park, Ontario	48° 30'	36	Spring	0.75	0.90	**1.21**	
Myodes gapperi	Algonquin Park, Ontario	48° 30'	36	Autumn	3.32	2.16	**0.66**	
Myodes gapperi	Keeley Creek, Lake County, Minnesota	47° 50'	13	Summer	8.85	7.85	**0.90**	6755
Myodes gapperi	Lake County, Minnesota	47° 50'	13	Summer	26.54	15.41	**0.59**	6747
Myodes gapperi	Lake County, Minnesota	47° 50'	13	Summer	12.08	13.29	**1.12**	6752
Myodes gapperi	Maine	43° 45'	23	Fall	23.00	2.22	**0.10**	(R. Boonstra)
Myodes gapperi	Powdermill, Pennsylvania	40° 10'	21	Spr/Au	11.88	7.78	**0.66**	(R. Boonstra)
Myodes glareolus	Boda, Sweden	62° 51'	28	Spr/Au	4.32	4.12	**0.96**	10068
Myodes glareolus	Finnish Lapland	68° 00'	16	Summer	2.57	2.59	**1.02**	9292
Myodes glareolus	Grimsø Wildlife Station, Sweden	59° 40'	11	Spr/Au	1.63	1.60	**1.01**	6999

Myodes glareolus	Karelia, Finland	63° 00'	21	Annual	2.03	2.21	**1.10**	10055
Myodes glareolus	Pallasjärvi, Finland	68° 00'	23	Spr/Au	6.33	5.18	**0.83**	9921
Myodes glareolus	Serpukhov, Russia	54° 53'	29	Annual	19.24	10.19	**0.53**	10044
Myodes glareolus	Monchegorsk, Kola Peninsula, Russia	67° 55'	19	Annual	9.24	12.30	**1.35**	10056
Myodes glareolus	Sotkamo, Finland	64° 09'	27	Annual	11.37	6.86	**0.61**	10045
Myodes glareolus	Strömsund, Sweden	63° 15'	12	Summer	1.88	1.83	**0.99**	7009
Myodes glareolus	Tataria, Russia	55° 00'	23	Annual	7.70	5.79	**0.76**	10046
Myodes glareolus	Tula Region, Russia	54° 11'	16	Summer	7.92	4.29	**0.55**	9294
Myodes glareolus	Umea, Sweden	63° 50'	18	Spring	4.19	3.75	**0.91**	6715
Myodes glareolus	Umea, Sweden	63° 50'	17	Autumn	1.16	1.38	**1.21**	6716
Myodes glareolus	Wytham Woods, UK	51° 42'	28	Summer	57.75	44.86	**0.78**	46
Myodes glareolus	Wytham Woods, UK	51° 46'	28	Winter	61.00	27.19	**0.45**	47
Myodes glareolus	Zvenigorod, Russia	55° 40'	31	Annual	16.29	9.78	**0.61**	10054
Myodes rufocanus	Cascade Mountains, Oregon	44° 00'	10	Spring	5.50	4.27	**0.80**	6702
Myodes rufocanus	Cascade Mountains, Oregon	44° 00'	13	Autumn	7.62	5.92	**0.79**	6703
Myodes rufocanus	Central Kola Peninsula, Russia	58° 45'	30	Summer	13.47	13.51	**1.01**	9894
Myodes rufocanus	Kilpisjärvi area, Finland	69° 00'	11	Summer	8.55	7.45	**0.89**	9854
Myodes rufocanus	Kilpisjärvi area, Finland	69° 00'	44	Spring	5.11	6.38	**1.25**	(H. Henttoneni)
Myodes rufocanus	Kilpisjärvi area, Finland	69° 00'	47	Fall	10.99	9.85	**0.90**	(H. Henttoneni)
Myodes rufocanus	Pallasjärvi, Finland	68° 00'	23	Spr/Au	1.39	2.68	**1.95**	9920
Myodes rufocanus	Monchegorsk, Kola Peninsula, Russia	67° 55'	19	Annual	19.09	16.95	**0.90**	10057
Myodes rufocanus	Monchegorsk, Kola Peninsula, Russia	67° 55'	19	Annual	15.23	18.31	**1.22**	10058
Myodes rufocanus	Umea, Sweden	63° 50'	18	Spring	0.92	1.08	**1.19**	6717
Myodes rufocanus	Umea, Sweden	63° 50'	19	Autumn	0.40	0.63	**1.61**	6718
Myodes rutilus	Heart Lake, Northwest Territories	60° 55'	19	Spr/Au	4.44	4.88	**1.11**	(W. Fuller)
Myodes rutilus	Kluane, Yukon	61° 21'	23	Spr/Au	8.55	8.32	**0.98**	(C. Krebs)
Myodes rutilus	Pallasjärvi, Finland	68° 00'	23	Spr/Au	2.62	2.40	**0.93**	9919
Myodes rutilus	Southeast Siberia	54° 20'	15	Spr/Au	14.72	9.97	**0.69**	(Koshkina)
Dicrostonyx groenlandicus	Cape Churchill area, Manitoba	67° 30'	13	Summer	15.33	15.24	**1.01**	9889
Lemmus lemmus	Hardangervidda Plateau, Norway	60° 15'	24	Annual	3.46	7.27	**2.12**	10014
Lemmus trimucronatus	Point Barrow, Alaska	71° 00'	21	Summer	16.24	19.18	**1.19**	9895

TABLE 2.1 ***(continued)***

Species	Location	Latitude	No. of years	Season	Mean	Standard. Deviation	CV	GPDB dataset number
Lemmus trimucronatus	Point Barrow, Alaska	71° 00'	20	Summer	44.44	69.34	**1.58**	9954
Microtus	Alojoki field plain, Finland	63° 05'	20	Spr/Au	3.99	5.18	**1.31**	9915
Microtus	Alojoki field plain, Finland	63° 05'	16	Spr/Au	2.68	4.19	**1.59**	9916
Microtus	Iesjavri basin, Norwegian Lapland	69° 40'	12	Spr/Au	7.04	6.54	**0.95**	9914
Microtus	Kauhava region, Finland	63° 06'	19	Spr/Au	4.80	4.61	**0.97**	9923
Microtus	Kauhava region, Finland	62° 45'	14	Spr/Au	6.98	6.61	**0.96**	9928
Microtus	Kauhava region, Finland	62° 45'	14	Spr/Au	6.91	6.73	**0.99**	9931
Microtus	Pallasjärvi, Finland	68° 00'	23	Spr/Au	12.10	10.39	**0.87**	9913
Microtus	Seinajoki region, Finland	62° 45'	12	Spr/Au	9.39	9.49	**1.03**	9925
Microtus agrestis	Grimsø Wildlife Station, Sweden	59° 40'	10	Spr/Au	0.80	1.36	**1.74**	6998
Microtus agrestis	Kilpisjarvi area, Finland	69° 00'	41	Spr/Au	7.69	8.52	**1.12**	9912
Microtus agrestis	Pallasjärvi, Finland	68° 00'	24	Spr/Au	2.29	4.02	**1.77**	9917
Microtus agrestis	Sotkamo, Finland	64° 09'	26	Annual	12.15	6.76	**0.56**	10047
Microtus agrestis	Umea, Sweden	63° 50'	18	Spring	0.53	0.58	**1.11**	6719
Microtus agrestis	Umea, Sweden	63° 50'	17	Autumn	0.48	0.83	**1.76**	6720
Microtus arvalis	Jelenia Gora, Poland	50° 55'	15	Sp/Su/Au	147.75	58.23	**0.40**	10075
Microtus arvalis	Koszalin, Poland	54° 10'	15	Sp/Su/Au	97.98	31.23	**0.32**	10074
Microtus arvalis	Leszno, Poland	51° 51'	15	Sp/Su/Au	207.50	210.19	**1.03**	10077
Microtus arvalis	Opole, Poland	50° 40'	15	Sp/Su/Au	418.10	352.89	**0.86**	10078
Microtus arvalis	Pila, Poland	53° 09'	15	Sp/Su/Au	302.79	355.15	**1.19**	10079
Microtus arvalis	Piutou-Charente, France	46° 15'	20	Summer	169.71	158.51	**0.95**	10037

Microtus arvalis	Poznan, Poland	52° 25'	15	Sp/Su/Au	288.25	378.19	**1.33**	10080
Microtus arvalis	Tula Region, Russia	54° 11'	17	Summer	7.28	3.81	**0.53**	9295
Microtus californicus	Brooks Island, California	37° 55'	5	Monthly	216.30	170.62	**0.83**	9955
Microtus californicus	Hopland Field Station, California	39° 00'	21	Bimonthly	25.04	35.92	**1.45**	6524
Microtus californicus	Hopland Field Station, California	39° 00'	22	Bimonthly	17.05	19.87	**1.18**	6525
Microtus montanus	Grand Teton National Park, Wyoming	51° 42'	18	Summer	2.30	2.31	**1.02**	1402
Microtus ochrogaster	Urbana, Illinois	40° 12'	25	Annual	11.72	19.13	**1.65**	(L. Getz)
Microtus ochrogaster	Urbana, Illinois	40° 12'	25	Annual	18.28	21.95	**1.21**	(L. Getz)
Microtus ochrogaster	Urbana, Illinois	40° 12'	25	Annual	6.44	13.91	**2.18**	(L. Getz)
Microtus oeconomus	Finnish Lapland	68° 00'	16	Summer	0.71	0.97	**1.38**	9293
Microtus oeconomus	Pallasjärvi, Finland	68° 00'	24	Spr/Au	1.27	2.23	**1.78**	9918
Microtus oeconomus	Monchegorsk, Kola Peninsula, Russia	67° 55'	19	Autumn	0.40	0.60	**1.51**	10059
Microtus oregoni	Cascade Mountains, Oregon	44° 00'	12	Spring	2.50	3.75	**1.53**	6698
Microtus oregoni	Cascade Mountains, Oregon	44° 00'	11	Autumn	5.36	4.80	**0.92**	6699
Microtus pennsylvanicus	Urbana, Illinois	40° 12'	23	Annual	11.83	15.16	**1.30**	(L. Getz)
Microtus pennsylvanicus	Urbana, Illinois	40° 12'	23	Annual	3.64	6.02	**1.67**	(L. Getz)
Microtus pennsylvanicus	Urbana, Illinois	40° 12'	23	Annual	35.04	28.57	**0.82**	(L. Getz)
Peromyscus maniculatus	Lake County, Minnesota	47° 50'	14	Summer	18.62	14.51	**0.79**	6756
Peromyscus maniculatus	Lake County, Minnesota	47° 50'	13	Summer	1.92	1.98	**1.05**	6748
Peromyscus maniculatus	Lake County, Minnesota	47° 50'	13	Summer	9.65	11.66	**1.23**	6753

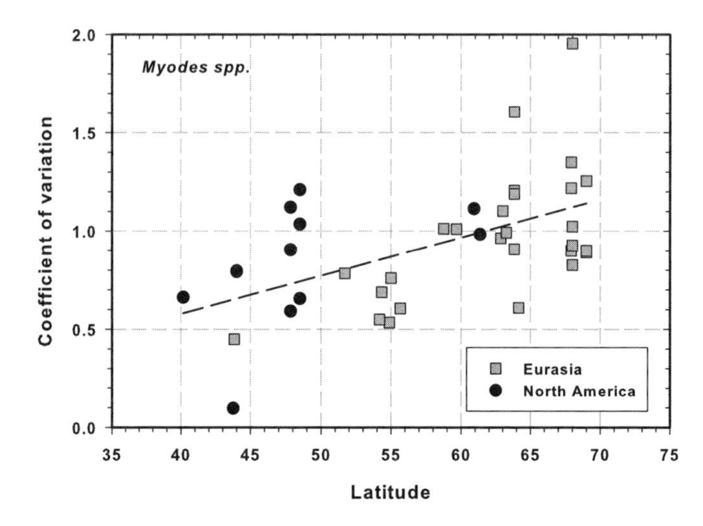

FIGURE 2.6 Coefficient of variation of density indices for *Myodes* (*Clethrionomys*) spp. from Eurasia and North America, in relation to latitude. Each point represents one population studied for at least 10 years at one location. The regression is CV = –0.1985 + 0.0194 latitude; $n = 39$, $r^2 = 0.29$.

a guideline that the coefficient of variation in population size for *Myodes* populations increases 8.6% for every five-degree increase in latitude.

Microtus species however do not show this pattern (figure 2.7). There is no suggestion that population sizes of *Microtus* species at more southerly latitudes fluctuate less strongly than those at more northerly latitudes. For this analysis I included *Arvicola terrestris* with *Microtus*, but the results are the same whether they are included or not. Averaged across all latitudes, *Microtus* species have 27% higher coefficients of variation than do *Myodes* species. For the nine species of *Microtus*, the average coefficient of variation is 1.18 ± 0.06 (SE), and for the four species of *Myodes* it is 0.93 ± 0.06, ($p = 0.005$).

This analysis has several weak points, and would be more focused if we had data on a series of populations of a single species spread from north to south. In North America *Microtus pennsylvanicus* is an obvious candidate for such studies because it has an extensive latitudinal range (figure 2.8). I obtained all the data sets on unmanipulated *Microtus pennsylvanicus* populations that spanned at least four years (table 2.2) and calculated the coefficient of variation for each site. Figure 2.9 shows that there is a weak pattern in these data, which support the general hypothesis of a latitudinal

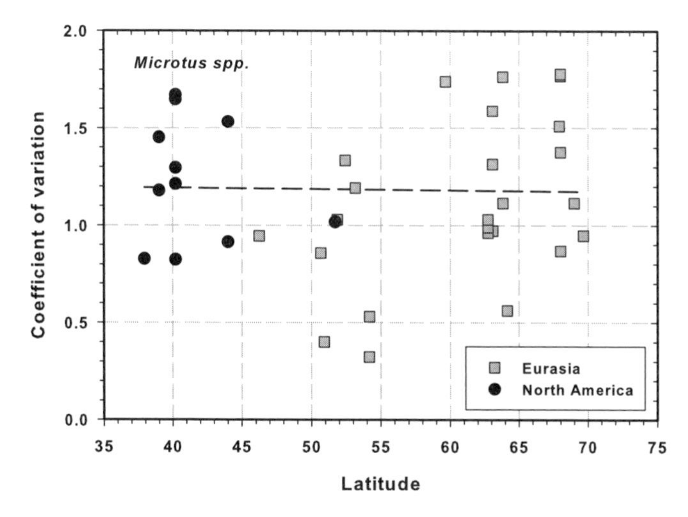

FIGURE 2.7 Coefficient of variation of density indices for *Microtus* spp. from Eurasia and North America, in relation to latitude. Each point represents one population studied for at least 10 years at one location.

gradient in fluctuations ($p = 0.07$). The most interesting points are those gathered in central Illinois by Lowell Getz and his students (points A, B, C in figure 2.9), because they are from different community types at exactly the same geographical site (A = alfalfa; B = bluegrass; C = tallgrass prairie). The scatter of points is more significant than the linear trend. The same point is shown by the two Yukon points D and I. The implication is that the variability of populations is driven more by habitat features than by latitude.

There could be several reasons why figure 2.9 shows only weak support for this widely held hypothesis while figure 2.7 shows none at all. There are unfortunately few data points for *M. pennsylvanicus* between latitudes 50° and 60° N, where there might be larger amplitude fluctuations. Many of these data also come from studies of short duration, and more long-term data are needed of the type already available in Eurasia, as shown in table 2.1. Other factors could bias the trend in the opposite direction to that postulated in hypothesis 1. Much of the data obtained near the southern edge of the geographic ranges could be affected by agricultural practices that subsidize predators and thereby reduce the amplitude of fluctuations (cf. Erlinge 1987).

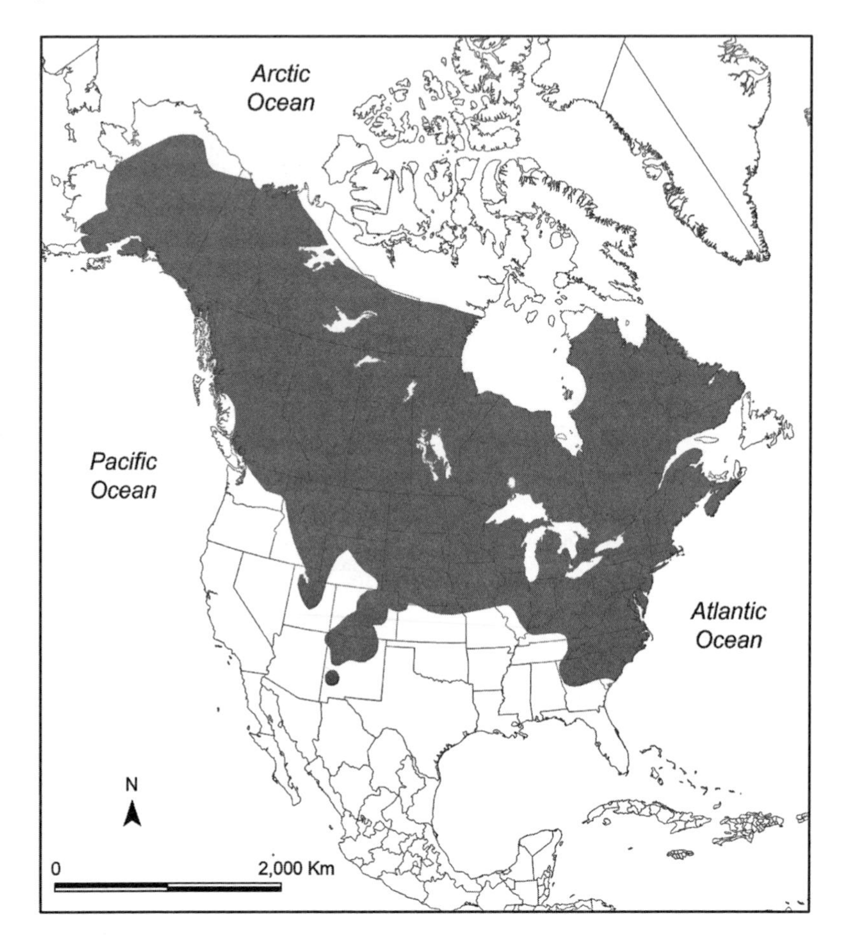

FIGURE 2.8 Geographical range of *Microtus pennsylvanicus*.

Hypothesis 2 (ecosystem complexity) and hypothesis 3 (landscape connectivity) are two alternative ideas that might explain the observed patterns of population variability. These ideas deserve further scrutiny, but on the face of the data we have, they do not seem to be supported. Landscape connectivity cannot explain the Illinois results of Getz, shown in figure 2.9, which span the entire range of variability found to date in *M. pennsylvanicus*. Ecosystem complexity is not a very attractive hypothesis for microtine rodents, which rarely show signs of competition (Galindo and Krebs 1985) and typically live in species-poor habitats. Nevertheless,

TABLE 2.2 **Coefficients of variation and latitude for the *Microtus pennsylvanicus* data in figure 2.9, along with sources of data. Location code refers to letter on the graph in figure 2.9.**

Location code	Location	Latitude	Years of data	Coefficient of variation	Reference
E	Bloomington, IN	39.10° N	5	1.16	Krebs et al. 1969
S	Kelowna, BC	60.07° N	10	2.05	Sullivan, unpubl.
B	Urbana, IL bluegrass	40.20° N	23	1.29	Getz et al. 2005
A	Urbana, IL alfalfa	40.20° N	23	1.67	Getz et al. 2005
C	Urbana, IL tallgrass	40.20° N	23	0.82	Getz et al. 2005
F	Toronto F, ON	43.65° N	4	0.50	Boonstra, unpubl.
G	Pinawa A, MB	49.00° N	5	0.67	Mihok, unpubl.
H	Pinawa B, MB	49.00° N	5	1.03	Mihok, unpubl.
I	Kluane Slims, YT	61.36° N	4		Krebs, unpubl.
D	Kluane, YT	61.35° N	23	1.64	Krebs, unpubl.
J	Grand Lake, NL	49.15° N	4	1.44	Thompson and Curran 1995
K	NB	45.88° N	5	0.63	Simmons et al. 1986
L	Prince George, BC	54.27° N	5	1.76	Sullivan et al. 1998
M	Cape Cod, MA	41.90° N	4	0.82	Tamarin 1977a
N	South Natick, MA	42.25° N	4	0.81	Tamarin et al. 1984
O	South Natick, MA	42.25° N	4	1.00	Tamarin et al. 1990
P	South Natick, MA	42.25° N	4	0.59	Tamarin, unpubl.
Q	South Natick, MA	42.25° N	4	0.55	Tamarin, unpubl.
R	Summerland, BC	49.66° N	10	1.99	Sullivan, unpubl.

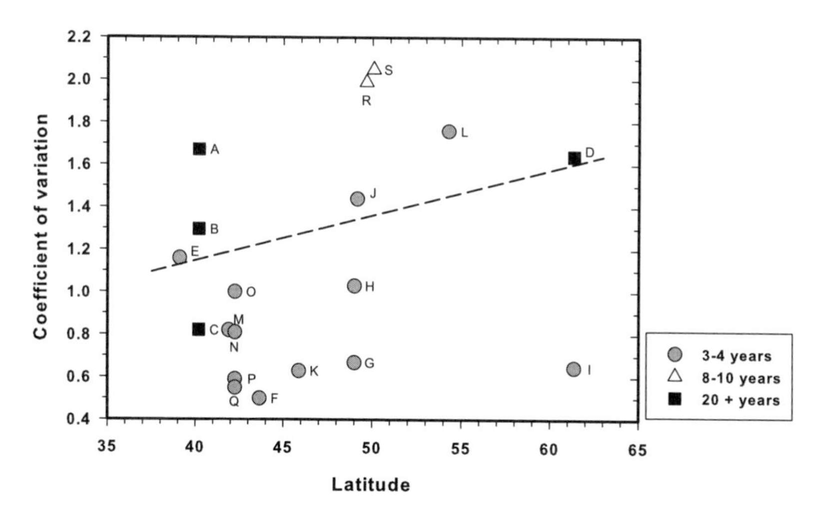

FIGURE 2.9 Coefficient of variation of density indices for *Microtus pennsylvanicus* from North America in relation to latitude. Each point represents one population studied for either 4 years (○), 8 to 10 years (Δ), or more than 20 years (■) at one location. Sites are identified by letters explained in table 2.2. The linear regression is weighted by the years of data; $R^2 = 0.12$, $p = 0.07$.

if ecosystem complexity can be identified in agricultural modifications of land use, hypothesis 2 might be supported. A recent example has emerged in northwestern Spain (41–42° N) where the common vole (*Microtus arvalis*) has become a serious agricultural pest coincident with lowland areas being converted to irrigated agriculture (Luque-Larena et al. 2012). Much could be learned from a detailed landscape analysis of rodent outbreak problems like this one.

Are Small Rodent Maximum Densities Higher in More Productive Habitats?

Another hypothesis that could explain the patterns shown in figures 2.6 and 2.7 suggests a bottom-up control of maximum rodent densities via plant productivity (hypothesis 4). The simple prediction is that maximum rodent density is linearly related to plant production. This idea could be made much more precise by identifying the major food plants of a particular rodent species and measuring their productivity, rather than that of the entire plant community. I can find no evidence that this kind of study has ever been done, and it remains an important idea. At the present I can only speculate on this hypothesis.

One explanation of the discrepancy between *Myodes* (*Clethrionomys*) and *Microtus* (figs. 2.6 and 2.7) could be that the primary productivity of *Microtus* habitats does not change greatly with latitude—or, more precisely, that the primary productivity of their food plants does not change. Thus, maximum density would not change with latitude, while minimum density might be much lower in more severe climatic regions. This speculation would include the idea that while primary production as a whole is higher in southern habitats, the fraction of this production that is usable food for *Microtus* is smaller than in northern habitats. For this speculation to hold, the variance in primary production of the food plants for *Myodes* would have to be higher in more northern habitats, while that of food plants for *Microtus* would have to be similar north and south. Data on this matter are insufficient. Berries are important in the diet of *Myodes* spp.; it is clear that there is an extreme variance in berry crops in northern ecosystems (Selås et al. 2002; Yudina and Maksimova 2005; Krebs et al. 2009). This type of mass seeding can affect population densities of rodents (Selås et al. 2002; Krebs et al. 2010). At present this is the best explanation for the discrepancy in the patterns shown by *Myodes* and *Microtus* in figures 2.6 and 2.7, but it must remain highly tentative until more precise data are collected.

The common vole (*Microtus arvalis*) of Europe shows a reverse latitudinal gradient in Poland, the Czech Republic, and Slovakia (Tkadlec and Stenseth 2001), which illustrates clearly that latitude cannot be the mechanistic explanation of gradients. While northerly populations in coastal Poland are relatively stable, southern populations in the Czech Republic are cyclic and more variable. Seasonality could be the key factor in this reverse gradient. Delattre et al. (1999, 2009) discuss a landscape model for the abundance of the common vole in France, which is discussed in chapter 12 (cf. figure 12.7).

Data on maximum abundance of brown lemmings in different parts of the North American tundra are available at a range of sites varying from extremely productive coastal plain areas near Point Barrow, Alaska (Pitelka and Batzli 2007), to much less productive tundra sites across the Canadian Arctic (Reid et al. 1995; Gruyer et al. 2010), providing estimates of maximum density over a range of two orders of magnitude. George Batzli (pers. comm. 2009) suggested that brown lemming maximum density would map directly onto the abundance of the grass *Dupontia fisheri* in the habitat. This grass is abundant in the coastal plain of northern Alaska, where brown lemmings reach peaks of more than 100/ha, and much less abundant on Herschel Island, where brown lemmings reach peaks of 7–8/ha (Krebs, unpublished). Adequate data are not available to test this interesting idea, and no one has directly addressed the general issue of what sets peak densities of microtine rodents.

Are Small Rodent Maximum Densities Lower in Predator-rich Habitats?

One alternative to the bottom-up model is the top-down control of maximum small rodent densities via predation and disease. Hypothesis 5 suggests that the maximum density of rodent populations is inversely related to the abundance of their predators and diseases. This implies a limitation of maximum rodent density by predation or disease and can be tested directly by removing or reducing predator abundance or reducing disease incidence at the population peak (Korpimäki et al. 2002).

The problem at present is that, while we have some experimental studies on the impact of predators on small rodent numbers (an issue we will discuss in chapter 9), we do not have the biogeographic data from a wide scale to test whether the abundance of rodent predators—including their functional and numerical responses—maps onto variation in maximum rodent densities. Again we are left in the unsatisfactory condition of having an interesting hypothesis but inadequate data to test it.

Conclusion

The conventional wisdom that small rodent populations fluctuate more strongly in northern ecosystems is correct for *Myodes* (*Clethrionomys*) species, but does not very well fit data from *Microtus* species. The reason for this difference is not clear. It could be associated with ecosystem complexity or landscape connectivity, as these interact with the effect of seasonality on *Microtus* populations (Tkadlec and Stenseth 2000). Seasonality must work through temperature and rainfall, and for voles its main effect should be on the length of the breeding season. More detailed field studies are needed to explore the validity of theoretical analyses of how seasonality may affect population fluctuations (Grenfell and Finkenstadt 1998; Stenseth et al. 1998).

We have almost no long-term data on population fluctuations of rodents in tropical environments, so it is impossible at present to extend any type of gradient analysis to tropical areas. A further shortage of data on primary productivity and, in particular, the relative abundance and productivity of food plants severely limits our ability to test bottom-up models of rodent abundance over broad gradients. We know that populations of the same rodent species have greatly different maximum densities in different ecosystems, but so far we can only guess what the key variables are to explain these differences.

The bottom line is that geographical ecology of rodent populations has come a long way in acquiring long-term data sets, particularly in Europe; but for much of the world, and even for North America, long-term quantitative data are in short supply. To try to test broad-scale hypotheses of the geographical ecology of rodents is still a challenge for researchers. If there is an agenda for rodent research in this century, it is to produce high-quality quantitative data over a long term for a variety of sites in temperate, polar, and tropical habitats.

CHAPTER THREE

Reproductive Rates in Fluctuating Populations

Key Points:

- Population fluctuations are partly driven by changes in reproductive output in strongly fluctuating populations.
- Age at sexual maturity is partly under social control and partly under seasonal control. It contributes strongly to changes in annual reproductive output.
- Litter size changes with age of females and season, but these changes contribute little to the contribution of reproductive rate to population growth rate.
- Length of the breeding season is the major determinant of annual reproductive output in rodents, and it may involve the breeding season extending into autumn and winter months.
- Hypotheses to explain changes in population growth rates for fluctuating populations of rodents must be able to explain the causes behind these shifts in reproductive output.

In this chapter I discuss the role that change in reproductive rates plays in determining population growth rates in a variety of small rodent species. The main focus is on female reproduction, under the global assumption that male fecundity is never by itself a factor limiting reproductive output in rodents.

Rodents are well known for their high rates of reproduction. Figure 3.1 shows the relationship between the maximum rate of population growth and body mass for rodents. Population growth is a joint function of fertility and mortality, but the maximum rate of growth (r_m) is calculated for nearly ideal conditions when losses to mortality are expected to be relatively low. A vole population with an adult weight of 30 g would be predicted to increase fivefold over a year, a 45 g vole about four times, and an 80 g lemming about three times. These rates of increase are relatively low, given that a typical rodent female in the breeding season can have a litter of five to seven every three weeks, which over a breeding season of six months (eight litters) would produce 24 female offspring on average, far more than a four- to fivefold increase over one year. One way to look at the reproductive component of the potential rate of population growth is to quantify litter size in relation to body mass in small rodents. Figure 3.2 shows this regression for the microtine rodents in the database compiled by Ernest (2003). There is a significant positive regression but the spread is large, and over the range of many microtine rodents of 0–100 g body mass, the regression is not significant because of all the spread.

Life history traits of many rodent species have been tallied by Ernest (2003), and an analysis of these data for microtine rodents shows no further

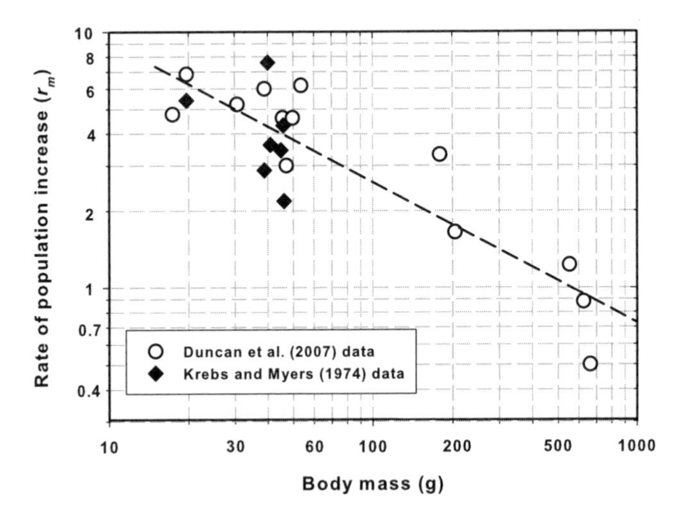

FIGURE 3.1 The relationship between the maximum rate of population growth (r_m) and body mass for 21 species of rodents weighing less than 1,100 grams. The regression line is $\log(r_m) = 1.5126–0.5503 (\log[\text{body mass}])$, where $r^2 = 0.80$, mass is in grams, and r_m is the instantaneous rate of population increase per year. Data from Duncan et al. 2007, and Krebs and Myers 1974.

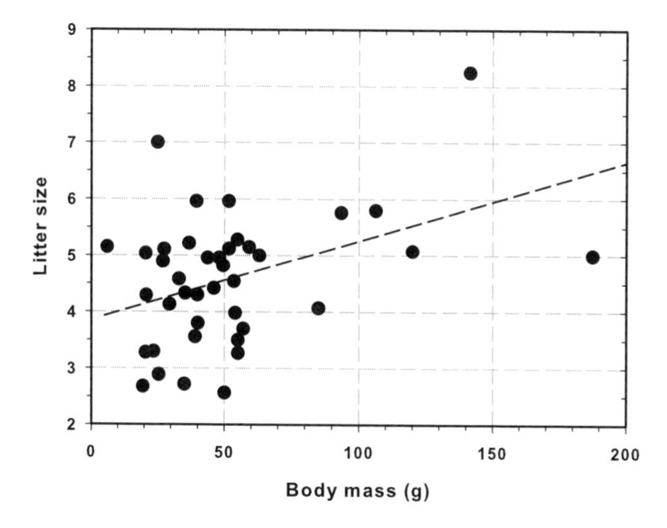

FIGURE 3.2 The relationship of average litter size to average female body mass in 40 species of microtine rodents. The regression line is litter size = 3.8551 + 0.140 body mass in grams; $r^2 = 0.18$, $p = 0.009$. No transformation of either axis improved the fit of this regression. Data from Ernest 2003.

significant correlations between body mass and reproductive indicators. For example, age at sexual maturity is not related to body size in microtines. The message is that these averages in reproductive performance are hiding many short-term fluctuations in reproduction, and we turn now to looking at these effects.

Methodological Issues

Reproductive performance in any small mammal population can be broken down into a series of parameters and measured separately to determine their contribution to total population output. Figure 3.3 shows the components of reproduction in small mammals. The key to understanding reproductive output lies in two directions. First, we need a precise estimate of the numbers of breeding females in the population, and second we need to measure the four components of reproductive output shown in figure 3.3: age at sexual maturity, litter size, pregnancy rate, and length of the breeding season.

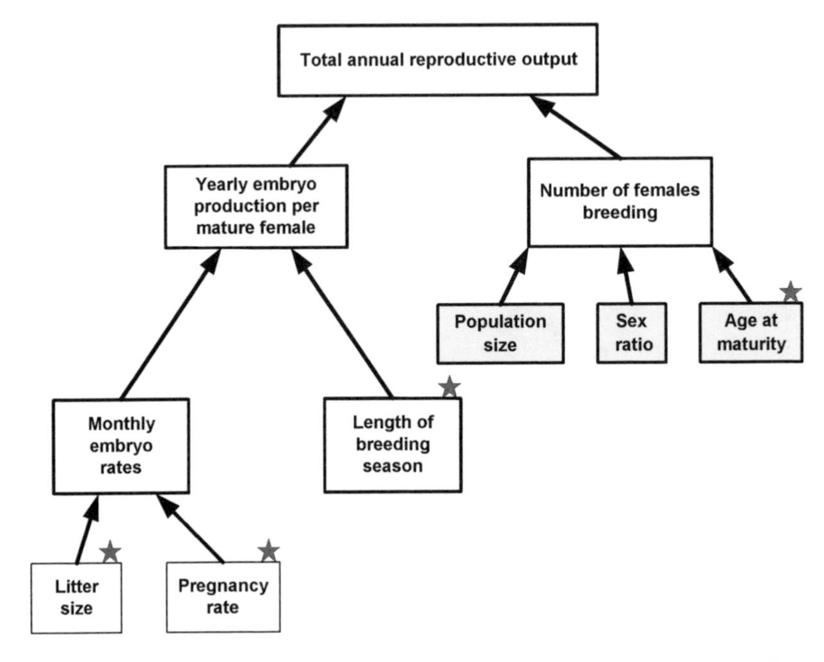

FIGURE 3.3 The components of reproduction in polyoestrous small mammals. The four components marked with stars comprise the key reproductive parameters discussed in this chapter.

Age at sexual maturity has been identified as the most important component affecting population growth rates for populations like those of small mammals that are characterized by early maturity and high reproductive rates (Oli and Dobson 2003). Models of population cycles can be driven most simply by phase-specific changes in the age at maturity and juvenile survival (Oli and Dobson 1999). These theoretical analyses of life history traits all point to the importance of estimating for small rodents changes in the age at sexual maturity.

For kill samples, age at maturity is in principle easy to obtain from eye lens weight (Hoffmeister and Getz 1968; Lidicker and MacLean 1969; Hagen et al. 1980). In general the eye lens grows rapidly during the first one to two months of age, and this is often the period in which small rodents reach sexual maturity. More difficulty ensues for individuals that do not reach maturity during their first summer of life: for older animals the determination of age from lens weight is much more difficult because lens growth slows to the extent that distinguishing eight-month old individuals from those 20 months old becomes impossible (Hoffmeister and Getz 1968). In *Myodes* (*Clethrionomys*) species with rooted molars, it is possible to age individuals by their molar root development (Abe 1976).

But in live-trap samples, age is not easy to estimate, and in practice a substitute variable—size at sexual maturity—is used in place of age. If individuals can be marked in the nest, direct observations of age can be made. Alternatively, in a detailed live trapping program in which juveniles are caught just after weaning, age can be estimated with relative precision if one knows the weaning period.

Weight at sexual maturity can be conveniently summarized by a logistic regression, first described by Leslie et al. (1945). Figure 3.4 illustrates an example for *Microtus pennsylvanicus* females from southern Manitoba. Leslie's approach is limited by the requirement of relatively large sample sizes to fit the regression cleanly; and because weight at maturity changes seasonally, the method can be applied most easily to relatively dense populations with large numbers of captures. Ergon et al. (2009) describe a

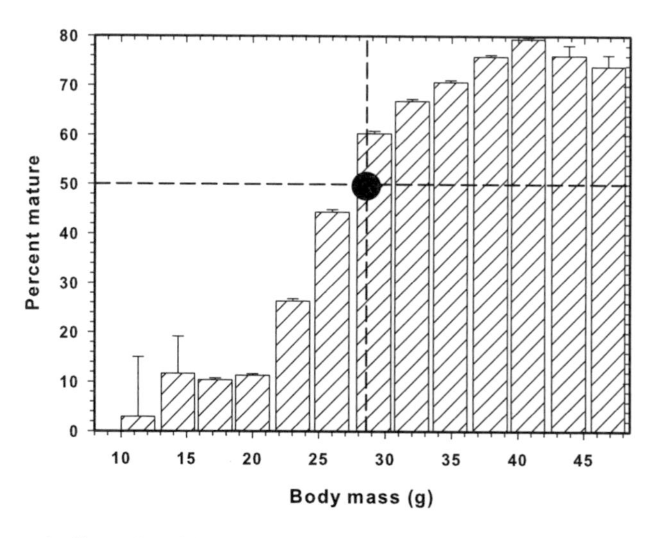

FIGURE 3.4 An illustration of the calculation of the median body mass at sexual maturity for female *Microtus pennsylvanicus* in Manitoba for 6,925 voles live-trapped from 1968 to 1979. The logistic regression for these data estimates the median body mass at sexual maturity to be 28.58 grams, with 95% confidence limits of 28.21 to 28.95 grams. Data courtesy of Steve Mihok (pers. comm.).

more elegant method of estimating the time of maturation from mark-recapture data on individuals. Their approach has the additional advantage of enabling one to estimate the survival costs of reproduction.

Litter size is perhaps the simplest of the reproductive parameters to determine, and there is a considerable body of data on this component (Keller 1985). Embryos can be counted in utero during necropsies, and depending on the stage of pregnancy at which the counts are taken and the rate of post-implantation loss, these counts are a good estimate of the litter size at birth. Alternatively, if nests can be found, litter size at birth can be determined directly (Morris 1989; Lambin 1994a).

Pregnancy rate is the most difficult of the four reproductive parameters to estimate, although in principle it seems simple. One has to measure the size of the pool of mature adult females in the population, and estimate the fraction of these that are pregnant at any instant in time. The crux of the problem is in determining the size of the pool of potentially mature adults. To do so, one needs to know the adult's size or age at maturity, and if this is used as a constant (for example, if all females over 25 g are seen as potentially mature), spurious results may be presented. Authors do not always state how they have determined the pool of potentially mature females, and this can make it difficult for one to compare the results from several papers. Necropsy samples for pregnant females will fail to detect visible swellings in the uterus for about four to five days of a 20- to 21-day gestation period, and consequently the maximum observed pregnancy rate will be about 75% even when 100% of females are pregnant. For live trapping studies, one needs to decide what physical characteristics will define a pregnant female. If this decision is based on body mass, it will mean that many females in early pregnancy will fail to be detected as pregnant, giving even more bias to an estimate of pregnancy rate. Consequently the best data must come from kill samples of females.

The length of the breeding season is measured from the first pregnancies in the spring (or start of the favorable season) to the last birth in the autumn (or end of the favorable season). While large changes in length of the breeding season are easy to pick up, subtle changes are much more difficult because of sample size problems. Individual variations in the start and stop of breeding may result in a few females beginning to breed long before the bulk of the population, or breeding long after most other females have stopped in the autumn. Again, one may have to make operational definitions such as "I will define the start of the breeding season as that time when 20% of adult females are breeding," or some similar measure. Because the start of breeding is often synchronous in small rodents, it is usually much easier to determine the start of breeding than its cessation.

The bottom line is that assessing the components of reproduction for polyoestrous mammals is not as simple as one might assume, and it is important to specify exactly which criteria are being used to define categories of reproduction.

Hypotheses Regarding Reproductive Changes

The central question we ask is: What role do changes in reproduction play in causing changes in rodent populations? We need to proceed in three steps to answer this question. First, we need to determine which, if any, of the four components of reproduction change with the population growth rate. If a component does not change with the density and population growth rate, it cannot be causing density changes. Second, once we have determined the important components, we must then ask whether they are quantitatively strong enough to affect population growth. To answer this question we must construct demographic models. It is quite possible for a reproductive component to change with population density but to have such a small effect on overall growth rate that it can be effectively ignored as a driver of density changes. Third, we need to determine which ecological factors cause the reproductive changes, and how they operate. We begin with the first step, to find out which components change with density and growth rate. To define more clearly the issues involved for rodents, I highlight four hypotheses.

Hypothesis 1. Rates of population growth are determined solely by changes in reproductive output in rodents. This hypothesis has two alternative hypotheses:

- *Alternative hypothesis 1A*: Rates of population growth are determined solely by changes in mortality rates.
- *Alternative hypothesis 1B*: Rates of population growth are determined jointly by changes in reproduction and mortality.

If reproductive rates are not associated with changes in density, we can ignore this demographic variable and concentrate on mortality. Note that this conclusion should be interpreted as meaning not that reproductive studies are not needed, but only that they are not necessary for understanding population dynamics. We know already that alternative hypothesis 1A is not correct (Oli and Dobson 1999), so we can proceed to discuss the contribution made by each of the four reproductive parameters.

Hypothesis 2. *The age at sexual maturity is inversely related to the rate of population increase in fluctuating populations.* This idea follows from early work by Cole (1954) on life history theory. The alternative hypothesis is that this parameter is not affected by the rate of population change.

Hypothesis 3. *The length of the breeding season is positively related to the rate of population growth in fluctuating populations.* Smith et al. (2006) have shown that this hypothesis could be significant for rodents, particularly if there is a time lag in the dynamic. The alternative hypothesis is that this parameter is a minor contributor to population changes.

Hypothesis 4. *Changes in litter size and pregnancy rate are of minor importance in driving the population growth rate in rodents.* These variables may change seasonally, and as such are part of seasonal patterns, but their demographic impact is minor, as was argued by Norrdahl and Korpimäki (2002) and by Keller and Krebs (1970). The alternative hypothesis is that these reproductive components are key drivers.

These hypotheses could be stated in a different manner, but the key point is that we need to test them by the comparative method of determining how changes in population growth rates associate or do not associate with each of the four reproductive parameters. We cannot test hypothesis 1 against 1B until the next chapter, when we consider the role of mortality factors in population fluctuations, so we will proceed on the assumption that either 1 or 1B is correct.

Does the Age at Sexual Maturity Change with Population Growth Rates?

In a few cases the observed changes in age at sexual maturity are so large that reproductive output is curtailed, and by inference population growth is lost. The best examples are from populations in which juvenile females mature or fail to mature in the year of their birth. The classic studies were by Kalela (1957), on *Myodes* (*Clethrionomys*) *rufocanus* in Finland, and Koshkina (1965), on *Myodes* (*Clethrionomys*) *rutilus* in the Soviet Union. Figure 3.5 illustrates how the proportion of juvenile females reaching maturity in their first summer of life relates to the subsequent population

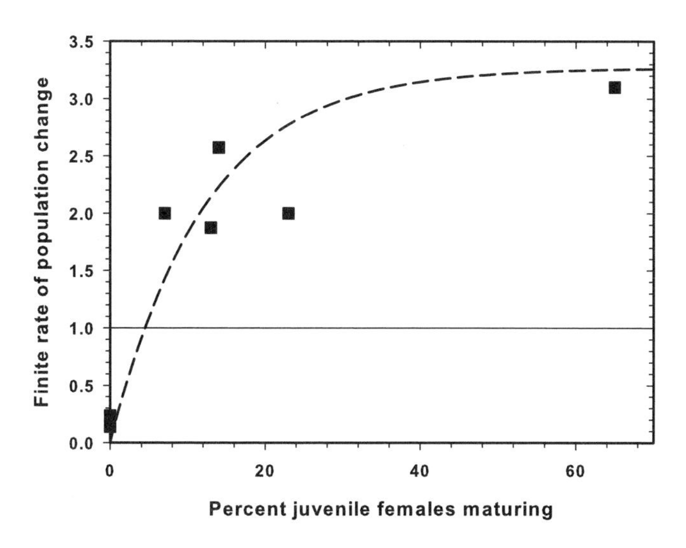

FIGURE 3.5 The percentage of female juvenile *Myodes* (*Clethrionomys*) *rutilus* maturing in the summer$_t$ of their birth in relation to the annual finite rate of population change (N_{t+1}/N_t) of the subsequent year, 1958–64, at Salair taiga in the Soviet Union. The horizontal line separates increasing populations above the line from decreasing populations below the line. Data from Koshkina 1965.

growth rate. This pattern described so well by Kalela (1957) has been observed in the brown lemming and collared lemming by Krebs (1964) in the Canadian Arctic, by Bujalska (1970, 1973) in *Myodes glareolus* in Poland, and by Gilbert et al. (1986) in *Myodes rutilus* in the Yukon. Bujalska (1973) and Gilbert et al. (1986) both showed by removal experiments that it was the presence of adult breeding females that restricted maturation in young-of-the-year voles. A series of elegant laboratory experiments by Drickamer (1984) have shown that urinary chemo-signals from adult mice are the mechanism by which reproduction is suppressed in juveniles.

Data on *Microtus* species are far less clear, because of the necessity of using weight at maturity as a proxy for age at maturity. Nevertheless, the same patterns are clear when detailed data are available. Figure 3.6 shows the median mass at sexual maturity for female *Microtus pennsylvanicus* from two live-trapping grids in southern Ontario in relation to the rate of increase of the population. Clearly, when population densities are increasing rapidly, voles mature at a smaller size and presumably at lower ages than when populations are decreasing. A similar pattern was described by Keller and Krebs (1970) for kill-trapped populations of *M. pennsylvanicus* and *M. ochrogaster* in southern Indiana in which the size at maturity of

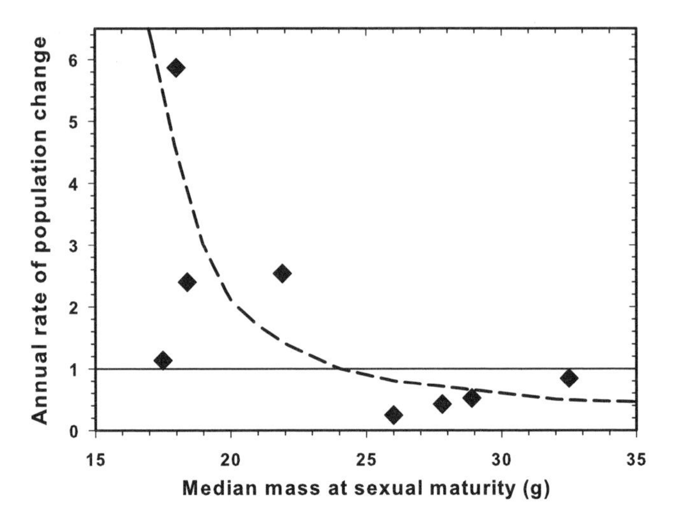

FIGURE 3.6 The median mass at sexual maturity for female *Microtus pennsylvanicus* in relation to the annual population growth rate (λ), southern Ontario, 1978–82. The horizontal line separates populations increasing (above the line) from those declining. The dashed line is a descriptive spline curve. Data courtesy of Rudy Boonstra.

females was larger in high-density, relatively stable populations than in low-density, rapidly increasing populations the following year. I have not been able to find any data that is at variance with the conclusion that age and size at sexual maturity in voles is closely associated with population density changes so that increasing populations have females that mature at younger ages and smaller body sizes than in stable or declining populations. Consequently I accept hypothesis 2 as correct.

Does the Length of the Breeding Season Change with Population Growth Rates?

In an ideal world, a population could increase its reproductive output by extending the breeding season with no change in litter size, age at maturity, or pregnancy rate. Many studies have found that the length of the breeding season is highly variable from year to year in microtine rodents. Hamilton (1937) suggested that a longer breeding season was one of the main causes of an increase in population size of *Microtus pennsylvanicus* in New York state. Norrdahl and Korpimäki (2002) found in bank

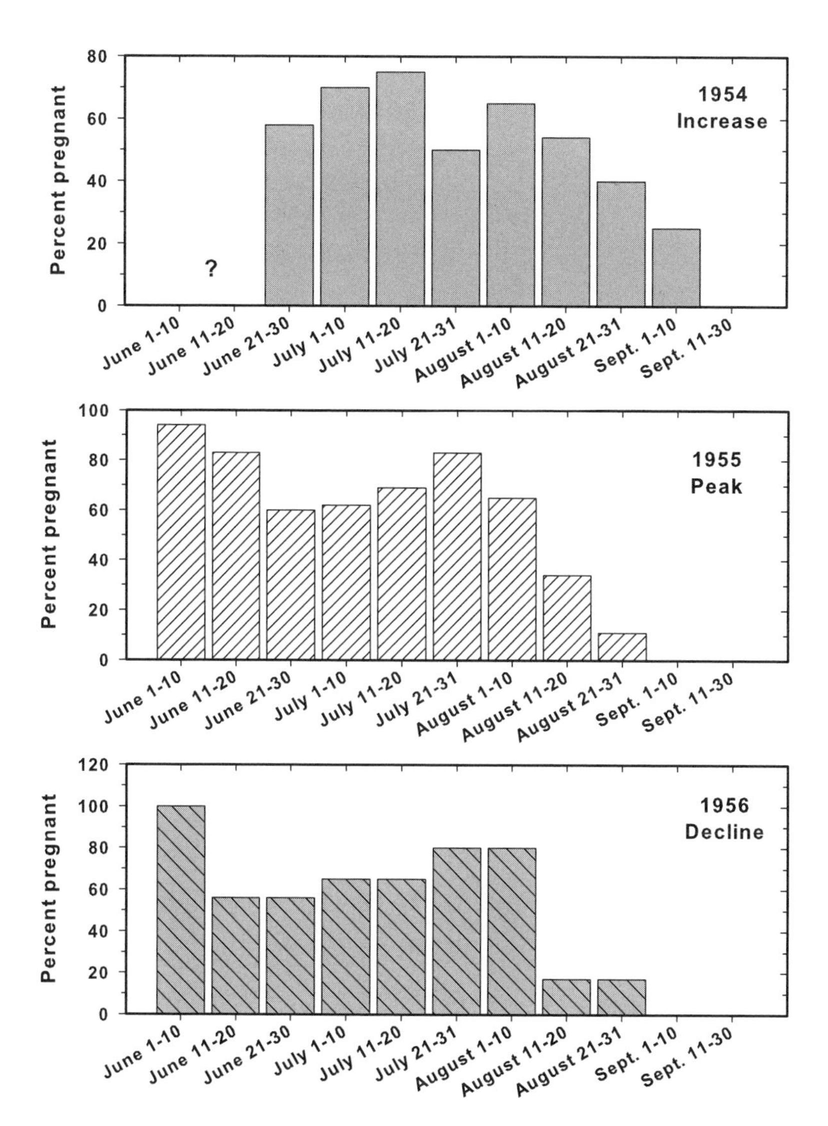

FIGURE 3.7 The prevalence of pregnancy in the grey-sided vole *Myodes* (*Clethrionomys*) *rufocanus* in Finnish Lapland, summers of 1954 to 1956. During the year of low density and population increase (1954), breeding began earlier in the spring and continued later into the autumn. Data from Kalela 1957, table 5.

voles that the summer breeding period began later in years of population decline.

Winter breeding is a characteristic of some species of microtine rodents, with the most spectacular example being lemmings, which breed under the snow (Hansson 1984). Winter breeding occurs in some winters but not all, and this raises the question of what controls the length of the breeding period—an issue we take up in chapter 11. Some species never breed during the winter months, and the important changes for these species could be when breeding begins in spring and when it ends in autumn. The most detailed data come from *Myodes* (*Clethrionomys*) *rufocanus* in northern Finland (figure 3.7). On the basis of the juveniles' appearance, Kalela (1957) states that breeding began earlier in 1954 than in 1955 or 1956. The end of the summer breeding period is clearly indicated in figure 3.7, showing that it had ended already in late summer in 1955 and 1956 when the population was at a peak or declining. Exactly the same sequence of changes in the length of the breeding period was observed in the brown and collared lemming by Krebs (1964). The conclusion is that the length of the breeding season correlates well with the rate of population growth so that increasing populations have longer breeding seasons. Because the breeding season remains short at least for one season after a population decline, the correlation is apparently less with immediate population density than with population growth rate.

The central question of what factors determine the start of the breeding season is still poorly understood (Ergon 2007). The assumption has always been that for rodents in colder climates, the start of breeding will be dictated by environmental changes in the timing of snow melt and subsequent vegetation growth. But in more temperate climates, snow will not be a factor, yet general climatic conditions will determine the start of vegetative growth. The tradeoffs that individual females must make between starting too early or too late are affected by their lack of perfect knowledge of future environmental conditions. Whatever the causes of the large variation in the timing of the start of the breeding season, that timing is an important determinant of subsequent population growth.

Does the Pregnancy Rate Change with Population Growth Rates?

The third parameter that defines reproductive output is the percentage of mature females that are pregnant throughout the breeding period.

Three caveats must be taken into account when estimating pregnancy rate. First, the rate must be calculated only for females that are mature, which typically means all females above a given body weight. If size at maturity changes seasonally, a constant expected weight at maturity cannot be used. Second, the pregnancy rate typically ramps up at the start of the breeding season and ramps down at the end, so that valid estimates should be made only for the central core of the breeding season, omitting the starting and ending phases (e.g., Nelson et al. 1991). Third, litter size can vary with parity of the female (e.g., Tkadlec and Krejcova 2001), and one needs to take into account this source of variation in the statistical analysis of litter size.

Kalela (1957) completed one of the earliest analyses of changes in pregnancy rate in a fluctuating population of *Myodes* (*Clethrionomys*) *rufocanus* in northern Finland. Figure 3.7 shows that during the main part of the summer breeding period there was no clear effect of population growth rate on the proportion of females that were pregnant. Basically all adult females were pregnant all the time during the main breeding period. Keller and Krebs (1970) found the same pattern of essentially equal pregnancy rates during the main breeding period for three years in fluctuating populations of *Microtus pennsylvanicus* and *M. ochrogaster*. They found

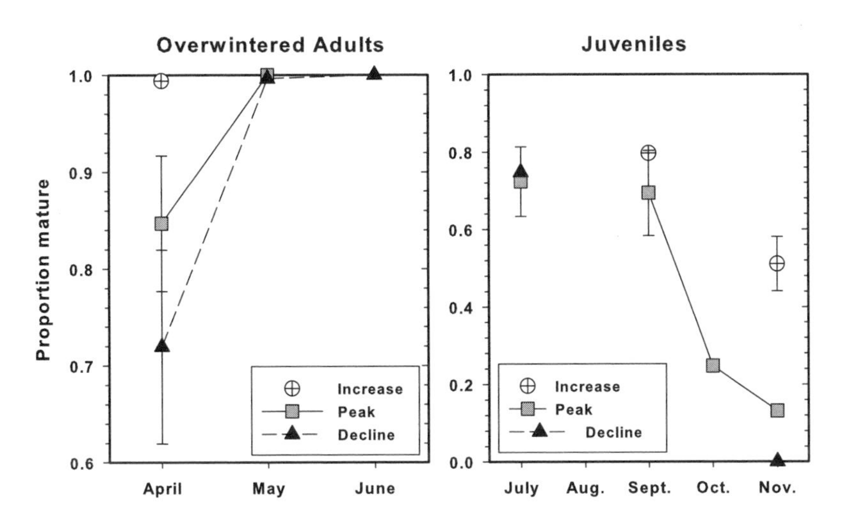

FIGURE 3.8 The proportion of female field voles (*Microtus agrestis*) classified as being in breeding condition during the three phases of the population cycle in southern Finland, 1984–92. Data from Norrdahl and Korpimäki 2002.

more evidence of winter breeding during the phase of population growth than in the decline phase.

Norrdahl and Korpimäki (2002) reported on changes in reproduction in an extensive study of cyclic populations of two *Microtus* species and the bank vole from 1984 to 1992, covering three vole cycles in southern Finland. Figure 3.8 shows their data for the proportion of females that were mature in the three phases of the cycle for the field vole *Microtus agrestis*. During the height of the breeding season, they could detect no differences between the proportion of females that are mature in different phases of the population cycle. There were differences in the length of the breeding season, shown in figure 3.8. Breeding started earlier in the spring of the increase phase of the cycle, and extended longer into the autumn for juveniles in the increase phase. They found winter breeding in the increase phase for both the field vole and the sibling vole (*Microtus rossiaemeridionalis*), but bank voles were never observed breeding in winter. The conclusion appears to be that voles and lemmings breed at a maximal rate during the height of the breeding period, and that this does not depend on either density or the population's rate of increase.

Does Litter Size Change with Population Growth Rates?

One way to increase reproductive output would be to increase litter size. Life history theory predicts that the age-specific pattern of reproduction should be triangular, so that for rodents litter size should progressively increase with age and size, and then decrease again in very old animals. This pattern is far from clear in microtine rodents (Tkadlec and Krejcova 2001), perhaps because of confounding effects. Litter size is potentially affected by three variables: body size, parity, and season of the year (figure 3.9). Consequently, if one wishes to determine whether population density or population growth rate is correlated with litter size, one must correct the raw data counts for these three variables (Hasler 1975). The basic problem in doing so was well stated by Pelikan (1979): one needs large sample sizes for these comparisons. Pelikan (1979) suggested that for comparisons of monthly means, one needs about 40 pregnant females to attain a precision (95% confidence limit) of 0.5 embryos. If a comparison is to be made between years, concentrating only on the main months of breeding (e.g., May to August), Pelikan recommends a sample size of 60 females. These statistical requirements are a function of the innate variability of

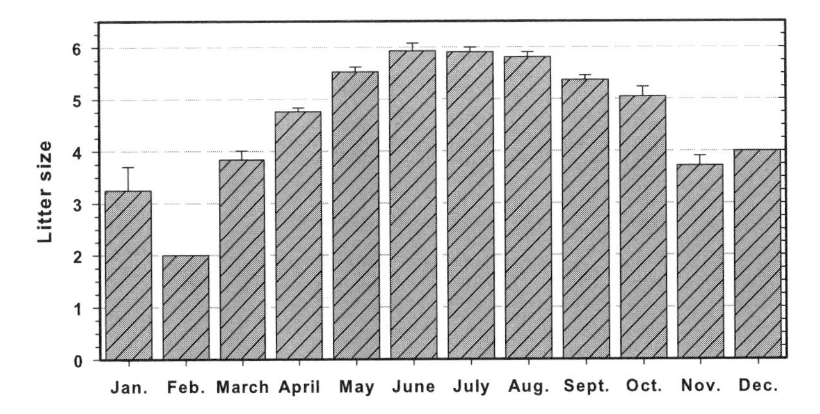

FIGURE 3.9 The seasonal change in litter size (± S.E.) in the common vole *Microtus arvalis* from Czechoslovakia. Sample sizes were small in the winter months. Litter sizes ranged from 1 to 11 during the summer months. Data from Pelikan 1979.

litter size. Pelikan (1979) reported an average standard deviation of litter size of 1.5 for *Microtus arvalis* (CV = 0.29). Anderson and Boonstra (1979) reported an average standard deviation of 1.0 to 1.2 for *M. townsendii* (CV = 0.20–0.25). Many studies of litter size in rodents have been unable to meet this sample size requirement; thus they risk making type 2 errors in their analysis, because of low statistical power.

Litter size changes with body mass, as illustrated in figure 3.10 for *Microtus townsendii*, but typically the variance about regression is large. Nevertheless, if no correction is made for differences in body mass of females, there is a risk of concluding that litter size has changed when in fact only the weight distribution has changed.

There is much less agreement about whether primiparous females (having their first litter) differ in litter size from multiparous females (having their second or third litters), assuming that both groups of females are of equivalent mass. Anderson and Boonstra (1979) found no significant difference between the litter sizes of primiparous and multiparous *M. townsendii* females. Keller and Krebs (1970) found that litter sizes of primiparous females were almost always lower than those of multiparous females, but these differences were usually confounded by differences in body size. In laboratory studies of the common vole *Microtus arvalis*, Tkadlec and Krejcova (2001) found a reverse pattern, in that the youngest and smallest females had larger first litters than did older and larger females. Norrdahl and Korpimäki (2002) found seasonal changes in litter

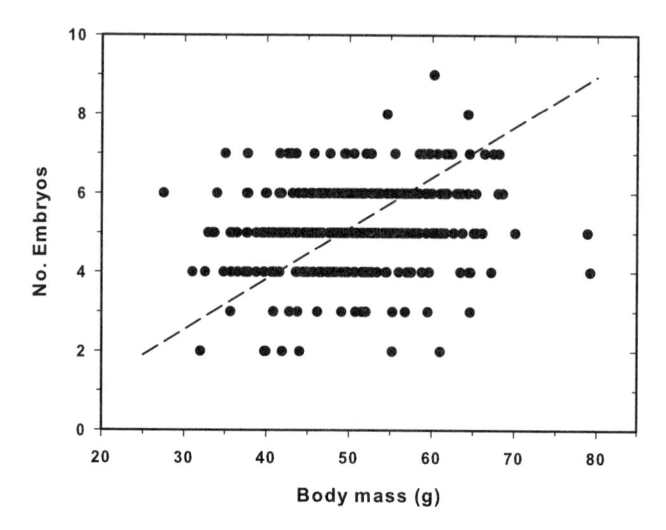

FIGURE 3.10 The relationship between body size and embryo counts in the vole *Microtus townsendii* in southern British Columbia. The functional regression is number of embryos = −1.302 + 0.1282 (body mass), r = 0.19, n = 353; body mass is mass corrected for total uterus weight including embryos. Data from Anderson and Boonstra 1979.

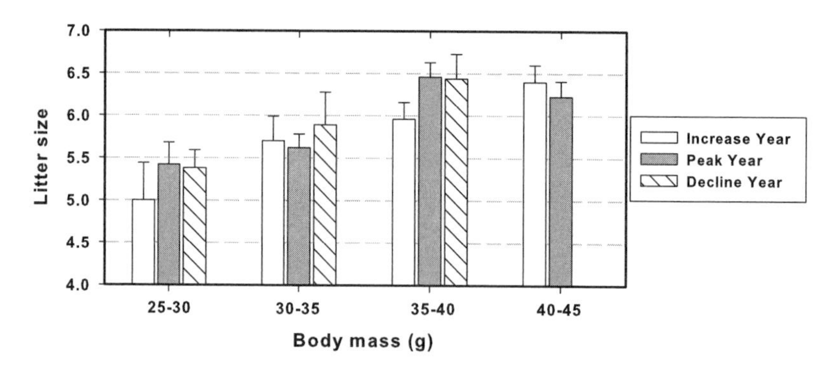

FIGURE 3.11 Changes in litter size in the grey-sided vole *Myodes* (*Clethrionomys*) *rufocanus* in Finnish Lapland, in relation to body size and phase of the cycle. Data from Kalela 1957.

size in *Microtus agrestis* but no significant changes in litter size in relation to the population cycle. Hasler (1975) concluded in his review of reproduction in microtine rodents that the first litter is significantly smaller than the second litter, but that age, size, and season often confounded comparisons of primiparous and multiparous litter sizes in field data.

The general conclusion is that litter sizes change dramatically with body size, season, and probably parity, but do not appear to change with population density or population rate of increase. This is illustrated in figure 3.11, from the classic paper by Kalela (1957) on the gray-sided vole in Finnish Lapland.

Synthesis and Modeling

To measure the impact of changes in the four reproductive parameters of figure 3.3, we need to construct a demographic model. Simple age- or stage-structured models are one approach, pioneered by Leslie (1945) and explored in detail by Caswell (2001). Figure 3.12 contains a simple life-cycle graph that has been used by Oli and Dobson (1999) and by Norrdahl and Korpimäki (2002) to explore the impact of reproductive changes on the rate of population growth in microtines.

Norrdahl and Korpimäki (2002) presented the summary of this model in figure 3.13 for their analysis of the demography of *Microtus agrestis*, *Microtus rossiaemeridionalis*, and *Myodes* (*Clethrionomys*) *glareolus* in southern Finland. The reduction in reproductive output clearly reduces the population's potential rate of increase in the peak and decline phases of the population cycle. The main reproductive effects causing this shift are changes in the age (size) at sexual maturity and changes in the length of the breeding season. Norrdahl and Korpimäki (2002) point out that changes in mortality rate are of greater significance in the model seen in figure 3.13, an issue we will take up in the next chapter.

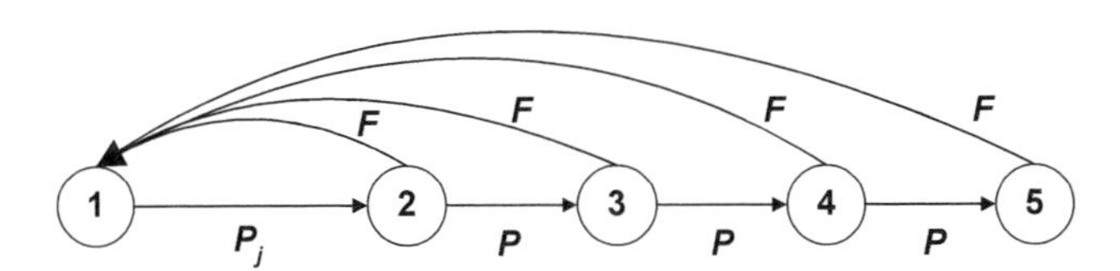

FIGURE 3.12 Schematic of the partial life-cycle model for exploring changes in the rate of population growth as a function of changes in mortality and reproduction. Individuals in age class 1 are newly born young, and they survive to age class 2 at maturity with probability P_j. per unit of time. Individuals in the subsequent age classes all survive at the adult survival rate P per unit of time, and their fecundity is symbolized by F offspring per unit of time. Survival P and fecundity F do not need to be constants, but can be adjusted for different age classes to make the model more complex and to explore how changes in life-cycle parameters affect the population growth rate. After Caswell 2001.

A similar analysis was done by Oli and Dobson (1999), using estimates for *Microtus pennsylvanicus* from data collected by Boonstra and Rodd (1983). They showed that one could generate model population fluctuations similar to those observed in natural populations by varying the age at sexual maturity and juvenile mortality rates, while holding all other reproductive parameters and adult survival constant. These results are complimentary to those of Norrdahl and Korpimäki (2002).

By contrast, Smith et al. (2006) showed with another model that delayed density-dependent changes in the length of the breeding season could generate population cycles. This model was based on the observations of Ergon (2003) that for *Microtus agrestis* in northern England, the start of the breeding season in year *t* was correlated with the density of the vole population in the spring of year *t*–1 (figure 3.14). Since any delayed-density-dependent factor can by itself generate population cycles, this is an important observation—and it focuses attention on the common

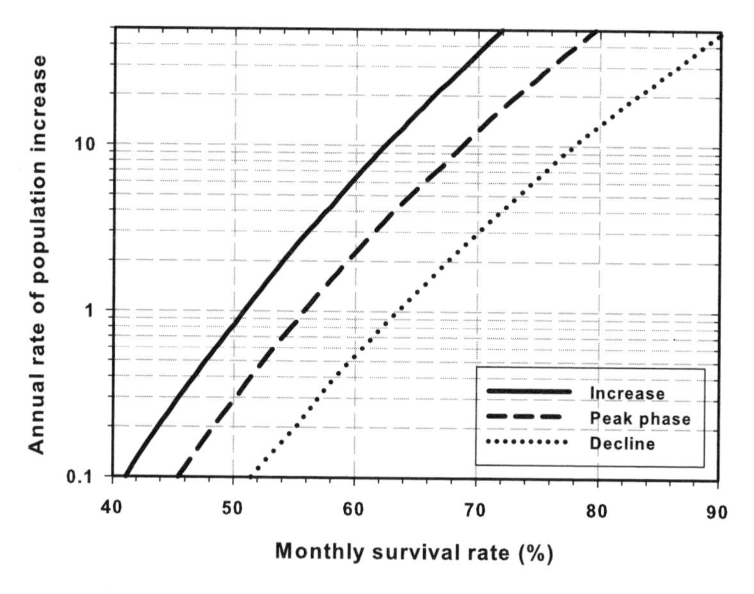

FIGURE 3.13 Changes in the annual rate of population growth (λ) as a function of changes in reproduction and in monthly adult survival rates for the vole model of Norrdahl and Korpimäki (2002), who used the measured values of the reproductive rate parameters for the three phases of the population cycle in a simple age-structured model to simulate these results. The length of the summer breeding season and the age at sexual maturity were the main variables changing the rate of population growth in this model, but changes in the mortality rate had a greater effect on λ.

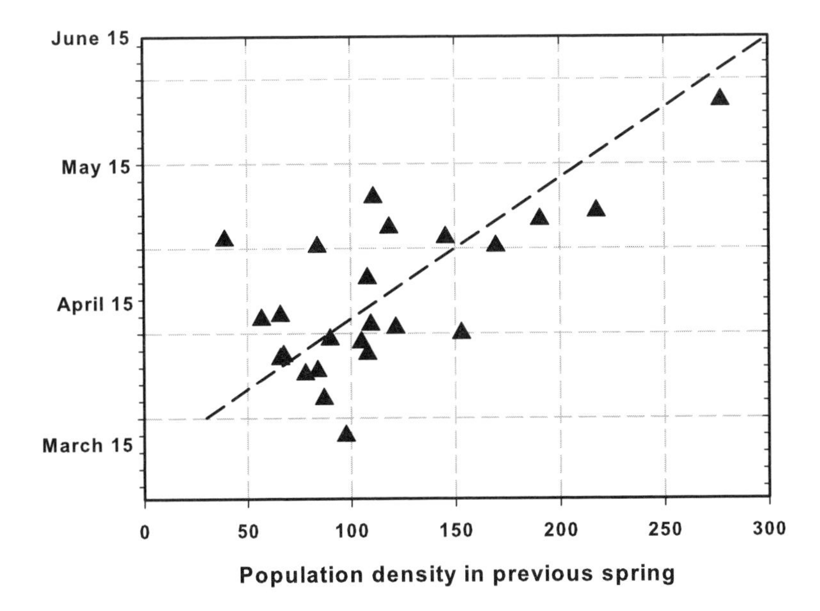

FIGURE 3.14 Relationship of the timing of the start of the breeding season for *Microtus agrestis* at Kielder Forest in northern England to the density of the population the previous spring. The start of breeding was defined as that date when 50% of the female voles had produced their first litter. For the functional regression, $n = 24$ and $r = 0.65$. This delayed density-dependent relationship by itself could produce a population cycle (Smith et al. 2006). Data courtesy of T. Ergon.

observation that the onset of breeding is often delayed in the year after a peak population, and that the length of the breeding season is also often shortened in that year (Krebs and Myers 1974; Boonstra et al. 1998). The question of what factors determine the start and cessation of breeding in rodents is still an important issue for further research.

Reproductive Rates in Declining Populations

The reproductive slowdown in lemmings and voles during the decline and low phases of the cycle is a product of lower fertility rates, smaller litters, lengthened time to maturation, and shortened breeding season. (Schaffer and Tamarin 1973; Krebs and Myers 1974; Erlinge et al. 2000). At first glance this behavior would appear to clash with the general idea from life history theory that when the death rate is high, individuals should reproduce at a maximal rate. Blachford (2011) has analyzed this conflict and shown that individuals are in fact maximizing fitness with these strategies. His analysis is as follows.

The period of time during a cycle in which reproductive rate is curtailed is longer than the lifespan of individuals, and is characterized by high adult mortality rates. Consequently the slowdown in reproduction is not a case of individuals delaying reproduction until a more opportune time; nor is it a case of individuals waiting, in a declining population, to have offspring later, when those offspring will comprise a larger fraction of the population (Fisher 1930; Ratcliff et al. 2009). Rather, it is a case of multiple cohorts of individuals aiming to produce fewer (or far fewer) offspring than they could.

The reproductive slowdown has thus far been assumed to be maladaptive, as is revealed by choices of phrase (e.g., "impairment" in Charbonnel et al. 2008, or "deterioration" in Sheriff et al. 2009), and in the reasoning that animals should invest in reproduction rather than survival if their chances of survival are very low (e.g., Norrdahl and Korpimäki 2000). A leading explanation of the reproductive slowdown has been that stress, while adaptive overall during the decline phase for reasons of individual survival (via energy mobilization and increased vigilance), has the inadvertent side effect of "impairing" reproduction (Boonstra et al. 1998; Wingfield and Sapolsky 2003). In contrast, Blachford (2011) showed that in these cycling systems the reproductive slowdown is adaptive.

Simulation of fitness metrics indicates that reproductive effort should be decreased at very high mortality rates (Blachford 2011). This result differs from the rule of thumb from classical life history theory that an individual should invest its resources in reproduction rather than survival if survival is unlikely. But that rule does not apply to declining populations, or to populations in which the ratio of juvenile to adult survival plummets (Charnov and Schaffer 1973). Blachford (2011) showed by simulation that slowdown of reproduction in a population with a high death rate is consistent with classical theory but occurs under conditions that previously attracted little attention. His simulation models also predicted stronger selection for smaller litter sizes when there is a premium on short-term survival. This can happen when numbers are so low that persistence is more important than rate of growth in numbers. Reproduction increases the probability of death in small rodents, so that in severe conditions less reproduction can be favored. The reproductive slowdown is known to extend into the low phase of the cycle after predator numbers have dropped substantially (Krebs 1964; Boonstra et al. 1998). Mortality rates can remain high enough to favor reproductive slowdown for some time into the low phase (see Blachford 2011). The bottom line is that cyclic rodent reproductive responses are adaptive at all phases of the cycle and do not conflict with life history theory.

Stress has been a leading explanation of the reproductive slowdown, because it is known to suppress reproduction (Boonstra et al. 1998; Wingfield and Sapolsky 2003). Stress is also known to take time to subside, and to have transgeneration effects. Both of these traits are consistent with slowdown lingering beyond the severe conditions of the peak and decline phases, and into the low-density phase. Blachford (2011) suggests that the reproduction-slowing effects of stress may well be adaptive in cyclic population systems and not merely an inadvertent side effect of physiology for short-term benefit to an individual's survival. The role of stress in population fluctuations is discussed in more detail in chapter 11.

Conclusion

Reproductive rates change dramatically when rodent populations are rising as compared with when they are falling, and these changes are seen most clearly in cyclic populations in which the phases of population growth and decline are clearly delineated. Only two of the four major components of reproductive output show substantial changes in relation to population density and population growth rate. The age or size at sexual maturity and the length of the breeding season are the key parameters of change. Litter size and pregnancy rate do not appear to change in relation to population changes, although they are significantly affected by body size and seasonal effects. The observed changes in reproductive effort over the cycle, formerly considered an exception to the predictions of life history theory, are now known to be adaptive. Clearly, reproductive changes are an important component of population change, but not the whole story. We turn now to consider the components of mortality.

CHAPTER FOUR

Mortality Rates in Fluctuating Populations

Key Points:

- Population fluctuations are partly driven by changes in mortality rates within strongly fluctuating populations.
- The loss of juveniles before weaning is very great, but there are few data on how these losses relate to population changes.
- As juveniles reach sexual maturity they often disperse, and these movements can be misinterpreted as mortality if the study area is small.
- The mortality rate of adult males is often of less significance than the mortality rate of adult females. High rates of loss of adult females can nearly eliminate effective reproduction because the females die before they can wean a litter.
- Changes in juvenile mortality are an important cause of changes in the population growth rates of small mammals, and we need to ask both why animals die and why they die at a particular age.

In this chapter I discuss the role that changes in mortality rates play in determining population growth rates in a variety of small rodent species. The main focus is on nestling losses, juvenile mortality after weaning, and adult mortality (figure 4.1). In addition, mortality of fertilized embryos can occur in utero, and we need to consider this component of early loss to get a complete picture of the life cycle of small rodents.

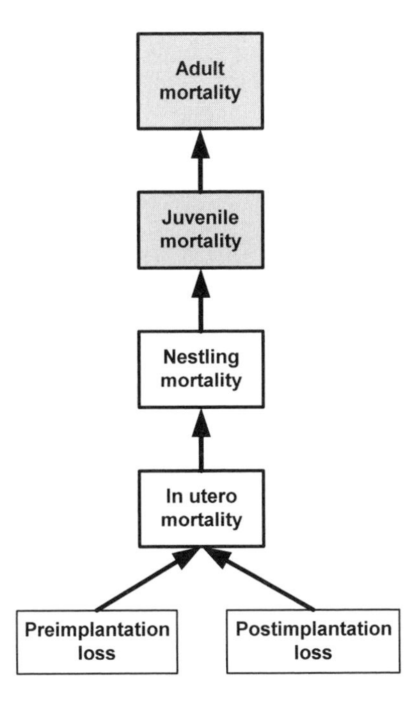

FIGURE 4.1 The components of mortality in small mammals. Recruitment indices measure the combined effect of the earliest four component mortalities in the life cycle.

The conventional view from life history analyses is that animals with high rates of reproduction are particularly susceptible to changes in the juvenile mortality rate and the rate of reproduction (Oli and Dobson 2003). Of second importance should be the mortality rate of breeding females, and relatively little importance should be given to the mortality of adult males, except in those few species that are monogamous.

Prenatal Mortality

Once a mammal releases ova and they are fertilized, processes of loss begin to operate. During gestation, two periods of loss can be operationally identified: preimplantation loss and postimplantation loss (Allen et al. 1947; Keller and Krebs 1970). A measure of these two components can be obtained by three counts:

$$\% \text{ preimplantation loss } = 100\left(\frac{\text{no. of corpora lutea} - \text{no. of implanted embryos}}{\text{no. of corpora lutea}}\right)$$

For example, if in a breeding female there were nine corpora lutea in the ovary upon autopsy and eight implanted embryos, the preimplantation loss for that individual would be 1/9, or 11%. Similarly, postimplantation losses arise from the discrepancy between the total number of implantations and the number of healthy implanted embryos:

$$\% \text{ postimplantation loss } = 100\left(\frac{\text{no. of failed embryos}}{\text{total no. of implanted embryos}}\right)$$

Failed embryos are difficult to detect in the early stages of embryo death (Brambell and Mills 1948); the main visual criterion is a difference in size between the implanted sites of healthy embryos (typically the majority) and the shrunken swellings of dead ones. If much loss occurs immediately after implantation, it may be impossible to separate these two components, so that one could calculate overall prenatal loss as

$$\% \text{ total prenatal loss } = 100\left(1 - \frac{\text{no. of viable embryos}}{\text{total no. of corpora lutea}}\right)$$

It is clear that these types of data are only approximate, since not all females that are sampled will be at full term, and some will not yet show visible pregnancy sites (as typically happens from five to seven days after fertilization). Some females may lose their entire litter before or after implantation, and some of these losses will not be detected as part of overall loss before birth (Keller and Krebs 1970). Some of the best data available are for the European rabbit *Oryctolagus cuniculus* (Adams 1960).

Table 4.1 gives some of the available estimates of prenatal losses for small rodents. These data must be taken with the qualifications given in detail by Beer et al. (1957), Keller (1985), and Loeb and Schwab (1987). In particular, postimplantation losses in some species increase toward the end of gestation, yet most of the samples on which the data in table 4.1 are based were gathered throughout the gestation period. It is possible that postimplantation losses in some species could be three times the values indicated in table 4.1, and more data are needed to confirm these estimates. As an overall figure for small rodents, one could assume a preimplantation

TABLE 4.1 **Estimates of preimplantation and postimplantation losses in small rodents**

Species	Preimplantation losses			Postimplantation losses			Reference
	Sample size	% females suffering loss	% ova lost	Sample size	% females suffering loss	% ova lost	
Microtus breweri	102		14.2	102		8.1	Tamarin 1977b
Microtus ochrogaster	148	19.7	6.6	151	14.0	7.0	Keller and Krebs 1970
Microtus ochrogaster	182		10.0	182		1.6	Rose and Gaines 1978
Microtus pennsylvanicus	152	30.3	9.0	159	14.5	6.4	Keller and Krebs 1970
Microtus pennsylvanicus	251		7.3	251		8.4	Beer et al. 1957
Microtus pennsylvanicus	57		6.4	57		2.8	Tamarin 1977b
Microtus californicus				82		7.2	Greenwald 1956
Microtus californicus				73		3.9	Hoffmann 1958
Microtus californicus						4.7	Lidicker 1973
Microtus montanus	110		5.9	110		3.5	Hoffmann 1958
Clethrionomys gapperi	107		7.7	107		1.3	Beer et al. 1957
Clethrionomys rufocanus[1]	439	36.0	12.9				Fujimaki 1981
Peromyscus maniculatus	288		8.5	288		2.9	Beer et al. 1957
Peromyscus maniculatus				1188		2.42	Loeb and Schwab 1987

[1] Fujimaki (1981) did not separate pre- and postimplantation loss, but suggests that most of the loss he observed was preimplantation loss.

loss of about 8.4% and a postimplantation loss of 4.6%, thus giving a total loss before birth of about 13% of the shed ova. In the worst case, this might rise to a 15 to 25% loss of shed ova.

The causes for prenatal losses are multiple and far from clear in rodents, a situation similar to the lack of understanding of these kinds of losses in humans (Jongbloet et al. 2007; Lummaa 2003). Genetic factors, seasonal effects, and physiological stress clearly underlie some of the observed levels of prenatal mortality. There is little evidence that the amount of prenatal loss is affected by population density or by the phase of the population cycle. Kalela (1957) found no change in prenatal losses in fluctuating *Clethrionomys rufocanus* populations in Finland. Hoffmann (1958) found a slightly increased amount of prenatal loss in declining *Microtus montanus* populations. Keller and Krebs (1970) found a constant level of prenatal loss that was independent of population density. Tamarin (1977b) could find no year-to-year variation in prenatal losses in *Microtus pennsylvanicus* or *M. breweri*. These data led Keller (1985) to conclude that as yet there is no evidence that prenatal losses are associated with changes in population density.

Nestling Mortality

Since mortality is a key component of population change, two questions are central to the discussion of nestling mortality: How much mortality occurs while young voles and lemmings are in the nest? And is this mortality rate associated with the rate of population change or the phase of the population cycle? For our purposes we will assume that nestling loss occurs in the interval of zero to 14 days of life. Some rodents will remain in the nest slightly longer than this, but exploratory movements typically occur at two weeks of age. Nestling mortality has two possible components. If the female is killed, all the young are lost. Alternatively, some or all of the young may be killed, but not the female. Clearly these two scenarios have different implications for the potential for population increase, even though in both cases nestlings are lost. Partitioning out exactly what happens at this stage of the life cycle has proven to be difficult.

It has long been known for small mammals that nestlings suffer a high rate of attrition. Millar (2007) reviewed the data available on nest mortality in small mammals and concluded that on average, about half of all nestlings do not survive to three weeks of age. Two methods have been

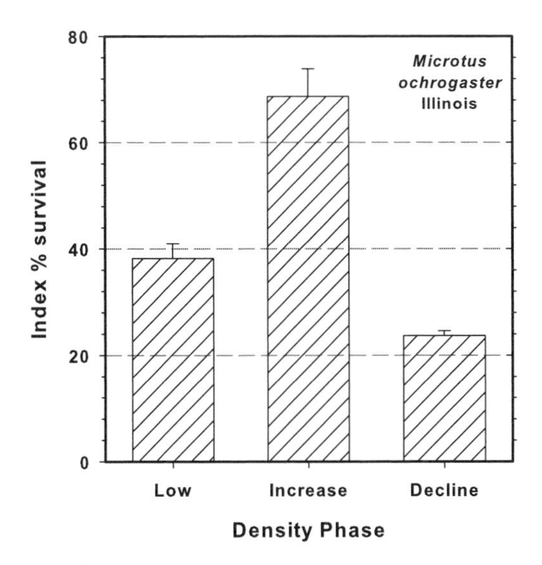

FIGURE 4.2 Nestling survival in *Microtus ochrogaster* in central Illinois, in relation to the phase of population fluctuations. Data from Getz et al. 2000. Live trapping twice a week over five years gave detailed data on the number of pregnancies and the number of new juveniles recruited. Error bars are 1 S.E.

used to estimate nestling mortality. At the population level, if one knows the number of pregnancies in adult females, the litter size at birth, and the number of juveniles that emerge from nests and enter live traps, one can calculate an index of nestling survival. This index can be given either as the fraction of nestlings surviving or as the number of nestlings surviving per pregnancy. Errors inherent in this approach involve knowing the number of pregnancies, the loss once juveniles leave the nest but before they are live-trapped, and variable litter sizes. Getz et al. (2000) is an excellent example of this approach. The second, more direct method is to locate nestlings in their nest after birth and mark them individually, so that they can be counted again after they emerge from nest burrows. Boyce and Boyce (1988a) and Lambin (1994a) are excellent examples of this second approach. The second method is clearly preferred, although both approaches appear to converge on the same conclusion that nest mortality ranges from less than 30% to more than 70%, and may reach 100%.

Figure 4.2 illustrates data from an intensive study on *Microtus ochrogaster* by Getz et al. (2000) that is one of the best examples of the first method. The data are consistent with the hypothesis that the rate of population growth is directly related to nestling mortality. Figure 4.3 illustrates similar data, collected much less intensively on *Microtus californicus* by Krebs (1966), that supports the same hypothesis.

Table 4.2 summarizes the estimates of nestling losses compiled by Millar (2007), and adds a few more recent estimates. The available data are sparse, as Millar (2007) noted, and there is a critical need for more data, particularly that gathered by the second method. The conclusions that can be drawn from the existing data are three. First, nest mortality can be extremely high, so that in some cases virtually all the reproductive output is lost in this stage of the life cycle. Second, the causes of nest mortality are a mixture of predation, weather, social interactions including infanticide, and possibly parasitism and disease. The primary mechanisms of nest mortality would appear to be predation and social factors, but for no species can we partition the losses adequately. Ylönen et al. (1997) are almost alone in reporting from field data on bank voles that 37% of both sexes were infanticidal, thus raising the possibility that a significant

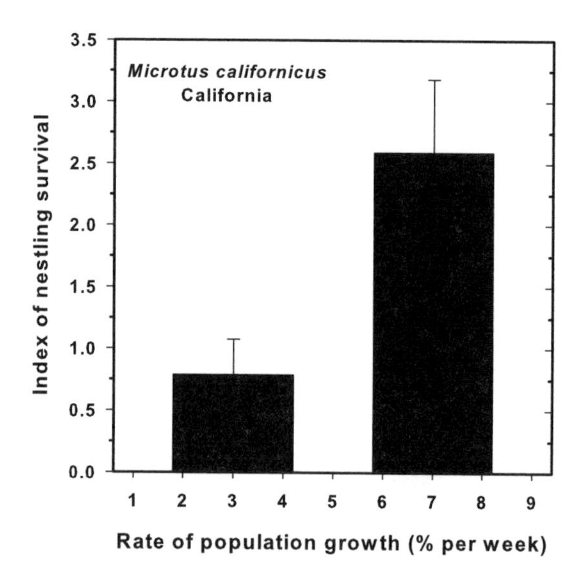

FIGURE 4.3 The index of nestling survival for *Microtus californicus* in a control population increasing at 3% per week (1962–63) and at 7% per week (1964). Data from Krebs 1966, table 10. The index is a crude measure of the number of juveniles recruited in live traps divided by the number of lactating females. Large differences suggest higher losses of juveniles during the first few weeks of life, but cannot separate nest losses from losses in the immediate period after leaving the nest and before being live-trapped. Error bars are 95% confidence limits.

TABLE 4.2 **Nestling mortality estimates (scaled to mortality rate per 21 days) from field studies in which individuals were monitored at least every week in natural populations. Results from experimental tests are given separately.**

Species	Nest mortality rate (%)	Observation or treatment	Reference
Microtus arvalis	63	Solitary females	Boyce and Boyce 1988a
	37	Groups of females	
Microtus ochrogaster	62	Low density	Getz et al. 2000
	31	Increasing density	
	76	Declining density	
Microtus ochrogaster	71	Inside predator exclosure	Getz et al. 1990
	81	Outside predator exclosure	
Microtus pennsylvanicus	57	Low-elevation site	Innes and Millar 1990
	80	High-elevation site	
Microtus townsendii	70	Low kin neighborhood	Lambin and Yoccoz 1998
	56	High kin neighborhood	
Clethrionomys gapperi	37–82	Variation among years	Innes and Millar 1990
Clethrionomys gapperi	48	No effect of experimental food addition	Kasparian and Millar 2004
Dicrostonyx groenlandicus	33	Inside predator exclosure	Reid et al. 1995
	62	Outside predator exclosure	
Peromyscus maniculatus	35–85	Variation over 16 years	Millar and McAdam 2001
Peromyscus maniculatus	55	No effect of protein supplement	McAdam and Millar 1999
Peromyscus maniculatus	61–83	Spring loss < fall loss	Goundie and Vessey 1986

fraction of nest mortality was socially induced. In the past we have largely treated nest mortality more or less as an unknown constant, but the extreme variation from year to year in long-term studies like those of Millar and McAdam (2001) gives a strong signal to investigate these losses as a potential key factor determining the rate of population growth. Third, the available data are consistent with the hypothesis that the rate of population growth is directly related to the rate of nestling survival in microtine rodents. Nestling survival is important to understanding the population dynamics of rodents.

The survival of nestlings and newly weaned juveniles in the bank vole is directly density-dependent. Koskela et al. (1999) raised litters of *Myodes glareolus* in the laboratory in Finland, and added two newborns to some litters. They then introduced these females with their litters to outdoor enclosures at two densities. They found a strong difference in female reproductive success between low-density enclosures (four offspring per pregnancy) and high-density enclosures (two offspring per pregnancy). Adding two pups to a litter did not increase reproductive success over that of unmanipulated litters. Of the 48 females with young used in these field experiments, 19 produced no surviving offspring, and 16 of these were in the high-density enclosures. Koskela et al. (1999) could not determine exactly when in the reproductive cycle these losses occurred, or how many might have been due to infanticide.

Juvenile Mortality

Once young have left the nest they are typically called juveniles, and they remain in this class until they reach sexual maturity, when they are classed as adults. The juvenile stage of the life cycle is of variable length because the age at sexual maturity changes with the season of the year, with population density, and with social structure. In *Microtus arvalis*, juvenile females may become sexually mature at 13 to 14 days of age (Frank 1954), so that in this situation there is virtually no juvenile stage. But more often the juvenile stage will last for three to five weeks in rodents, or much longer if a period of no reproduction occurs, such as in winter or in a dry season. In these cases it is important to separate juvenile survival into two components: survival when breeding is occurring, and survival in the nonbreeding period.

We are concerned here with two questions: First, what is the range of juvenile losses in small rodents? Second, how do juvenile losses relate to population growth rates or cyclic phases? The problem with answering these simple questions lies in issues of methodology for estimating survival rates in mark-recapture studies. While these methods are now well established and resourced (e.g., Program MARK, White 2008), many of the available data predate these methods. Many studies have relied on absolute census methods using minimum survival rates based on actual captures of individuals. These methods approach an accurate estimate of survival rates only when the probability of capture is very high, but they always have the disadvantage of having no method of estimating confidence limits. Mark-recapture studies also suffer from the inability to capture juveniles exactly when they leave the nest, and typically there is a one- to two-week interval of early juvenile life outside the nest when little

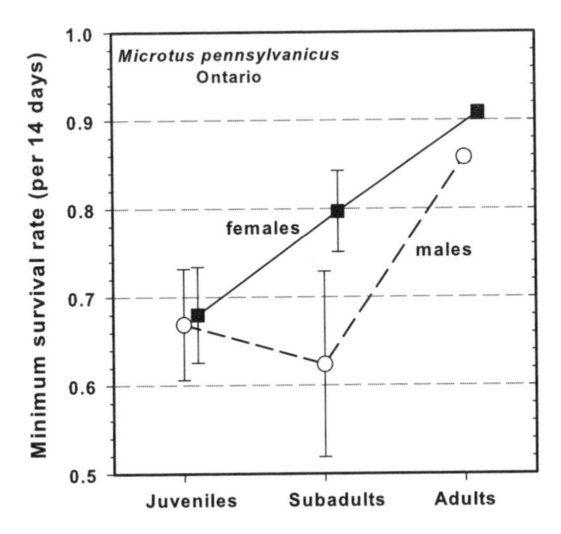

FIGURE 4.4 Minimum survival rates per 14 days (± 1SE) for *Microtus pennsylvanicus* averaged over four years, Toronto, Ontario. Only breeding season data are used. Of the three age groups, juveniles survive most poorly in both sexes (subadult males disperse so survival rate on the trapping area is biased low). Data from Boonstra and Rodd (1983, table 1). Densities each breeding season reached approximately 200 to 300 per hectare. Estimates of survival based on 228 and 196 juveniles (male:female), 482 and 558 subadults, and 1,321 and 1,983 adults.

data are available even when intensive live trapping is being carried out. Nevertheless, the patterns that have emerged from several intensive studies are suggestive, and we proceed with the full knowledge that the data are far from perfect estimators of juvenile survival.

Figure 4.4 illustrates from data in Boonstra and Rodd (1983) the general principle that survival rates of juvenile voles are lower than those of subadult and adult voles. In most cases this simply represents a continuation of the pattern of low survival in the early age classes of small rodents, whether in the nest or as juveniles before reaching sexual maturity. Getz et al. (2000) found the same pattern in *Microtus ochrogaster* in Illinois. They estimated the 3.5 day survival rate of adults as 94.9%, while that of juveniles was estimated at 65.7%. They also reported that juvenile males and females had identical survival rates, as did reproductive adult females and adult males.

How do juvenile survival rates change with population density and cyclic phase? The general pattern for cyclic populations has been for juvenile survival to decline first, followed by adult survival only later in the late

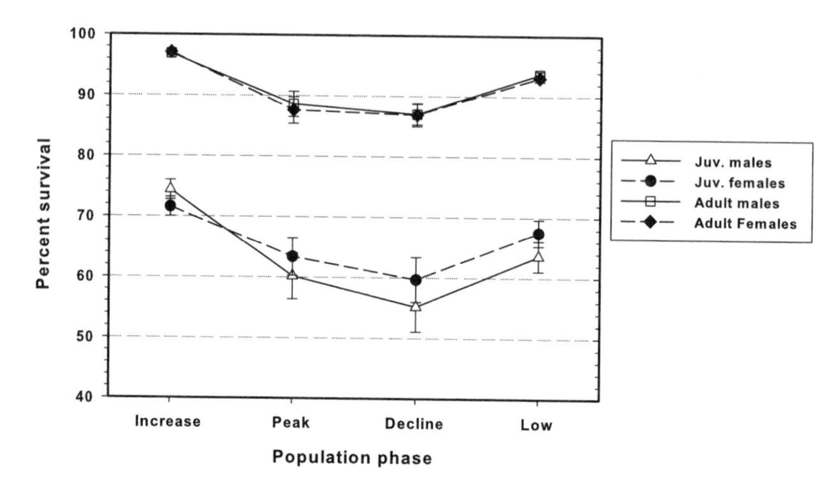

FIGURE 4.5 Minimum survival rates (± 1SE) per 3.5 days for *Microtus ochrogaster* in Illinois, averaged over the four phases of population change. Data for the reproductive season only from Getz et al. (2000, table 2).

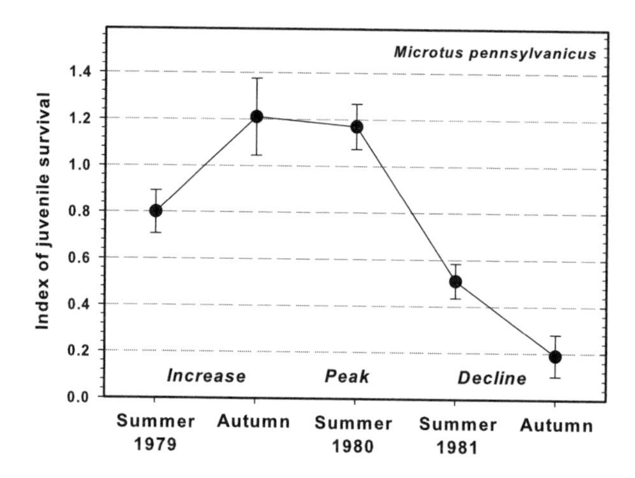

FIGURE 4.6 The index of early juvenile survival in *Microtus pennsylvanicus* in Ontario. The index is the number of juveniles caught in pitfall traps and Longworth traps per lactating female. Data from Boonstra (1985, table 1). Boonstra (1985) suggests that most of this loss occurs around or shortly after weaning. Given an average litter size of about 5 (Keller 1985) these data imply a loss of approximately 76% to 96% of juveniles produced by the adult female population. Error bars are 95% confidence limits.

peak and decline phases. Figure 4.5 shows data for *Microtus ochrogaster* from Getz et al. (2000). In this population the decline phase is associated with particularly low survival in juveniles and poor survival in adults as well. That this is not always the case is shown by the data in Boonstra (1985). For *Microtus pennsylvanicus* in Ontario, the survival rate of adults of both sexes during the breeding season was similar in all of the cyclic phases, and the main cause of the collapse was poor juvenile survival. Boonstra (1985) used pitfall traps to capture juveniles more effectively, and concluded that the poor juvenile survival was centered on the time of weaning and might have been related to maternal condition and the nutrition provided in lactation (figure 4.6).

Subadult and Adult Mortality

The subadult stage in small rodents is a transitory stage in which individuals are becoming sexually mature. In most cases, mortality rates in this stage of the life cycle are calculated on the basis of body weight rather than sexual status, and this confounds the dispersal of individuals with survival. In most small mammals, males are the sex that disperses (Lambin et al. 2001) and consequently male "survival" in this life history stage is completely confounded with movements off the study area. The use of radio telemetry can overcome this limitation (Gaines and McClenaghan 1980; Madison and McShea 1987: Wolff 1993). There is a great amount of literature on dispersal in small rodents (Clobert et al. 2001) and I will discuss this aspect of rodent ecology in chapter 5.

Adult small mammals in general do not disperse, so consequently we can measure their survival relatively easily. Two patterns of adult survival are evident in small rodents. The first pattern is of constant adult survival independent of population density and cyclic phase. A good example is shown by adult females in figure 4.7 from Boonstra (1985). The second pattern is of variable adult survival with the expectation of the highest survival in increasing populations and the worst survival during a decline phase. Krebs and Myers (1974) summarized several studies that showed this pattern. Beacham (1980b) reported another case for *Microtus townsendii*. Gaines and Rose (1976) found that the adult survival rate in *Microtus ochrogaster* in Kansas was directly related to the rate of population growth (figure 4.8), thus supporting the second pattern. There does not seem to be any universal pattern in adult survival over population fluctuations in

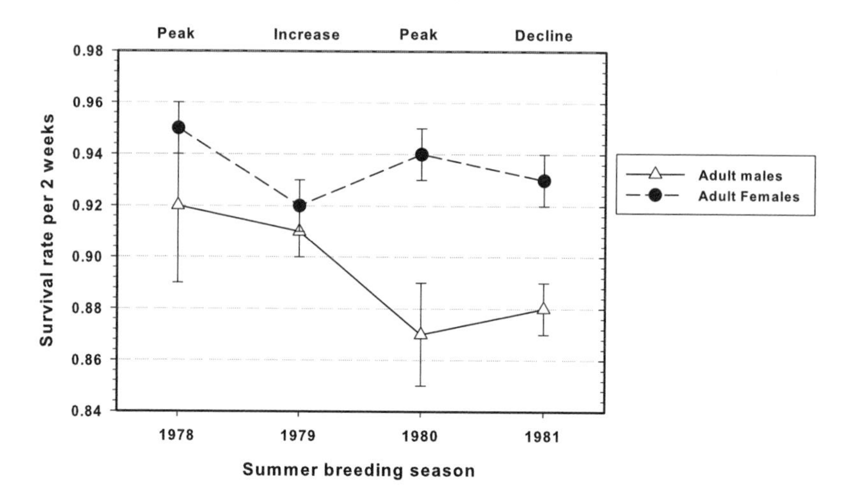

FIGURE 4.7 The survival rate per 14 days (± 1 SE) estimated by the Jolly-Seber model for *Microtus pennsylvanicus* in Ontario. All the survival estimates are high, and adult female survival is constant over a density range from about 50/ha to more than 250/ha. Data from Boonstra 1985, figure 3.

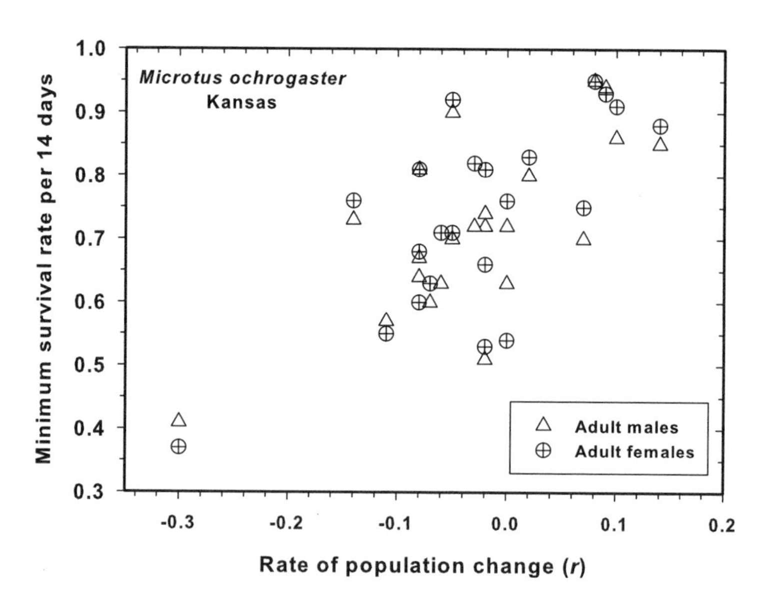

FIGURE 4.8 The survival rate of adult *Microtus ochrogaster* in Kansas over three years is directly related to the rate of population growth (r = instantaneous rate of population change per week) with Pearson's correlation coefficient equal to 0.70 for both adult males and females. These data support the view that adult survival follows the density cycle.

small rodents, although there are more cases of the second pattern than the first.

Which components of mortality are most significant in causing changes in small mammal numbers? We have discussed this issue with respect to reproductive changes in chapter 3. The same partial life cycle analysis shown in figure 3.12 can be used to investigate the relative importance of mortality changes in determining population growth rates. Oli and Dobson (2003) conducted an extensive analysis of the relative importance of different life history variables in determining population growth rates in mammals. They analyzed age at sexual maturity, age at last reproduction, juvenile survival, adult survival, and fertility rate in a range of mammal species (n = 142 species) ranging from large to small in body size. They recognized a continuum of life histories ranging from "fast" to "slow" (Dobson and Oli 2008). Fast life histories are typical of small rodents, and

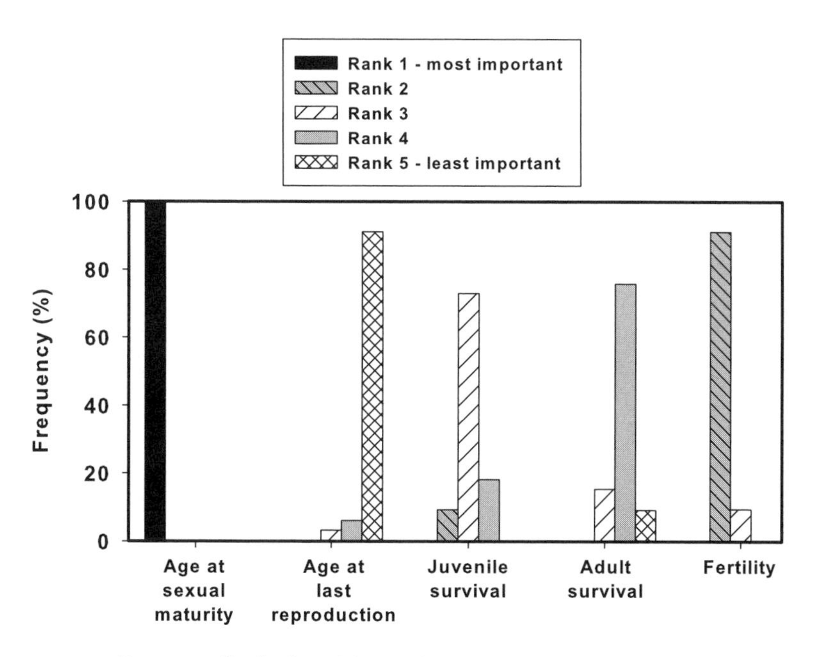

FIGURE 4.9 Frequency distribution of the elasticities of 33 populations of small mammals. For each population, the relative importance of each life history variable in contributing to increased population growth was ranked in a descending order such that a life history variable with the largest relative influence would receive a rank of 1 while a life history variable with the smallest relative influence on the population growth rate would receive a rank of 5. From Oli and Dobson 2003.

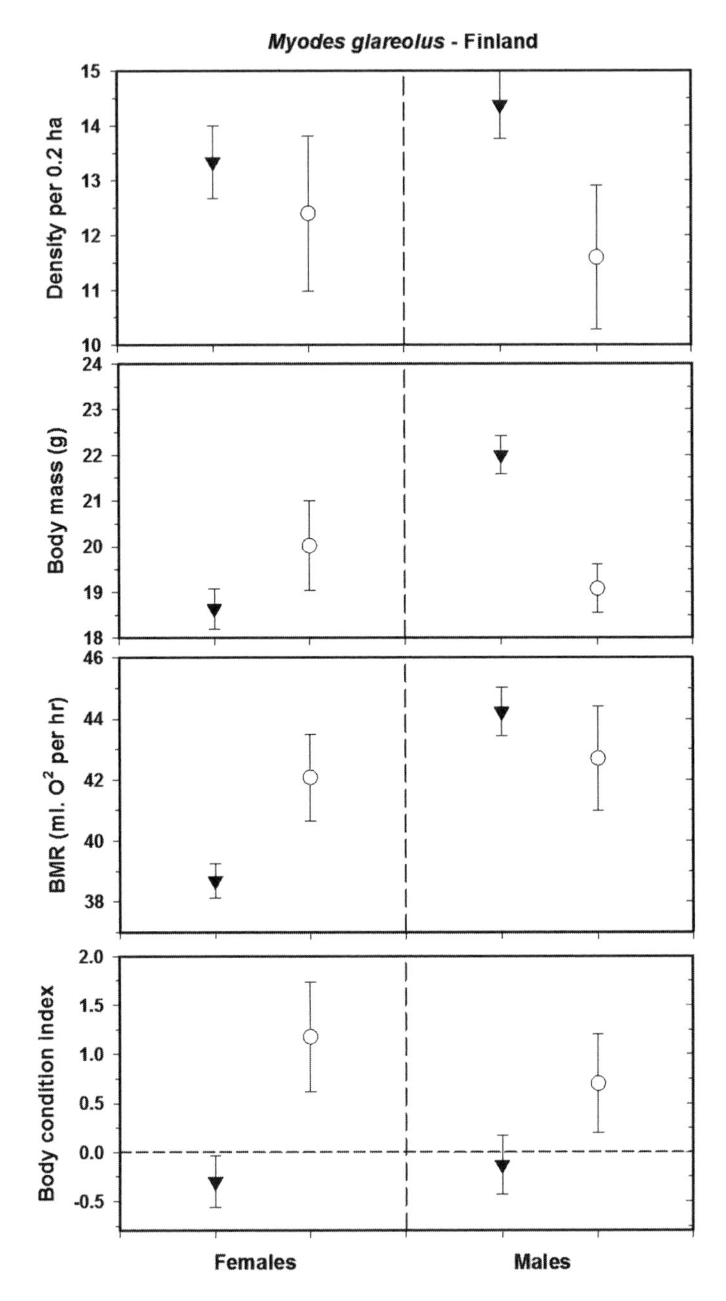

FIGURE 4.10 Mean (± SE) of bank voles (*Myodes glareolus*) that survived overwinter (open symbols) and those that died (black symbols) in relation to population density, body mass, basal metabolic rate, and body condition (from the regression of body mass on head width). Modified from Boratynski et al. 2010.

are characterized by an early age at sexual maturity, a high fertility rate, and a short lifespan. Figure 4.9 shows the results of their analysis for small mammals. Age at sexual maturity was always the most significant variable affecting population growth rate (as we noted in chapter 3), and fertility rate was second. Only then did juvenile survival enter in importance, and adult survival was rarely an important component leading to high population growth. One caveat of the Oli and Dobson (2003) partial life cycle analysis is that it includes in the fertility component prenatal losses and early juvenile mortality, so that fertility is defined as effective fertility: the number of young that leave the nest at approximately two weeks of age. Thus, when we conclude from this analysis that fertility is a critical variable, we must remember that the term "fertility" includes some early-stage juvenile mortality, as well as litter size and the length of breeding season.

The physiological characteristics of individuals that live and those that die have been too rarely studied. In northern voles the winter is a stressful period, and overwinter survival can be a critical variable in population changes. Boratynski et al. (2010) followed the fate of 136 bank voles (*Myodes glareolus*) released in autumn into large outdoor enclosures in central Finland. They asked whether individuals with higher metabolic rates would survive better during the winter, and figure 4.10 illustrates their results. Females that survived over winter were on average heavier, and had higher basal metabolic rates and body condition values than the females that did not survive (figure 4.10). Males that survived had lower body mass and slightly higher body condition than males that did not survive. The physiological correlates of behavior in relation to survival in voles and lemmings are an area of great promise, but of relatively little data at present (Lantova et al. 2011).

Conclusion

Mortality must, of course, be a critical component of population changes in small rodents, since it is not constant or independent of age in any mammal. In fluctuating small rodent populations the key mortality variable is juvenile loss, including loss of young in the nest and in the stage of independence after they leave the nest and before they reach sexual maturity. Adult female survival is particularly critical only when populations are declining rapidly, and when females cannot complete pregnancy and lactation before they die. The resulting population traces are thus related to the interplay of changes in reproductive performance and in the probability of death.

What we have ignored so far are movements and dispersal, and the possibility that some changes in population size are the result of source-sink dynamics, a topic to which we turn in chapter 5.

CHAPTER FIVE

Immigration and Emigration

Key Points:

- Populations can be sources if emigration exceeds immigration, or sinks if immigration exceeds emigration.
- Emigration is an essential part of the life cycle of small mammals that are territorial, and male dispersal is the general rule, presumably to prevent inbreeding.
- If emigration and immigration are stopped by a barrier fence, populations may explode to high numbers; this is the fence effect.
- The costs of movement for small mammals in additional mortality are difficult to study. The behavioral ecology of immigration and emigration is more clearly understood than are the population consequences of such movements.

Most studies of population changes in small rodents assume that movements are occurring into and out of the population under study, but that these movements are effectively equivalent to births and deaths. In an equilibrium world immigration into and emigration out of an area are equal, so that they cancel each other out and can be ignored. This simple view was challenged by many ecologists who recognized that habitats are heterogeneous, and this could cause a spatial component to population dynamics. Lidicker (1973) introduced the idea of "dispersal sinks" from his studies of the California vole, and the idea was generalized by Pulliam (1988), who coined the terminology of "source-sink dynamics."

These ideas were already implicit in the early work of Huffaker (1958) on insect populations. In this chapter we will investigate what studies of rodent movements, immigration and emigration, have contributed to our understanding of their population dynamics.

Source-Sink Dynamics

A simplified view of source-sink dynamics is illustrated in figure 5.1. If reproductive output exceeds deaths, the population must grow in size, export the excess or surplus individuals, or both. An important point made by Pulliam (1988) was that sink populations could be quite dense, so that population density alone is not a sufficient variable on which to classify habitats as sources or sinks. To the simplified view in figure 5.1 needs to be added the fact that a particular population in a habitat may be a source population in one year and a sink population in the following year, so that both temporal and spatial dynamics need to be considered.

In most studies of small rodents, the survival rate is measured by knowing which individuals stay on the study area. Animals that disappear may have died, or may have emigrated to a new area outside the study zone. Thus, measures of survival should be called *apparent survival* (Runge et al. 2006). If all individuals that disappear are assumed to have died, apparent survival will be equal to true survival, but this situation is unlikely to be common in nature. If only one study area is available for analysis, Runge et al. (2006) suggest that a valuable way to differentiate sources and sinks is to determine the self-recruitment rate for a local population, defined as

$$R = \Phi_A + \beta\Phi_J \quad \text{(eq. 1)}$$

where R = self-replacement rate ($R < 1$ defines a sink, $R > 1$ a source),
Φ_A = adult survival rate,
β_A = number of juveniles per adult, and
Φ_J = juvenile survival rate.

These rates are typically defined on an annual basis, but any biologically relevant time period could be used. Estimating whether a local population is a source or sink demands high-quality data, and Runge et al. (2006)

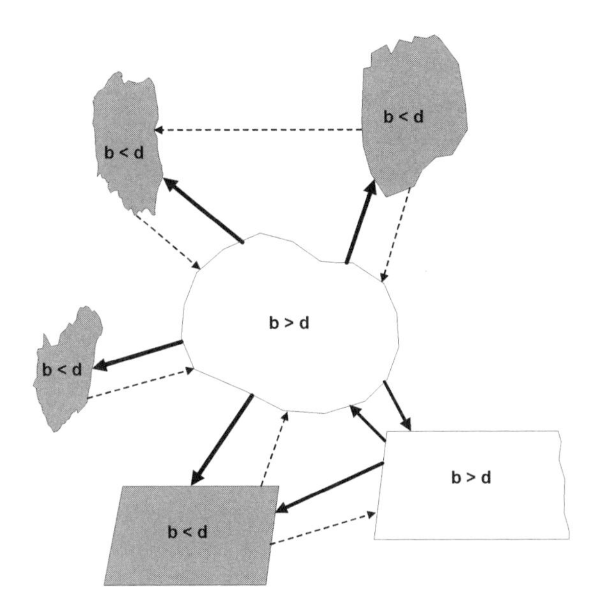

FIGURE 5.1 Hypothetical landscape of sources and sinks. Sources, in which births (b) exceed deaths (d), are indicated in white. Sinks, in which deaths exceed births, are indicated in gray. Widths of arrows indicate the number of individuals moving between patches in the landscape. Without immigration, sink populations would go extinct. Sinks can be thought of as poor habitats, sources as good habitats.

list many studies that have used inadequate methods. Not all small rodent population studies find evidence of source-sink dynamics, and it is far from clear how often this population metaphor is appropriate. For example, Hanley and Barnard (1999) studied *Peromyscus keeni* in four distinct forest habitats in southeast Alaska over four years. They found no differences in survival or densities among the habitats, which ranged from floodplain forest to upland old growth, and no evidence of source-sink dynamics. By contrast, Paradis (1995) found in the Mediterranean pine vole living in an agricultural landscape that immigration was the major factor causing population changes, thus suggesting a source-sink structure in a small-scale, patchy agricultural landscape. Clearly, movements of individuals can in some cases be a major factor in population changes, and we need to consider the demographic consequences of movements in more detail. In table 5.1 Krebs et al. (2007) provide some hypotheses of how population density affects social behavior resulting in dispersal events.

TABLE 5.1 **Predicted density-dependent effects on various aspects of rodents' social behavior. From Krebs et al. 2007, p. 178.**

	Low population density	High population density
Territoriality	Large territories wth individuals widely spaced, mutual avoidance, low aggression, and available vacant space	Small territories with considerable overlap and high aggression
Dispersal and philopatry	Distant dispersal by all males, dispersal by females close to natal site, bequest by some dams of maternal sites to daughters	Delayed emigration with sons and daughters remaining on natal site, extended families, cooperative and communal breeding of females
Age at sexual maturity	Maturity of sons and daughters at young age	Delayed sexual maturation for both sexes, cooperative and communal breeding for some species
Infanticide	High incidence for males, low for females	Low incidence for males, high for females

The next problem involves how we should classify dispersal movements in relation to population changes in field populations.

A Classification of Dispersal Movements

In a seminal paper, Lidicker (1962) suggested that emigration might regulate population size below the carrying capacity of the habitat. This was a radical idea at a time when population changes were considered to be driven by births and deaths because it was assumed that immigration equaled emigration so that the two canceled each other out of any dynamics. Lidicker (1962) considered the carrying capacity to be set by the food supply and, given the confusing definitions of "carrying capacity" as outlined in Dhondt (1988), it seems better to avoid the term if possible in discussions of dispersal. In an important review, Lidicker (1975) expanded on his views and introduced two important concepts (figure 5.2). He defined *presaturation dispersal* as emigration from a population increasing in size and not yet at maximum density ("carrying capacity"). By contrast, *saturation dispersal* is emigration from a population already at high density. These definitions immediately raise the behavioral ecological issue of why small animals emigrate at all. Lidicker (1975) suggested that presaturation emigrants were high-quality individuals of either sex in good condition and that "chose" to leave to seek better home sites, or because they were sensitive to increasing density. Saturation dispersers, meanwhile, were ani-

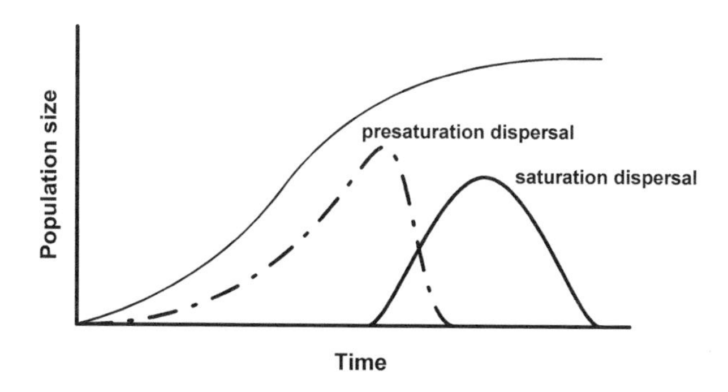

FIGURE 5.2 Presaturation and saturation dispersal in relation to the population growth curve, as defined by Lidicker (1975).

mals leaving out of desperation with poor chances of survival in a habitat too crowded to support them. These definitions raise the issue of how one can operationally define a dispersing individual.

Methods of Defining Dispersers

Several methods have been suggested to define a disperser in the field (Stenseth and Lidicker 1992). If trappability is high, one way would be to define dispersers as individuals that exceed a specified body mass at first capture ("adults"). Dueser et al. (1981), for example, used 30 g at first capture as a cutoff point for defining dispersers of *Microtus pennsylvanicus*. This criterion was criticized by Tamarin (1984) and Danielson et al. (1986), who pointed out that misclassification in live trapping studies could be significant. Porter and Dueser (1990) responded that if the cutoff body mass criterion is chosen carefully and trappability is high, it is possible to make relatively few errors.

An alternative experimental method of defining dispersers is to use removal areas in which all colonizing animals are regularly removed. This was used by Krebs (1966) on populations of *Microtus californicus* to measure the rate of emigration in relation to population growth rates on a control area. The number of adult colonists entering the removal area every two weeks was astonishing (figure 5.3), and a total of 1,758 adult individuals were removed from a 0.8 ha area between November 1962 and November 1963. Not all the colonists in the removal area were dispersers but enough were to suggest that dispersal was common and that empty

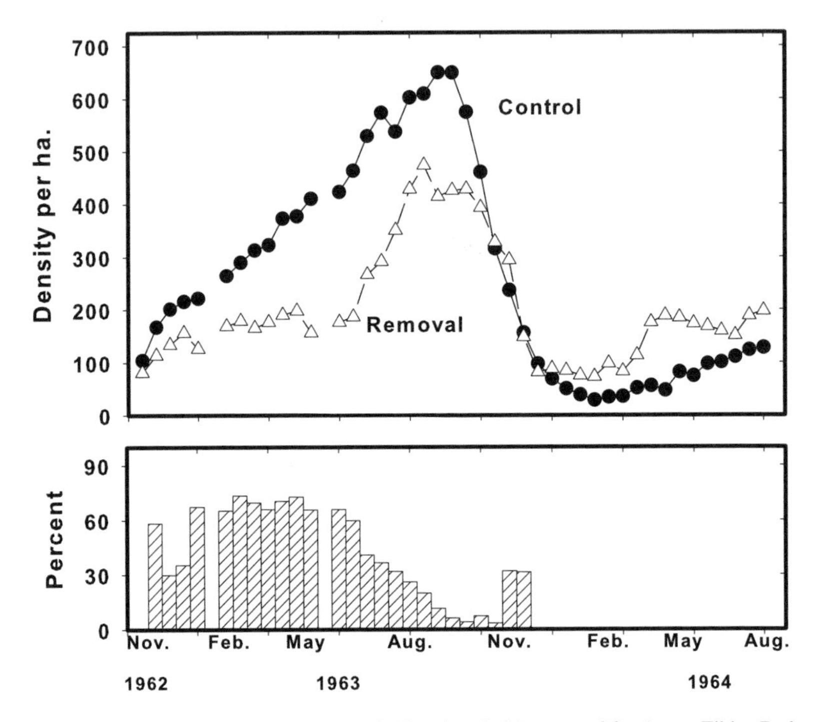

FIGURE 5.3 Population fluctuations in the California vole *Microtus californicus* at Tilden Park in Berkeley, 1962–64. Until November 1963 the removal area was cleared of all voles every two weeks and was rapidly recolonized. The lower graph shows the numbers of adult colonists to the removal area as percentages of the concurrent population on the unmanipulated control area. Every two weeks, up to 60 individuals colonized the 0.8 ha removal area during the stage of population growth. Data from Krebs 1966.

areas were filled rapidly even when control populations were increasing. The main result of this study was to show that "surplus" individuals exist in large numbers to colonize relatively empty but favorable sites.

The next step in studying dispersal in field populations was to use enclosures that stopped both emigration and immigration. The question being asked with this design concerned how small rodent populations would respond in the absence of these two processes. The answer produced the fence effect (Krebs et al. 1969): fenced populations increased to abnormally high densities, exhausted the food supply, and starved. Figure 5.4 illustrates one example from *Microtus ochrogaster* in Indiana. Figure 5.5

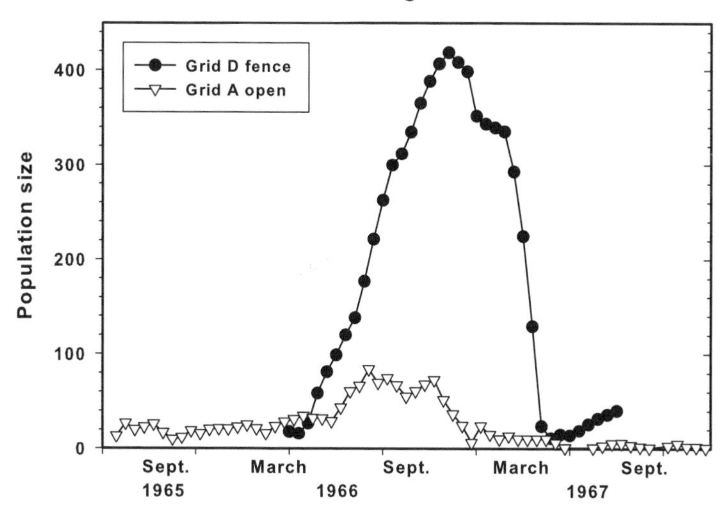

FIGURE 5.4 The fence effect in *Microtus ochrogaster* in Indiana. The fenced population with no immigration or emigration increased to 420 individuals on 0.8 hectares, while the open control grid reached a peak of 84 individuals. Data from Krebs et al. 1969.

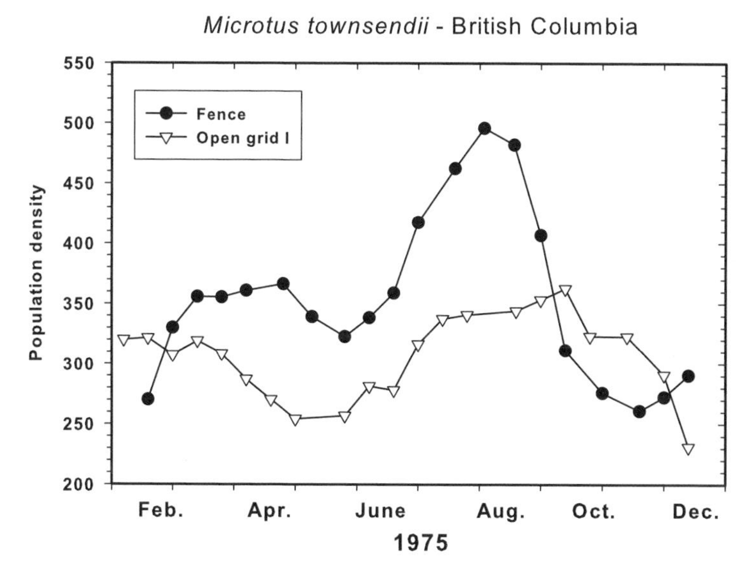

FIGURE 5.5 The fence effect in *Microtus townsendii* in British Columbia. Population density per 0.8 hectare on y-axis. The fenced population with no immigration or emigration increased to a density of about 440 per hectare, while the control populations reached a maximum of about 625 per hectare during the year of study. Data from Boonstra and Krebs 1977.

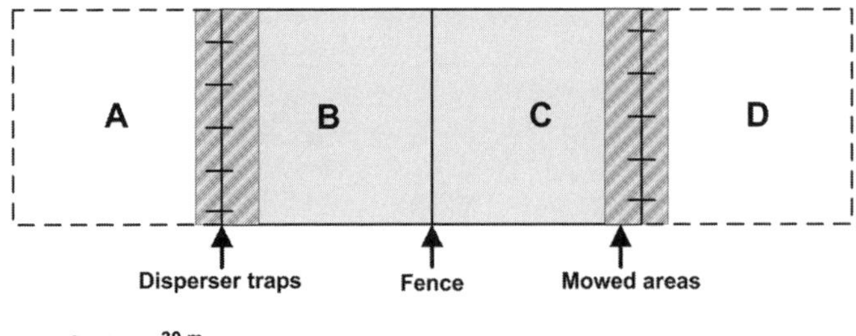

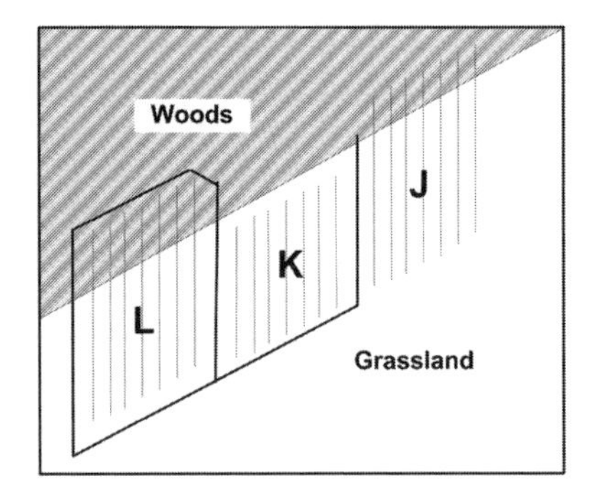

FIGURE 5.6 Two designs for fenced grids that provide a dispersal sink for voles. (A) Beacham (1980) design. Each grid is a 7 × 7 checkerboard of traps, spaced at 7.6 meters. The shaded areas were mowed to make the habitat unsuitable for voles, and were provided with traps at the outer edges; grids A and D were open-habitat. (B) Tamarin (1984) design. Each grid is 0.8 hectares in size, and trapping lines are spaced at 7.6 meters. Both these designs eliminated the fence effect.

shows another example from *Microtus townsendii* in British Columbia (Boonstra and Krebs 1977). The fence effect was completely unexpected, and it provided a strong demonstration that dispersal was necessary for normal population control in these *Microtus* species.

If the plots used in these study areas contained source populations, the relevant variable was expected to be emigration rather than immigration. Attention then turned to trying to provide exit doors or exit traps in fences to see whether this would eliminate the fence effect. Gaines et al. (1979) did this experiment with *Microtus ochrogaster* and found that the fence effect disappeared when one-way exit doors were used. Tamarin et al. (1984) used an elegant design with a fence open at one end into woodland that acted as a dispersal sink (figure 5.6), and also found no fence effect when emigration was permitted. Beacham (1980a) achieved the same effect with dispersal sinks inside the fence. The consensus seems to be that study areas are source populations, and that emigration is the key to the fence effect.

Ostfeld (1994) questioned many of the findings regarding the fence effect. Many of his comments are well taken: there should be more replication in all studies of populations, and not all rodent species living in all habitats may necessarily show a fence effect. But his analysis is flawed in citing five enclosure studies that do not show a fence effect when only one of the five had an unfenced control population for comparison. His own studies on *Microtus pennsylvanicus* were done in enclosures that were too small for demonstration of natural population dynamics: this small enclosure problem has been a continuing source of misleading data in studies of small rodents. His comments on food limitation in declining populations will be discussed in chapter 8. The useful message in Ostfeld (1994) is that more studies of potential fence effects would be desirable. Wolff et al. (1996) showed that barrier strips in enclosures did not produce a good separation of individual animals that were dispersing, but instead caught individuals making exploratory movements or individuals at the edge of their home range. The question remains: Are barrier strips in enclosures an effective way of allowing natural dispersal to occur (and thereby of defining dispersers), and might the utility of barrier strips be species-specific?

Who Disperses and Why?

The behavioral ecology of dispersal has been extensively summarized in Wolff and Sherman (2007), and there is an extensive literature on dispersal that we can only touch on here. Two age-related types of dispersal are recognized by behavioral ecologists—natal dispersal and breeding dispersal. In natal dispersal a young animal leaves its birth area before it reproduces. In breeding dispersal an adult animal leaves its home range after it has successfully reproduced, and moves to a new breeding site. Natal dispersal is most common in small animals, and has a strong sex bias toward males dispersing (Wolff 1993).

Four types of mating systems occur in rodents (figure 5.7). Female territoriality seems nearly universal, and the most common system is promiscuity, in which male home ranges overlap partly or entirely. Monogamy is rare, and even when it occurs it is not universal at all densities and seasons (e.g., *Microtus ochrogaster*: Getz et al. 1993). Because young in the nest

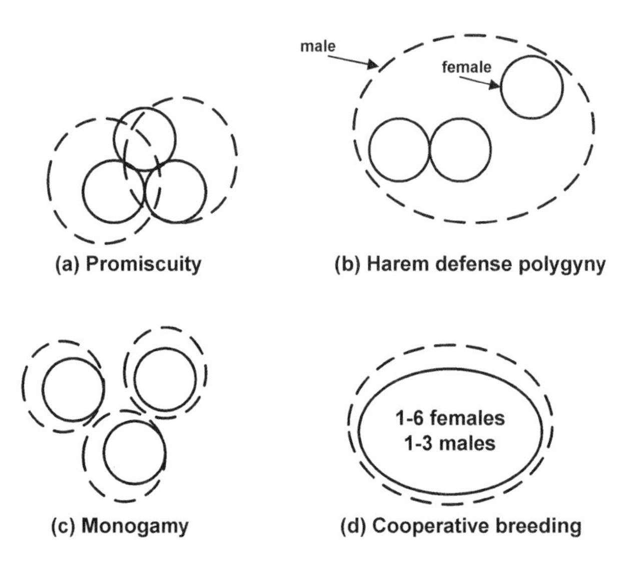

FIGURE 5.7 Four types of mating or spacing systems found in rodents. (a) Female territoriality and males with large overlapping home ranges, with promiscuous mating; this is the most common system. (b) Harem defense polygyny, in which one male defends exclusive mating access to several females. (c) Monogamy, in which one male guards and mates exclusively with one female. (d) Communal cooperative breeding, in which several related females breed cooperatively with one or more males. Male home ranges are represented by dashed lines, female territories by solid lines. After Wolff 2007.

are altricial, they are subject to infanticide by adult males or females that view them as potential competitors for living space. Territoriality in female rodents thus seems to have evolved as a tactic to protect their young from intruding infanticidal females (Wolff 2007). In rodents the defense of food supplies is not practical for an individual, and territoriality is not designed to defend food resources. Male spacing tactics are much more labile, and they depend on what other males in the population are doing. The key limiting resource for males is breeding females.

Much discussion has hinged on the question of to what extent dispersal is hardwired genetically. Wolff (2007) suggested that in rodents, natal dispersal of males is caused by the presence of opposite-sex relatives, and has evolved to prevent inbreeding. Females are philopatric because adult males do not remain in residence to permit inbreeding. Thus, for most rodents male natal dispersal and female philopatry are the basis of the social system.

The question of whether the social system can affect population growth rate depends on the intensity of infanticide and the amount of direct aggression between breeding individuals. Infanticide can be studied in the laboratory readily, and it is common in laboratory situations (Mallory and Brooks 1978), but it is nearly impossible to document and measure in the field. It remains a lacuna in our knowledge of rodent population dynamics. Much indirect evidence suggests that infanticide or direct aggression can affect the rate of population change through juvenile survival. Redfield et al. (1978) removed males from one live-trapped grid and females from another every second week for two breeding seasons to drive the sex ratio to 80% of the favored sex. Figure 5.8 shows the results on recruitment of juveniles into the trappable population. Clearly, in this species adult females control the recruitment of juveniles, via either infanticide or direct killing of independent young. In a more carefully controlled experiment, Wolff et al. (2002) found the same result for *M. canicaudus*.

Another impact of the social system on population dynamics arises when breeding adults suppress the maturation rate of their young. The best evidence for this comes from *Myodes* (*Clethrionomys*), as discussed in chapter 3 (see figure 3.4). The key issue, as Wolff (2007) has pointed out,

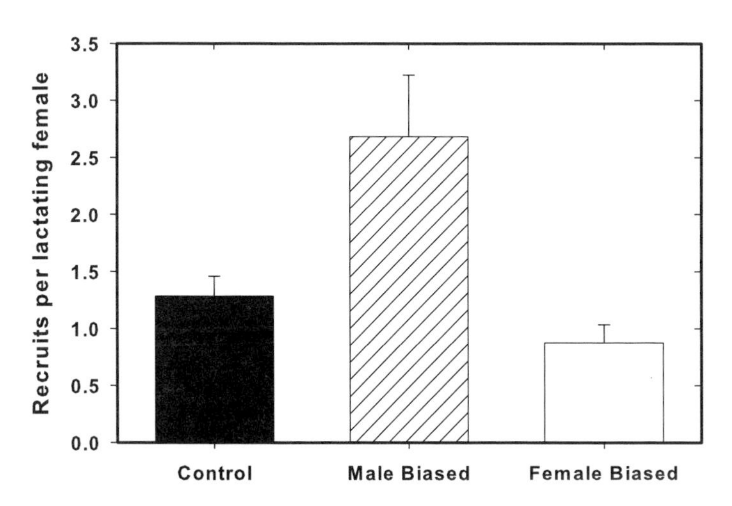

FIGURE 5.8 The average recruitment of juvenile *Microtus townsendii* for the summer breeding seasons of 1972 and 1973 on control (unmanipulated), male-biased (80% of adult females removed), and female-biased (80% of adult males removed) areas. Recruitment is measured as the number of juveniles live trapped at two to five weeks of age per pregnancy. Error bars are 95% confidence limits. Data from Redfield et al. 1978, table 6.

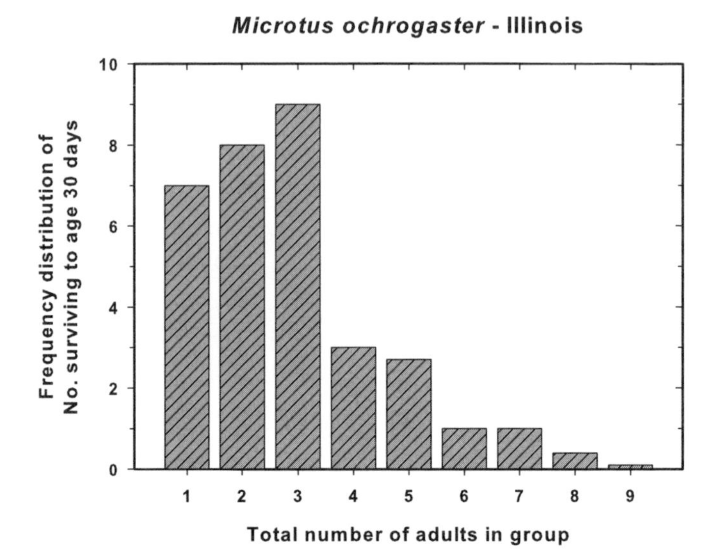

FIGURE 5.9 Reproductive success and size of social group in prairie voles *Microtus ochrogaster*. Reproductive success was measured as the number of young that survived to 30 days of age per adult female in the group. It was not known whether members of the group were related. Data from McGuire et al. 2002, figure 2.

is whether juvenile females interact with related or unrelated females. If the interactions are mainly with related females, kin groups can form with communal breeding, and this can accelerate population growth (Lambin and Krebs 1993). By contrast, if the interactions are with unrelated individuals, in some rodent species juvenile females may be reproductively suppressed, thus reducing the population growth rate. In addition, infanticide could be more common among unrelated females (Wolff and Peterson 1998). Unfortunately we have too little data on the social structure of females in wild populations to know how often breeding groups occur and whether or not the individuals are related. A notable exception is the prairie vole *Microtus ochrogaster* (McGuire et al. 2002). Figure 5.9 shows that in this species, maximum reproductive output was achieved at a group size of three, particularly when there were two males in the group. The prairie vole is often monogamous, in contrast to most other *Microtus* species, and it is not known whether the two males in a group were related. But the suggestion that kin group cooperative breeding may be beneficial has been found in *M. townsendii* (Lambin and Krebs 1993). In *M. arvalis*

Boyce and Boyce (1988a) found that group breeding had both positive and negative effects on relative fitness. Solitary *M. arvalis* females produced fewer offspring, but their offspring began to breed at an earlier age and also survived better than the offspring of group-breeding females. It is presumed, but not known, that these group breeding females were related.

But an important question from the viewpoint of population dynamics is whether these social behaviors are an important component of population change, or just a minor sideshow (Wolff 2007). For species in which females can breed in their mothers' territory, it seems unlikely that population growth will be restricted by social behavior. But if each breeding female requires its own exclusive breeding territory, some limitation on population growth is possible.

Two Hypotheses Regarding Dispersal

Two ideas were proposed during the 1980s about how social interactions could be involved in population fluctuations. Charnov and Finerty (1980) proposed a simple and elegant kin selection model for population fluctuations. At low density, nearby individuals would be related, and aggression would be low between individuals. As the population grew in density, Charnov and Finerty postulated, more unrelated individuals would disperse into the neighborhood, thus leading to more aggression, with consequences for mortality and reproductive output. Their model predicted that aggression should be elevated if the coefficient of relatedness is low, and vice versa. Several tests of the Charnov-Finerty model have been carried out in enclosures into which were introduced related individuals or, alternatively, completely unrelated individuals to test for differences in population growth rates (Boonstra and Hogg 1988; Ylönen et al. 1990; Pusenius et al. 1998). No differences between populations begun with related individuals and those begun with unrelated individuals have been detected in any of these studies (reviewed in Dalton 2000).

An alternative view to the Charnov-Finerty hypothesis was suggested by Lambin and Krebs (1991) on the basis of research on open populations of *Microtus townsendii*. Table 5.2 summarizes the different predictions from these two models. Lambin and Krebs (1991) pointed out that familiarity is the basis for kin recognition in small rodents, so that experiments that use relatives, like that conducted by Boonstra and Hogg (1988), could fail if the relatives are not familiar with one another. Familiarity may thus

TABLE 5.2 **Assumptions and predictions of the Charnov-Finerty and Lambin-Krebs models. From Lambin and Krebs 1991.**

	Charnov-Finerty model	Lambin-Krebs model
Assumptions	Kin groups result from low dispersal and inbreeding in low-density populations. Relatedness declines in increasing populations because of high dispersal rate. High relatedness induces higher breeding success because of kin selection.	Kin groups result from philopatry and lack of immigration, and disappear because of predation and immigration. High relatedness increases breeding density through reduced competition for space between closely related females in spring.
Predictions	Average relatedness is higher at low density than at high density. Competition for space is most intense at high density.	Degree of female relatedness fluctuates seasonally, increasing during the breeding season and decreasing in winter. Average relatedness is higher in spring of outbreak or peak years than in other years. Competition for space between females is least intense at high density, when average relatedness is high.

replace relatedness as the key parameter to be analyzed, and this feeds into a large literature on how individual rodents recognize relatives (Holmes and Mateo 2007). The field data reported by Lambin and Krebs (1993) supported the predictions of the Lambin-Krebs model (see table 5.2), but more studies are needed to test the model's generality.

A second model was suggested during the 1980s by Hestbeck (1982): the social fence hypothesis. Hestbeck suggested that when density is low, spacing behavior regulates populations through emigration. But as density increases, the effectiveness of spacing behavior to regulate populations is lost because neighboring groups fence one another and prevent emigration. Once emigration is blocked, population regulation is achieved via resource exhaustion. This hypothesis, like the Charnov-Finerty model, rests on the assumption that aggressive interactions between unfamiliar individuals are more severe than those that take place between familiar individuals. It also operates on the assumption that the social group size changes with density. It differs from the previous models in suggesting that at peak densities resources will be exhausted and starvation will reduce numbers. Hestbeck's hypothesis makes the interesting prediction that populations that occupy adjacent high-quality and low-quality habitats will show different population dynamics because there will be no social fence at the border between the two habitats and emigration will continue from the high-quality habitat. Hestbeck (1982) presages the source-sink habitat idea that would appear in the literature six years later.

Some studies have provided limited support for the predictions of the social fence hypothesis, such as the prediction that movements in a population are inversely density-dependent (e.g., Smith and Batzli 2006), but these results are often consistent with other hypotheses. Stenseth (1988) argued on theoretical grounds that the social fence hypothesis should lead to population stability rather than fluctuations. The social fence hypothesis is consistent with results showing a negative correlation between colonization success and population density, as has been found in water voles in Scotland (Fisher et al. 2009). The key behavioral question is: What happens to emigrants when they move into a strange area and become immigrants? The suggestion from all the field data is that emigration is almost always equivalent to death, unless there is a vacant area nearby. But some immigrants are successful, at least a low population density, and at those times emigrants could be at a selective advantage.

Mechanisms Underlying Dispersal

The analysis of the ecology of emigration and immigration leads directly to questions about the mechanisms underlying an individual's decision to disperse (Clobert et al. 2001). There is a nexus of interrelationships that complicate the attempt to dissociate the proximal causes of dispersal in any organism (figure 5.10). The three immediate questions are: (1) Is there a genetic basis for dispersal in small rodents? (2) Is dispersal condition dependent such that a measure of body condition can predict dispersal? (3) How does dispersal relate to population density? We have already discussed the third issue, and Lidicker's important distinction between presaturation and saturation dispersal is an important contribution to our understanding of this component (Gaines and McClenaghan 1980). To address the first issue, Roff and Fairbairn (2001) reviewed the genetic basis of dispersal in animals and asked the important question of whether dispersal traits were correlated with other components of fitness. They concluded that there was considerable evidence for substantial additive genetic variance for traits related to dispersal. Virtually all of this evidence comes from studies of insects, and in particular they reviewed experiments in which selection for traits related to dispersal in insects could

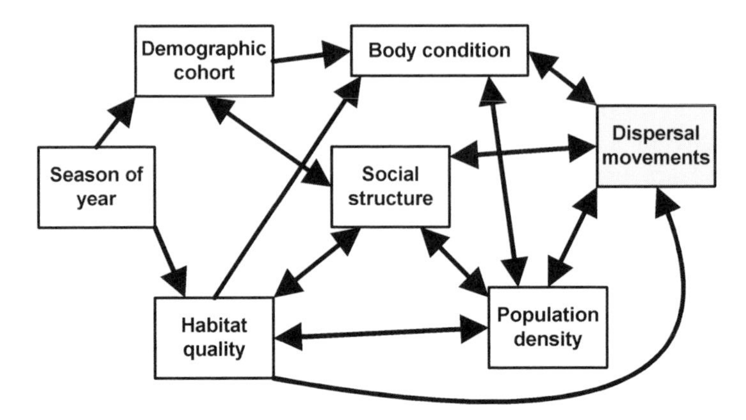

FIGURE 5.10 The nexus of interrelationships that illustrate the multifactorial nature of dispersal in small mammals and complicate simple explanations of the proximate causes of dispersal. Modified from Ims and Hjermann 2001.

be subject to natural selection. There are some data on the genetic control of dispersal in squirrels (Selonen and Hanski 2010), but unfortunately I cannot find any data of this type for voles and lemmings.

Most studies of dispersal in mammals tacitly assume that it is phenotypic and condition-dependent. Ims and Hjermann (2001) addressed this second issue, which is mentioned above, and they recognized that dispersal is a joint function of environmental conditions (habitat, food quality, density, social structure) and intrinsic conditions (fat reserves, body size, physiology), as illustrated in figure 5.10. Because of these complications, almost no one has attempted to make quantitative predictions about the expected dispersal rate under different conditions. In some mammals testosterone, along with body condition, stimulates male natal dispersal (Dufty and Belthoff 2001). The recently discovered hormone leptin may also play a role. Clearly we are a long way from understanding the nexus of interactions that result in dispersal movements in small rodents.

Conclusion

Simple field experiments on small mammals have revealed the existence of a large population of surplus individuals that apparently move out of their natal site and search for a suitable new area of habitat. The fence effect in some species shows graphically the significance of frustrated dispersal, and focuses our attention on the demographic consequences of immigration and emigration. In continuous habitats most of the surplus individuals are presumably lost to predators, starvation, disease, or social mortality due to direct aggression by territorial breeders. In patchy habitats they may become colonists of suitable vacant areas. The concept of "sources" of surplus individuals and "sinks" of immigrants may apply to small mammals, but in any fluctuating population an area that is a source in one year may become a sink in the following year, thus complicating interpretations of the dynamics.

Lidicker's distinction between presaturation and saturation dispersers has fostered many mark-recapture studies that try to operationally define and classify individuals. The proximate mechanisms underlying dispersal are not easily studied in small rodents, and the most promising approach now available entails detailed radio-telemetry studies of individuals.

CHAPTER SIX

Spatial Dynamics of Populations

Key Points:

- Spatial synchrony has been observed in many populations of small rodents that fluctuate in more or less regular cycles.
- Interspecific synchrony between species living in the same general geographical area is also commonly observed.
- If populations are nearby, spatial synchrony can be obtained by dispersal, but this is restricted to populations less than 5 to 10 km apart.
- For synchrony on a larger spatial scale, only two mechanisms have been suggested: mobile predators and large-scale weather events.
- Without landscape level studies, these explanatons for synchrony are difficult to evaluate.

One of the earliest observations on small rodent population fluctuations was that they tended to occur at the same time over large geographical areas of hundreds to thousands of square kilometers. In 1935 the Bureau of Animal Population at Oxford University, under the guidance of Charles Elton, began to circulate questionnaires to Hudson's Bay Company post managers and missionaries in Labrador and the Northwest Territories of Canada to map the relative abundance of lemmings, snowy owls, and arctic foxes. Figure 6.1 shows one summary of the lemming data obtained for a year of peak lemming numbers. This particular survey was carried out

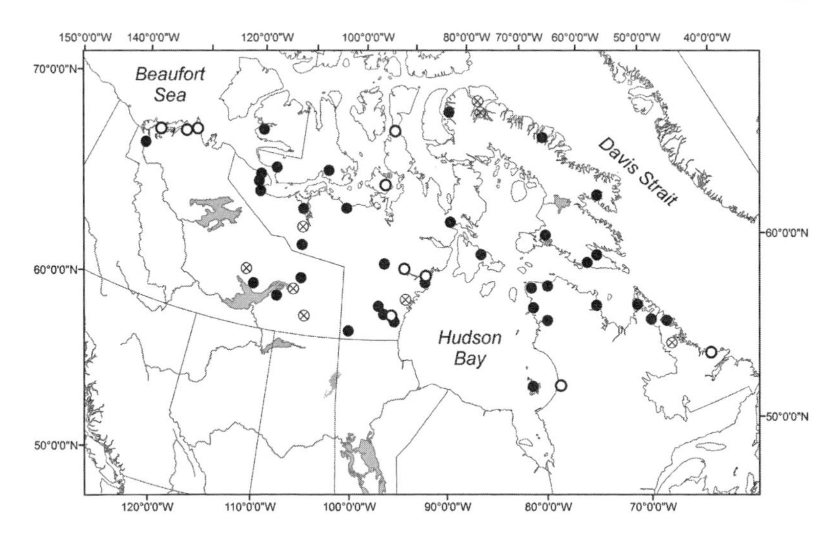

FIGURE 6.1 The Canadian Arctic Wildlife Enquiry results for lemmings in 1940–41. Lemming populations were at a peak in 1940–41 across most of the eastern and central Canadian Arctic. Questionnaires sent out to fur trading posts and government agents in arctic settlements recorded increase (black dots), decrease (white dots), or no change and not abundant (crossed dots). From Chitty and Nicholson 1942.

each year from 1933 to 1949. Much longer time series of qualitative information have been recorded in Fennoscandia (Angerbjörn et al. 2001). The information gained from these surveys helped to cement the observation that there was widespread synchrony in lemming population fluctuations in the Canadian Arctic. From these early observations, summarized in Elton (1942), an abundance of data and hypotheses has emerged to show a diversity of patterns in time and space.

Methods for Analyzing Synchrony

A fundamental problem in analyzing synchrony is a negative relationship between the length of the time series and the quality of the data. It is impossible to use all the elegant statistical methods for synchrony analysis without having at least 20 to 30 years or more of data. Almost no one has data of high quality for this length of time. Consequently, most analysts use questionnaires, hunter bags, or fur return data gathered by government agencies over long time periods, and treat these data as adequate indices of population size and density change. There has been a lively debate over the value of indices in population analysis (Anderson 2003; Engeman 2003), and I think the problem of data quality must always be uppermost in evaluating population studies of synchrony. Keeping this problem in mind, we will proceed on the worrisome assumption that the available indices are adequate for investigating synchrony.

The simplest method for looking at synchrony is to plot the time series and look for concordance between isolated populations. Figure 6.2 illustrates this approach for a few of the 225 time series of *Myodes* (*Clethrionomys*) *glareolus* analyzed by Saitoh (1987) and Saitoh et al. (1998). There is a strong tendency toward synchrony in these four populations, illustrated in figure 6.2, but the synchrony is not perfect. It can be tested by the cross-correlation coefficient, and it is significant statistically. An excellent synopsis of methods of measuring synchrony is given in Ranta et al. (2006, chapter 4). But the important questions are biological: What produces this synchrony, and how can we analyze it in more detail?

Three mechanisms are postulated as sources of synchrony among populations (Ranta et al. 1995; Ims and Andreassen 2005). Dispersal is one possible cause, but clearly it can be relevant only at relatively small spatial scales of a few tens of kilometers. Predation by highly mobile bird and mammal predators is a second possible mechanism (Ydenberg 1987). Weather effects are a third, and Moran (1953) showed theoretically that regional weather could synchronize local populations (Ranta et al. 1997).

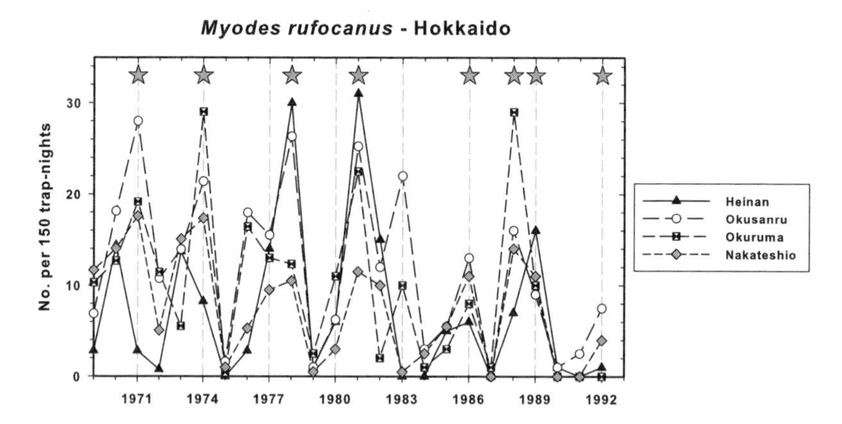

FIGURE 6.2 Synchrony in time series plots of the vole *Myodes* (*Clethrionomys*) *rufocanus* on four forest areas of northwestern Hokkaido. Density was estimated from snap trap lines over three nights of trapping in autumn. Stars indicate the peak years on most of these areas, which were up to 100 kilometers apart. Data from T. Saitoh.

Patterns of Synchrony

Synchrony among lemming populations has often been cited as one of the best examples of widespread synchrony of population change across several hundred to a thousand kilometers. Angerbjörn et al. (2001) analyzed 137 years of qualitative data on population fluctuations in the Norwegian lemming in Fennoscandia, and they found a high degree of synchrony between regions. Synchrony fell off slowly with distance (figure 6.3), but over a scale of 1,000 km it was still significant. So we can begin with the expectation of synchrony for independent fluctuating populations of small mammals, and see if it holds up.

To look at regional synchrony, Steen et al. (1996) trapped at 31 stations for five years along a 276 km transect in southeastern Norway (figure 6.4). Their results are broken down in figure 6.5. They found that regional synchrony for the bank vole in the boreal forest was strong, but only over a linear distance of 40 to 80 km, and they recognized that the relatively short duration (five years) of the study was limiting generalization. They suggested that the most likely hypothesis for synchrony in this species was dispersal. This result, supporting the dispersal hypothesis, led to a more detailed experimental study of the role of dispersal in synchronizing populations of voles, conducted by Ims and Andreassen (2005), who used enclosed populations of the tundra vole *Microtus oeconomus* in a patchy habitat to determine whether dispersal among the patches would synchronize population changes. Patches were separated by mowed strips that were not inhabitable by the voles. They repeated the experiment over two summer seasons and found a complete lack of synchrony among the populations in each patch. Their surprising results cast doubt on the ability of dispersal alone to synchronize populations at a local level. The explanation Ims and Andreassen (2005) offered was that dispersal in their populations was inversely density-dependent (figure 6.6). Since dispersal at high densities was so uncommon, local populations could not achieve synchrony within the summer breeding season.

The opposite conclusion about the role of dispersal in generating synchrony was reached by Diffendorfer et al. (1995), working in the grasslands of eastern Kansas. They fragmented a grassland in the pattern shown in figure 6.7, and live-trapped the prairie vole *Microtus ochrogaster* in each of these patches of variable size for eight years. Population fluctuations were identical in all of the patches (figure 6.8) with peak densities in 1985, 1987, and 1990. I do not understand why the studies of Ims and Andreassen

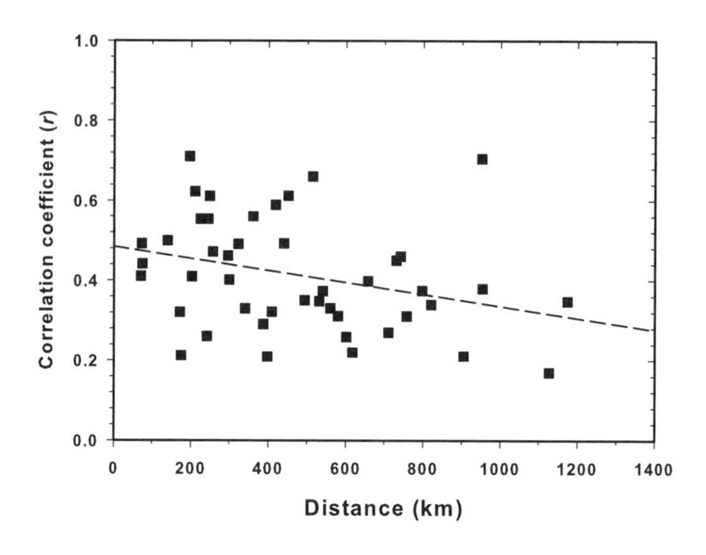

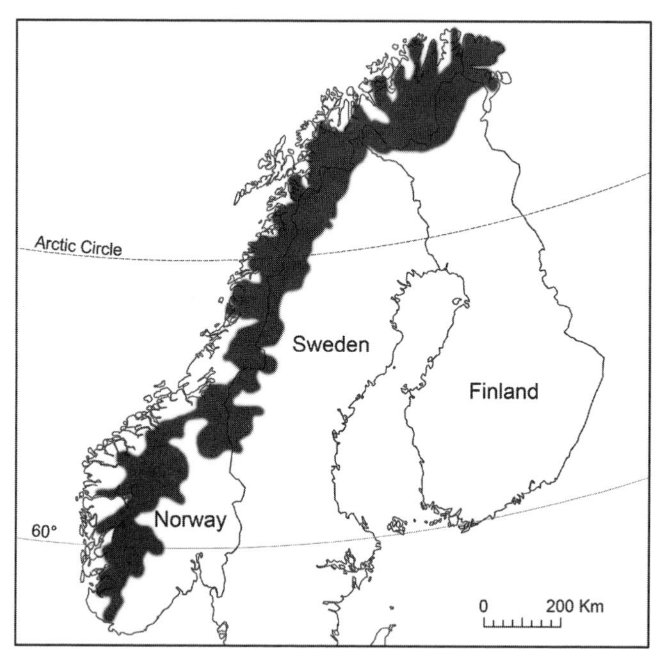

FIGURE 6.3 Synchrony of population changes in the Norwegian lemming in 10 regions (gray areas on map) of Fennoscandia, from 1862 to the present. Synchrony falls off with distance, as seems to be a common pattern, but for this species synchrony occurs over a very large spatial area. From Angerbjörn et al. 2001.

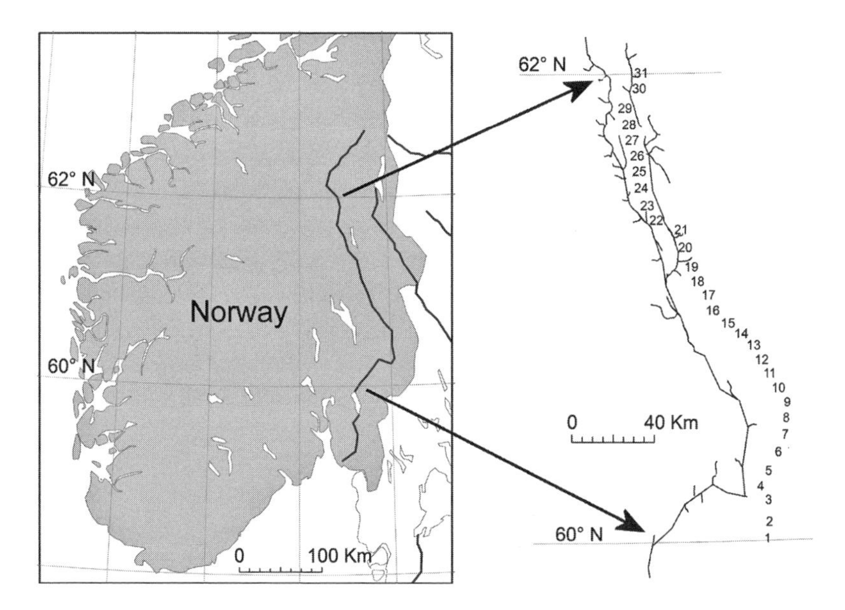

FIGURE 6.4 The spatial map of 31 trapping sites spread over 256 kilometers of southeastern Norway used by Steen et al. (1996) to investigate local synchrony in populations of the bank vole *Myodes (Clethrionomys) glareolus*. The transect was designed to remain in the same boreal forest habitat type from south to north. From Steen et al. 1996.

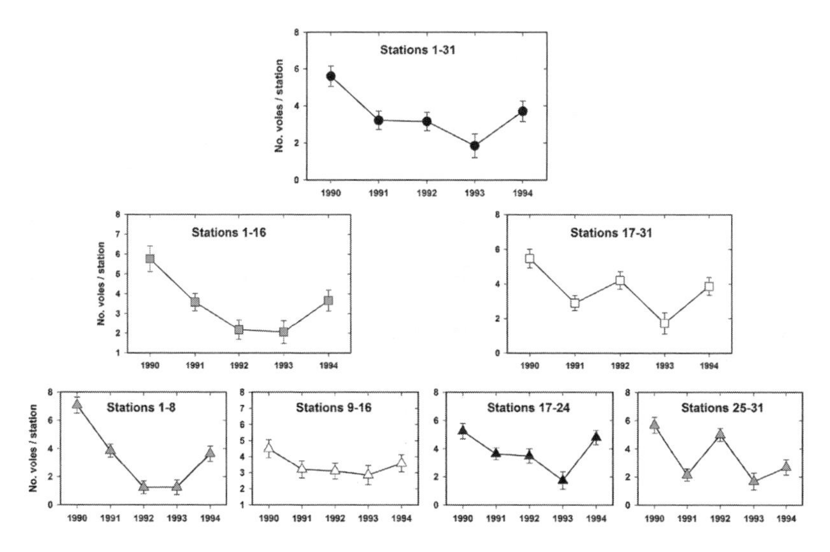

FIGURE 6.5 Fall population density indices of bank voles for different groupings of the 31 trap stations mapped in figure 6.4. Means ± 1SE. The top graph averages all 31 stations, the middle two graphs subdivide the transect in half, and the bottom four graphs subdivide it into quarters. From Steen et al. 1996.

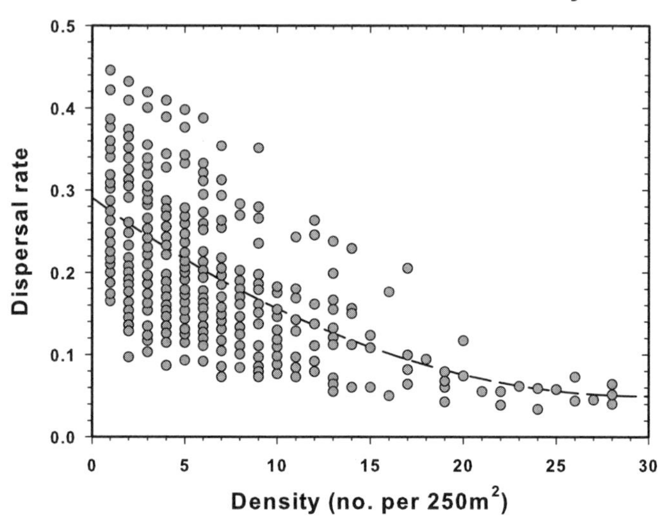

FIGURE 6.6 Dispersal rate of tundra voles between isolated patches of grassland in southern Norway. Dispersal rate is the fraction of individuals moving into an adjacent population during a week. From Ims and Andreassen 2005.

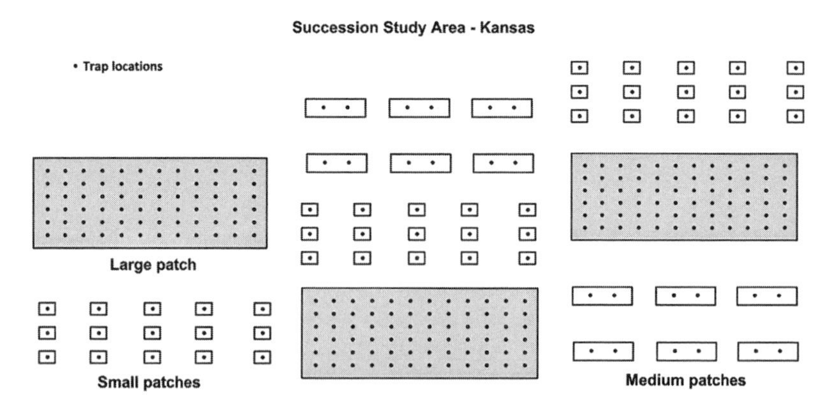

FIGURE 6.7 Diagram of experimental blocks of grassland used in the eastern Kansas study of small-mammal movements. The 40 small patches are each 4 × 8 meters; the 12 medium patches are 12 × 24 meters; the three large patches are 50 × 100 meters. A single large patch is a "large" block, a cluster of 6 medium patches is a "medium" block, and a cluster of 10 or 15 small patches is a "small" block. Blocks are separated by 16 to 20 meters. Dots indicate live traps. The total study area is six hectares. (From Diffendorfer et al. 1995.)

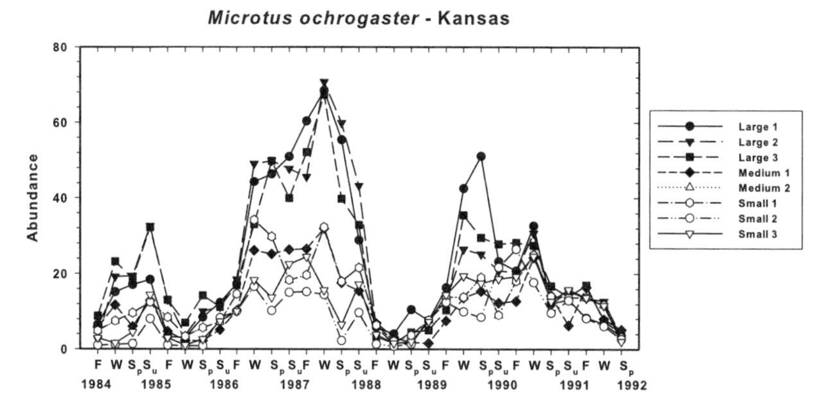

FIGURE 6.8 Abundance estimates for eight years for the prairie vole *Microtus ochrogaster* for the eight blocks on the experimentally fragmented grassland landscape shown in figure 6.7. Note the temporal synchrony among these populations. After Diffendorfer et al. 1995.

(2005) and of Diffendorfer et al. (1995) reached opposite conclusions. The reason could be species-specific responses to fragmentation, or the fact that the Ims and Andreassen study was confined to the summer months.

An extensive study of synchrony among local populations of the field vole *Microtus agrestis* in northern England has introduced another idea to the study of synchrony: traveling waves. The managed conifer forest in this area covers about 600 km^2 (figure 6.9). An extensive sampling program at 133 sites in the areas surrounding Kielder Forest from 1984 to 2004 showed that sites within 8 to 20 km were synchronous in population changes, but that sites farther away were somewhat out of phase (Lambin et al. 1998). At the extremes of the study area (figure 6.9) populations could be one or two years out of phase. Population peaks moved across the study area as a traveling wave at 12 to 15 km per year toward the northeast (Bierman et al. 2006). But since 2001 the cycle in Kielder Forest has broken down and become an annual fluctuation. Bierman et al. (2006) concluded that the major changes were in winter dynamics, thus implicating climatic factors; but the exact mechanism by which traveling waves were generated and then dropped out in this system remains unknown. Periodic traveling waves can arise in models of predator-prey and plant-herbivore interaction, and they can also be generated by dispersal when its timing is age-dependent (Kareiva 1990).

If dispersal in small rodents is a process that promotes synchrony only over short distances of up to a few kilometers, we are left with only

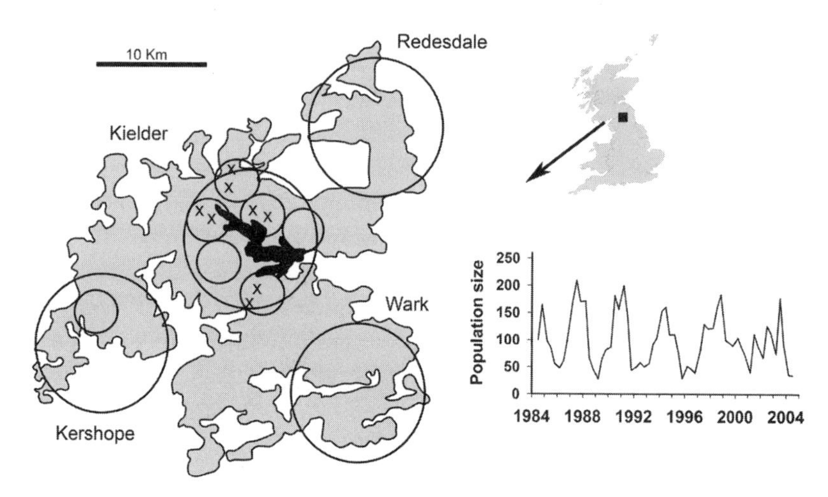

FIGURE 6.9 Variation in field vole densities in southern Scotland was sampled at three spatial scales in the forested area (grey shading). Seven different sampling sites were located within each of the circles, and each patch was indicated by a cross. Sampling was most intensive in six valleys around Kielder Reservoir (black shading). The black crosses indicate the approximate positions of grassland patches used as the smallest sampling scale. The graph shows the overall changes in vole density in Kielder Forest from 1984 to 2004. From Mackinnon et al. 2001.

two possible mechanisms to explain regional synchrony: predation and weather. These two mechanisms could be isolated by studies examining independent populations that operate under the same regional weather system but do not share predators. Islands offer one possible system for such studies, because at least small mammal predators like weasels might be absent from some islands but present on others or on the adjacent mainland. Krebs et al. (2002) tried to make use of islands to dissect the reason for synchrony in lemming populations in the central Canadian Arctic. Eight sites were sampled annually in early summer with snap traps for lemming abundance (figure 6.10). Six of the sites were small islands, and we searched for predator signs in the hope that mammalian predators (arctic foxes and ermine) would not be present because of the islands' isolation and small size. Collared lemming populations in the whole region fluctuated in spatial synchrony (figure 6.11). Interspecific synchrony was also present. When one species at a site was at peak abundance, all other species tended also to peak (Krebs et al. 2002). Because mammalian predators did not access the islands in those peak years, we favored the weather hypothesis for generating regional synchrony. But this conclusion was relatively weak, because we could not study predation in detail on any of the islands. It was also undercut by an analysis that indicated genetic similarity among collared lemmings at all the sites, which could have been achieved only by continuous immigration from the mainland (Ehrich et al. 2001). The gap between the genetic results and our ecological intuition—that dispersal to these islands across the ocean was impossible in summer and risky, thus probably rare, across the ice in winter—was left unresolved. Studies on small arctic islands could resolve these issues, especially if their small populations periodically go extinct and they must be recolonized.

If weather is the synchronizing agent in small rodent fluctuations, we need to find out exactly how it affects reproduction and survival in lemmings and voles (Reid and Krebs 1996). There are many possibilities. Late onset of snow could cause more exposure to cold temperatures in the autumn, and more exposure to avian predators. Icing events in winter could prevent a substantial amount of food supplies from being accessed. Deeper snow could encourage more winter breeding because of its thermal amelioration (Merritt 1984). A corollary of the weather hypothesis of

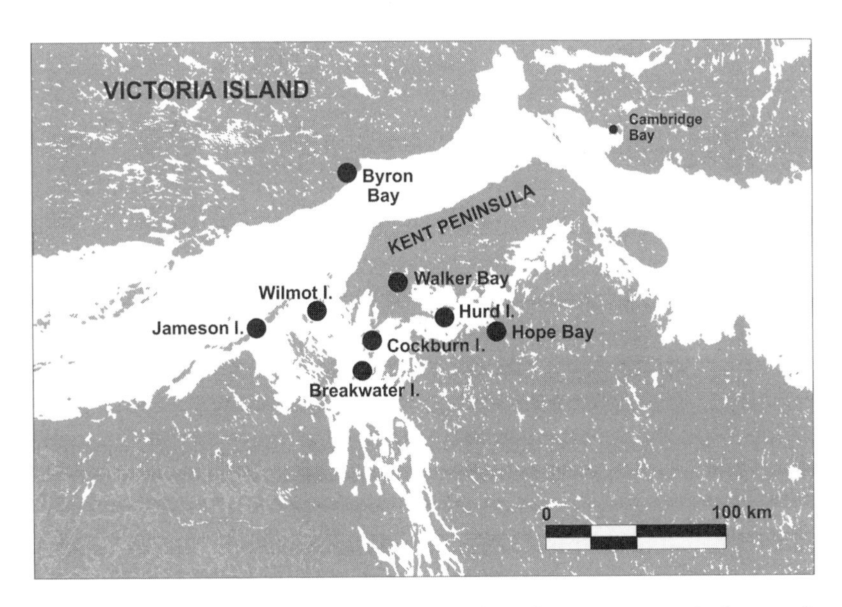

FIGURE 6.10 Eight sites sampled for lemming abundance from 1994 to 2000 in the central Canadian arctic. Five sites were islands and three were mainland sites 50 to 80 km apart. Movements of lemmings between islands and the mainland is most unlikely and all the area is under the same weather systems. (After Krebs et al. 2002.)

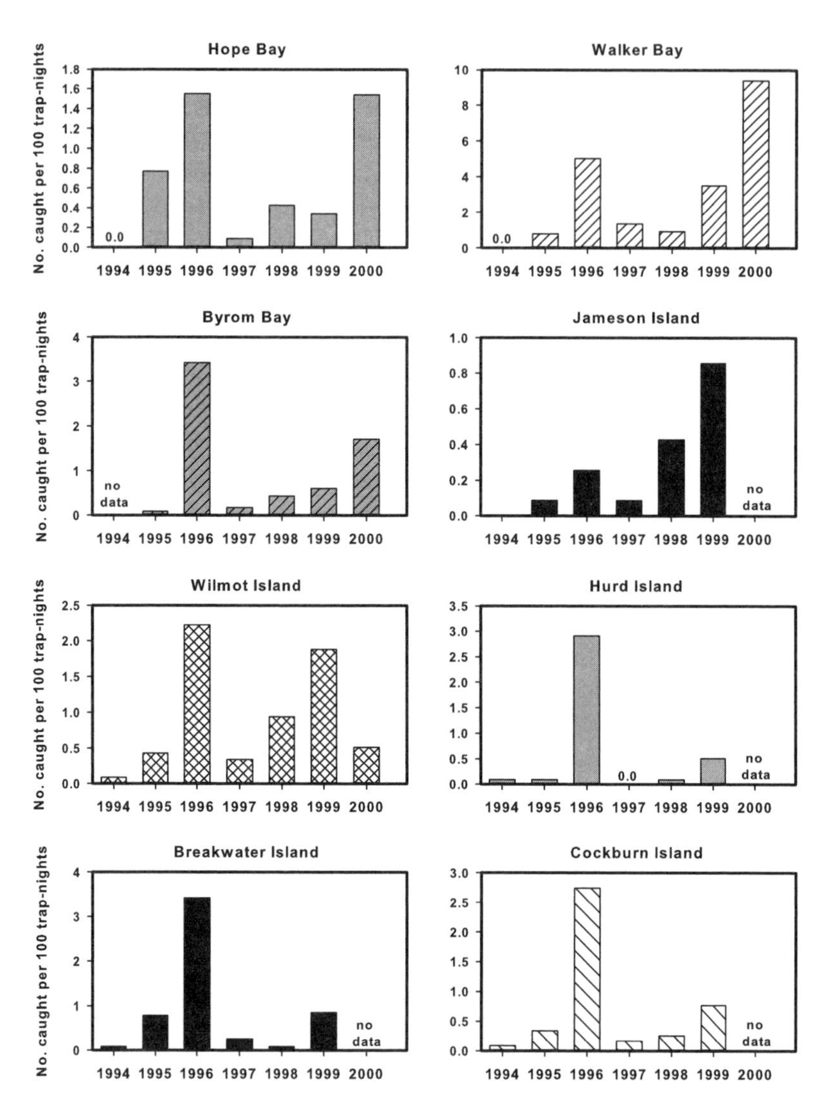

FIGURE 6.11 Collared lemming (*Dicrostonyx groenlandicus*) population indices for six island and two mainland sites in the central arctic. Except on Jameson and Wilmot Islands in 1999, lemming populations were in synchrony across this region from 1994 to 2000. Each estimate is based on data from 1,170 trap nights using snap traps. Data from Krebs et al. 2002.

synchrony is that arctic climate change in the coming decades will possibly alter patterns of synchrony.

Long-Term Patterns of Synchrony

Because most studies of synchrony are short-term (covering less than 20 years), small mammal ecologists have tended to assume that our recent observations are of a stable system. We can test this with very long-term data, and for small rodents these kinds of data are available only from Fennoscandia. Henden et al. (2009) analyzed the fox bounty data from 18 counties of Norway spanning 1880 through 1976. They assumed that the red fox and arctic fox populations in Norway depended on small rodents and fluctuated with the same periodicity as the rodent prey base. They divided the data into three time periods in which cyclicity and synchrony appeared to change. Figure 6.12 shows how between 1880 and 1910 all areas with cycles showed a four-year periodicity, while many areas in southern Norway were not cyclic. Spatial synchrony was well established in this time period, and populations up to 1,200 km apart fluctuated in phase. But from 1911 to 1932, some areas in southern Norway shifted, cycle length became more variable, and spatial synchrony decreased so that some counties were out of phase with adjacent areas. The bounty system was stopped from 1932 to 1947, so no data are available from that time. But from 1948 to 1976 some areas again became noncyclic, and for those that were cyclic the period length again centered on four-year cycles. The key point is that there is no single "typical" period or pattern in these long-term fox data, so analyses of relatively short time series must be careful to limit their generalizations.

The mechanism behind these shifts from cyclic synchrony over large areas to loss of cycles and only smaller-scale regional synchrony is not known. The suspected factor is weather, and the difficulties of testing weather hypotheses are well known (Berteaux et al. 2006). The time period from 1880 to 1910 was one of cold and uniform weather systems in Norway, and the weather began to warm by the 1930s (Henden et al. 2009). The suggested shifts might then be tied to the alternation of cold and warm climatic periods. The key again is to find out exactly how these weather changes affect the reproduction and mortality of small rodents.

There are no long-term data sets from North America that could be used to test some of the suggestions about synchrony that have come from

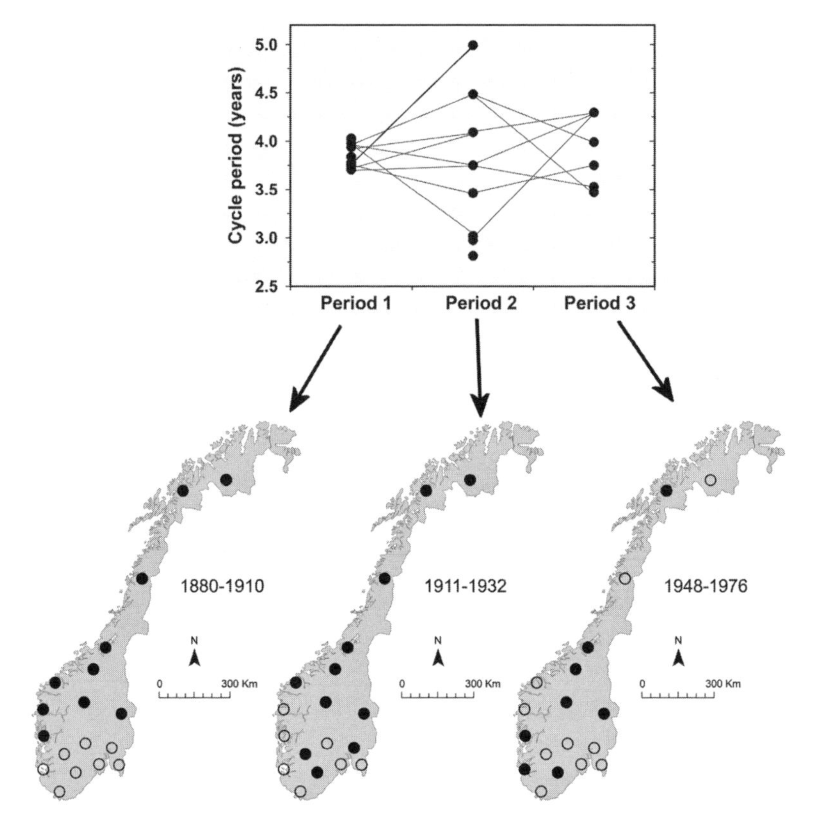

FIGURE 6.12 Cyclicity and cycle period length for arctic and red foxes in 18 Norwegian counties from 1880 to 1976. Upper panel: estimated period cycle length over three time periods in which cycle length changed. Lower panels: maps show the spatial distribution of cyclic (filled circles) and noncyclic (open circles) populations in Norway over the three time periods. (From Henden et al. 2009.)

Fennoscandian data. For the present it will only be possible to test such hypotheses only with suitable experiments or observations.

Genetic methods can also shed light on long-term synchrony, and while there is no direct demographic connection, we can ask whether populations separated in space are slightly or highly differentiated. We have evidence that if they are very similar, dispersal between local populations has occurred. Ehrich et al. (2001) sampled collared lemmings, *Dicrostonyx groenlandicus* along a linear transect on southern Victoria Island in the Canadian arctic and measured their genetic differentiation with mitochondrial DNA and four nuclear microsatellite loci. They found only weak differentiation over large geographical distances (figure 6.13), and only above about 50 km of distance did genetic similarity begin to fall off. These genetic results imply frequent and extensive exchange of individuals among local populations, so that synchrony of population fluctuations could occur by local and regional dispersal.

Others have found that genetic similarity falls off much more rapidly with geographical distance. Stacy et al. (1997) studied genetic differentiation in bank vole (*Myodes glareolus*) populations in the boreal forest in Norway using the same mtDNA sequence used earlier by Ehrich et al. (2001), and concluded that local populations of the bank vole typically extend over less than about 8.5 km. Genetic tools could be very useful for outlining the possible extent of dispersal and synchrony in small mammals, and this approach to the study of population structure has much to recommend it (Rousset 2001).

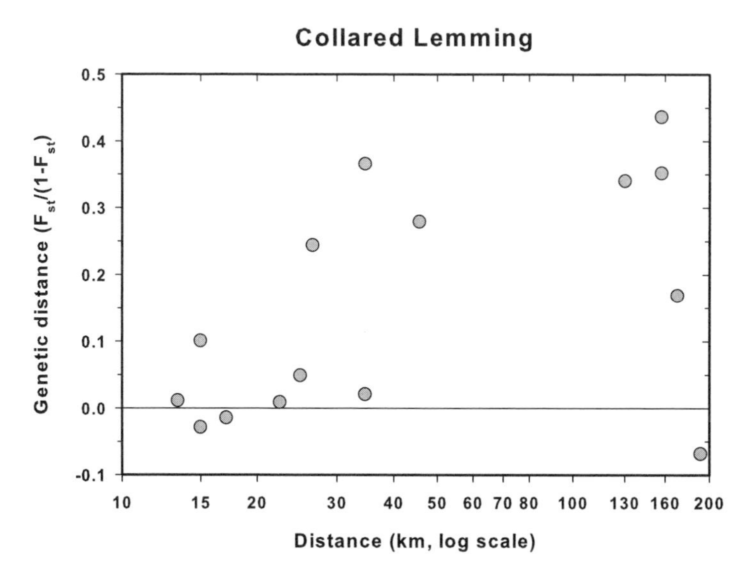

FIGURE 6.13 Differentiation with distance among collared lemmings at six localities on Victoria Island in the central Canadian Arctic. Pairwise differentiation estimates from mtDNA CR haplotypes are plotted against the logarithm of geographical distance. There is very little genetic differentiation between populations 30 kilometers apart, and even populations more than 200 kilometersapart are genetically similar, thus implying much dispersal among regional populations. From Ehrich et al. 2001.

Conclusion

There is no question that synchrony occurs in both space and in time between sympatric species of small rodents. Spatial synchrony comes and goes, however, and for now we can only keep mapping it and designing experiments with island populations or other experimentally isolated patch populations to see whether we can get more evidence regarding the roles of dispersal, predators, and weather in producing populations that fluctuate in phase with one another. The present consensus, if there is one, seems to be that local populations are synchronized by dispersal, and that regional populations are synchronized by weather. But mobile predators could also play a role in synchrony, and it is too early to ignore their contribution. Much more field research needs to be done on these questions before we can give strong, evidence-based answers.

CHAPTER SEVEN

How Can We Determine What Drives Population Changes?

Key Points:

- The major question in population dynamics is: What factors determine the rate of population change in any species?
- Two different ways of answering this question are to watch and wait, and to manipulate the population with adequate controls.
- Either approach requires clear hypotheses that include predictions, so that a decision can be reached on whether to accept or reject each hypothesis. This is rarely done in published papers.
- Any type of population change can be explained after the fact by any number of factors, particularly if the factor in question has not been measured.
- Multiple-factor hypotheses are useful if they include clear predictions about the mechanisms involved and the evidence that will confirm or reject them.
- The method of testing all our models must be the hypothetico-deductive system that is used in all the successful sciences.

If we observe a population that is fluctuating in abundance, one central question in ecology concerns what factors influence these fluctuations. This has been the global agenda of population ecology for the last 100 years, and much progress has been made in describing population changes, particularly those of higher vertebrates and insect pests. Rather less progress has been made in explaining the factors that cause the changes, and various reasons have been suggested for this lack of progress (Krebs 2006; O'Connor 2000; Peters 1991; Weiner 1995). Possibly no topic in the discipline has been more controversial than this question of how to do population ecology properly.

Small rodent studies, along with bird and insect studies, have been at the forefront of these controversies because they have provided the raw material for observations and experiments on species with short life cycles. In this chapter I will use the accumulated small rodent data to address this issue, to see if we can map a way forward.

What Is the Question?

There are many different questions we can ask about populations of any species, and we can first make two fundamental observations. The first observation is that abundance varies from place to place: there are "good" habitats where the species is, on the average, common, and "poor" habitats where it is, on the average, rare. This fundamental question of why abundance varies spatially has rarely been asked or answered for any small rodent or indeed for most animal species. The second observation is that no population goes on increasing without limit, and the population growth rate changes from positive to negative after some time interval. To explain this observation two different paradigms have been used (Krebs 2002; Sibly et al. 2003). Figure 7.1 illustrates these two approaches to the study of population problems, which are often confused.

The *density paradigm* uses population density as the independent variable for analysis. I have argued that this is a poor variable to use, because the key relationships between density and reproduction, survival, and movements are not repeatable in time or space (examples in Krebs 2002). Population density is affected by many habitat factors and is critically significant for managers involved in conservation programs, but the important point for understanding population change is that density is not a mechanism like predation or disease. It is essential, of course, to know the density in order to calculate population growth rates.

Population growth rates are defined most simply as a ratio:

$$\text{finite population growth rate} = \lambda = \frac{N_{t+1}}{N_t}$$

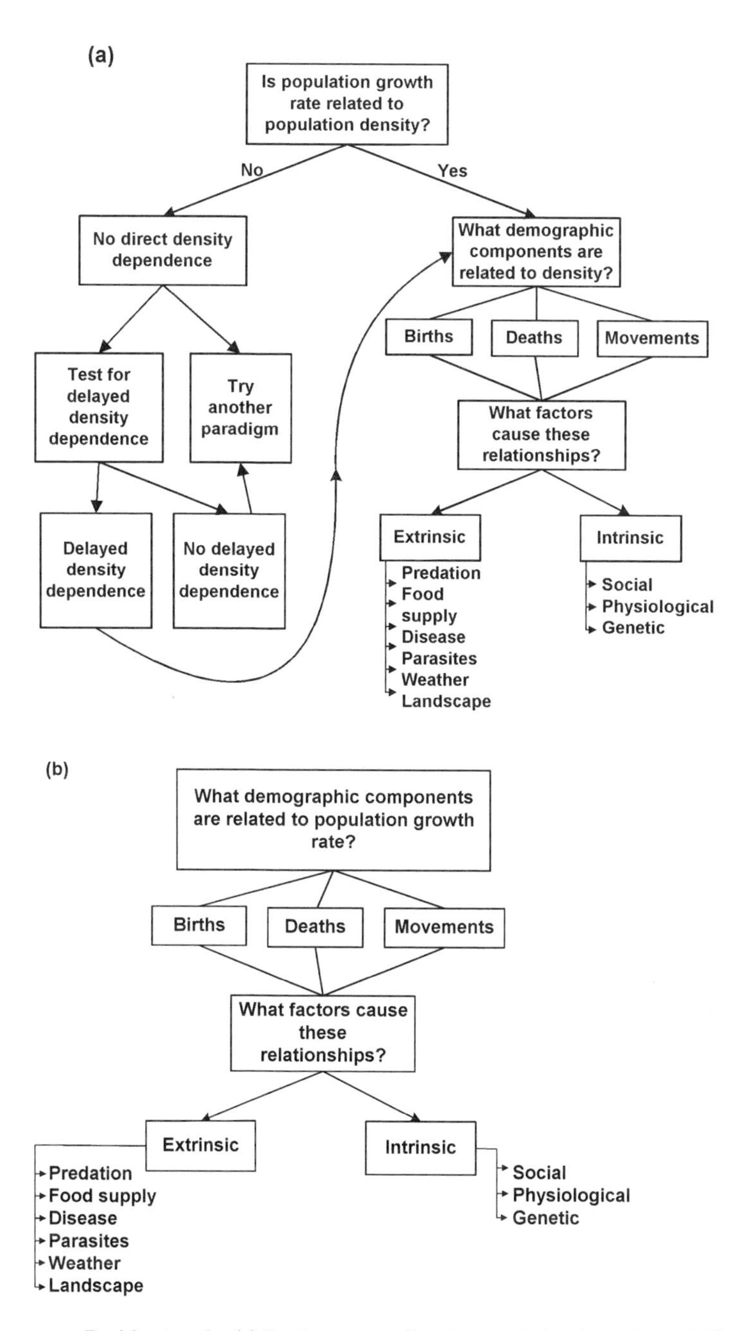

FIGURE 7.1 Decision tree for (a) the density paradigm for population dynamics, and (b) the mechanistic paradigm. The significant difference between these two approaches lies in the question of whether the independent variable is population density or population growth rate. From Krebs 2002.

where λ = finite population growth rate (rate per capita),
N_{t+1} = population density at time t+1, and
N_t = population density at time t.

For many analyses the population growth rate is expressed on a log scale as

$$\text{instantaneous population growth rate } = r = \log(\lambda).$$

The instantaneous population growth rate is sometimes abbreviated as *pgr* (Sibly et al. 2003) and may be expressed in natural logs or base-10 logs (which differ only by the multiplier 2.302585).

The key question thus becomes which variables determine population growth rates. This question must be answered in a hierarchical manner (figure 7.1b). First, we need to know for any system what the relationships are between population growth rate and measures of reproduction, mortality and movements. This problem has been the focus of theoretical population ecologys since the seminal works of Cole (1954) and Lewontin (1965). The relative importance of life history variables to population growth rate has centered on the sensitivity of the finite rate of population growth λ to changes in various life history variables. Oli and Dobson (2003), for example, subdivided the demographic variables into several measurable quantities: age at sexual maturity, age at last reproduction, juvenile survival, adult survival, and fertility. Fertility is a composite term defined as the number of young that join the juvenile population per female (litter size × sex ratio × probability of breeding × survival to juvenile age). In such analyses, emigration and immigration are ignored. These methods allow one to calculate elasticity, which provides a direct quantification of the relative importance of life history variables to the population growth rate. Caswell (2001) describes the methods in detail.

Given that we can determine which of the primary demographic variables change with population growth rate, we can now proceed to the focal point of the mechanistic paradigm. There are six extrinsic variables that might affect the growth rate, and three intrinsic variables. The next decision is how to proceed with nine possible variables as possible key influences. Traditionally, researchers have focused on two extrinsic variables, food supply and predation, as the factors most likely to be critical in affecting population growth rate. This has been mirrored in the dichotomy between the top-down (predation) and bottom-up (food, nutrients) paradigms of community ecology (Hairston et al. 1960; McQueen et al. 1986).

TABLE 7.1 **Characteristics of two schools in small mammal ecology that attempt to define the research agenda for understanding population fluctuations**

School	Positive attributes	Difficulties	References
Single-factor	1. Deems simple mechanistic hypotheses for any factor important	1. Cannot hold other variables constant in field experiments	Tamarin 1978a
	2. Is applied to a wide variety of species, and is therefore simple to reject if not correct	2. May apply to some species but not others	Krebs 1978
	3. Enables models of population dynamics to be constructed simply	3. Uses models perhaps too simple to be useful tests of reality	Dambacher et al. 1999
	4. Narrows the focus of investigation to a manageable work level	4. Ignores possible interactions of two or more factors	Hilborn and Stearns 1982
	5. Is compatible with Popper's method of falsification	5. Has problems with the falsification approach to science	Lakatos and Musgrave 1970
Multiple-factor	1. Can include a wide array of species and environments	1. May combine groups of species with different mechanisms	Lidicker 1973, 1978
	2. Brings together many possible interactions in a holistic approach	2. Makes reductionist testing of components difficult	Hilborn and Stearns 1982
	3. Is compatible with systems thinking in other sciences	3. Uses models that may be vague and intuitive	Loehle and Pechmann 1988
	4. Encourages inductive approaches to population regulation	4. Encourages explanations after the fact	Murray 2001
	5. Cannot be refuted by any particular case history	5. Makes no specific predictions to test	Popper 1963

But it is clear in figure 7.1 that there is potentially more to population control than food and predators, so a more complex approach is needed.

Population ecologists are caught in a dilemma because they are taught Ockham's Razor—the idea that, other things being equal, simplicity is the best starting point for scientific explanation—and on the other hand they know from natural history that many factors influence populations in the field. One consequence of this dilemma is that rodent ecologists have split into two camps, which for convenience I will call the multifactor school and the single-factor school. Table 7.1 lists some of the advantages and disadvantages of each approach. One way to illustrate this difference in approaches is to describe in more detail a good example of the two schools, which for convenience I will label the Chitty approach and the Lidicker approach, after two of the main scientists around whom the controversy has swirled. But first it is useful to discuss one approach that has not worked out as well as one might have hoped.

Approaches That Have Not Proven Useful

In an ideal world, laboratory experiments on rodent populations would quickly test the variables hypothesized to be relevant to understanding dynamic changes in births and deaths. This was the prevailing belief in the 1950s, when a large number of confined populations (particularly of *Mus musculus*) were studied. These studies—remarkable in their design, rich in their control of environmental conditions, and accurate in their vital statistics—have provided little or no understanding of real-world populations. Singleton et al. (2005), for example, pointed out how laboratory studies of *Mus rnusculus* populations were of little use in determining which factors generate outbreaks of house mice in the wheatlands of southeastern Australia. Laboratory experiments were useful in suggesting that spacing behaviour might be relevant to understanding population changes (e.g., Clarke 1955; Christian 1970). The reason for their failure to increase our understanding of population dynamics is in complete contrast to all other areas of biology, in which laboratory studies are essential to all progress. Why should this be? Laboratory populations simultaneously alter too many variables—dispersal, food quantity, food quality, predation rates, disease transmission, parasite transmission, amount of space, and seasonality—so that when we observe, for example, that laboratory populations of *Microtus* do not fluctuate cyclically, we do not know which variables to blame for this change in population dynamics.

Consequently, almost no one does laboratory population studies of small rodents any more in spite of the general scientific belief that laboratory studies are essential to progress in understanding (Krebs 1988, table 1) The lack of spatial heterogeneity in laboratory studies is possibly their critical shortcoming with respect to rodent population dynamics. Although this shortcoming can be overcome with sufficient knowledge, cleverness, and facilities (Crowcroft 1966), these conditions are rarely achieved. It may be that laboratory studies will provide essential detailed analyses of some of the features of rodent population dynamics that have been discovered in field populations. One example is the analysis of reproductive suppression in *Myodes* (*Clethrionomys*) females, reported by Bujalska (1970) from field observations and analyzed in enclosures by Saitoh (1981) and Bondrup-Nielsen and Ims (1986). But at this time only a few very specific questions can be usefully addressed in laboratory population studies of rodents.

Chitty's Approach

Dennis Chitty, working with Charles Elton at Oxford from the mid-1930s, attempted to determine the causes of population fluctuations in the field vole *Microtus agrestis* in Great Britain. In a 1952 paper based on his D.Phil. thesis, he described some demographic details of the population fluctuations the Oxford group had studied (Chitty 1952). In particular, Chitty felt that their investigations of the food shortage hypothesis (Summerhayes 1941), the disease hypothesis (Chitty 1954), and the predation hypothesis (Chitty 1938) had convincingly shown that none of these conventional factors could explain the changes in numbers they had observed. More details of this early work are described in his book *Do Lemmings Commit Suicide?* (Chitty 1996). Chitty then began to investigate the effects of the social environment on population changes in voles, and in the process he developed the self-regulation hypothesis (Chitty 1957). At the same time he became convinced that there must be generality in the study of population dynamics, so that hypotheses that could apply to voles might also apply to birds and insects.

Chitty first stated the self-regulation hypothesis in a classic paper (Chitty 1960): "On the assumption that vole populations are a special instance of a general law, the hypothesis is set up that all species are capable of regulating their own population densities without destroying the renewable sources of their environment, or requiring enemies or had weather to keep them from doing so" (p. 99). The key point Chitty recognized was that much population ecology was carried out on the assumption that the properties of individual organisms were constant. He pointed out that mortality rates could change for two quite different reasons, diagrammed in figure 7.2. The physiological attributes of individuals and the genetic composition of populations should not be considered constants in studies of population change, according to Chitty's view. Chitty believed that for voles and lemmings the extrinsic factors of food shortage, predation, and disease were present but not always operative or strong enough to

	First Hypothesis		Second Hypothesis	
Time	1	2	1	2
Death rates	$D_1 < D_2$		$D_1 < D_2$	
Organisms	$O_1 = O_2$		$O_1 \neq O_2$	
Mortality factors	$M_1 \neq M_2$		$M_1 = M_2$	

FIGURE 7.2 Two contrasting approaches to population dynamics. D = death rate, O = physiological and behavioral properties of individuals, M = mortality rate. The first hypothesis for understanding changes in population growth rate is the conventional approach to population dynamics, which assumes individuals with constant properties. The second hypothesis, suggested by Dennis Chitty, recognized that individuals could change to become more susceptible to extrinsic factors. The real world is somewhere between these two extreme suggestions.

limit population growth rates. The explanation for population fluctuations, he suggested, lies more in intrinsic factors of the changing social environment.

Dennis Chitty was a strong advocate of Popperian science, which advocates a falsifications approach to scientific hypotheses (Popper 1963). The strength of this approach has always been the rapid advancement of science (Platt 1964). The weakness is that in any specific test of a hypothesis, the investigator always has a set of background assumptions as well as the original hypothesis. Consequently, if an experiment results in a rejection of a hypothesis, it may be because the hypothesis was wrong and should be changed, or because one or more of the assumptions were not correct (Murray 2001). Other ecologists are less enamored of the Popper approach (e.g., Diamond 1986), and prefer a greater variety of approaches to ecological questions.

Chitty's approach has highlighted another controversial aspect of the study of population fluctuations in rodents. Should we assume that all fluctuations in different species and different habitats have a common explanation, and thus seek generality? Chitty himself (1960) originally restricted his ideas to rodents that showed cyclic fluctuations, on the assumption that the cycles were repeatable sequences of events that should have a common explanation and thus could be subject to standard tests. He argued that population cycles in a broad range of species from lemmings in northern Canada to field voles in Germany, with different food plants, predators, and diseases, were more likely to have a unified explanation based on a common social structure and the resulting interactions between individuals. This is, of course, an article of faith in the principle of uniformity, which we can justify only in retrospect when the causes are well understood. On this basis, for example, Chitty could dismiss the role of snow cover for population cycles because many species of voles that fluctuate in cycles live in temperate areas where snow is rare or nonexistent. The assumption of generality is a key one on which ecologists have varying opinions.

Lidicker's Approach

The multifactorial model has been an alternative approach to population questions for many years (Thompson 1929), and the comprehensive school of population ecology has long been prominent in insect ecology (Andrewartha and Birch 1954). Bill Lidicker has been a prominent advocate of the multifactorial approach to population studies (Lidicker 1978), the principles of which many others share (e.g., Holmes 1995), and he serves as an important advocate of it. The multifactor hypothesis will be discussed in more detail in chapter 12.

Lidlicker's (1978) multifactor model of population regulation has several basic premises. It suggests that it will not be possible to explain densities of natural population on the basis of one or a few of the factors listed in figure 7.1. It thus rejects the generality promised by the Chitty approach. It also implies that both extrinsic and intrinsic factors will be involved in causing changes in populations, and that the mechanisms by which they change in abundance are expected to vary in both time and in space. How much loss of generality this will involve is not clear. At the extreme a study of the field vole in Wales in 1990 could not be compared in a Popperian sense with the study of the same species in 2008, nor with a study in Scotland in 1990. This conjecture may in fact be correct, but we will not know whether it is or not without a great deal of field study.

One strength of Lidicker's approach is that he recognizes that both extrinsic and intrinsic factors may be involved in population regulation. His early suggestion that dispersal could affect population changes (Lidicker 1962) is a good illustration of this belief. He illustrated it by summarizing how studies on the California vole *Microtus californicus* fit into the multifactor model (Lidicker 1973, 1978). He recognized six factors that are

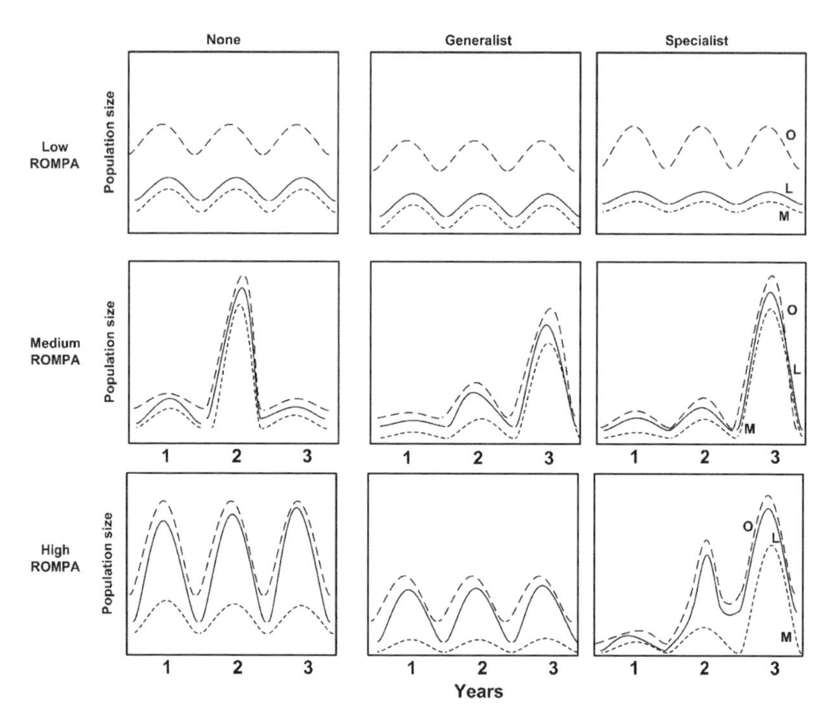

FIGURE 7.3 Some predictions from Lidicker's trophic/ROMPA interaction model (TRIM) for microtine rodents. N = number of voles, L = numbers averaged over the entire landscape (solid lines), O = numbers in optimal patches (dashed lines), and M = numbers in marginal patches (dotted lines). ROMPA is the ratio of optimal to marginal patch areas from the vole perspective. Types of predation are listed along the top. After Lidicker 2000.

important in causing three- to four-year cycles in the California vole in the coastal Mediterranean climate of central California:

1. seasonal rainfall pattern (determines vegetation growth),
2. photoperiod and food supplies (affect litter size),
3. interaction of winter vole density and subsequent summer food supply,
4. physiological effects of high densities in summer,
5. presaturation dispersal, and
6. mammalian predation.

Lidicker pointed out that, although at any given time a single item like predation can be the limiting factor for population changes, the important point is that no single factor acts alone.

A further important addition to understanding population changes in rodents has been the introduction of landscape ecology ideas and habitat heterogeneity as important components in understanding patterns of population change (Cockburn and Lidicker 1983; Ostfeld et al. 1985; Lidicker 2000). This approach has resulted in a landscape model for vole populations, illustrated in figure 7.3. Lidicker (2000) calls this the TRIM model (trophic/ROMPA interaction model, where ROMPA is the ratio of optimal to marginal patch areas in the landscape). "Optimal habitat" in this model refers largely to the quantity and quality of food resources. The TRIM model is basically a two-factor, food supply–predation model, with little recognition of intrinsic factors other than movement between source and sink habitats. It is largely descriptive, which is good in providing a framework in which to fit various studies; but it is impossible to test, because it forbids nothing from happening.

That landscape affects populations is clear from studies of experimentally fragmented populations. Figure 7.4 shows the average population density of the prairie vole *Microtus ochrogaster* in the patchy system illustrated in figure 6.7. The smaller the patches the larger the density of these voles. The explanation suggested is that another competing species,

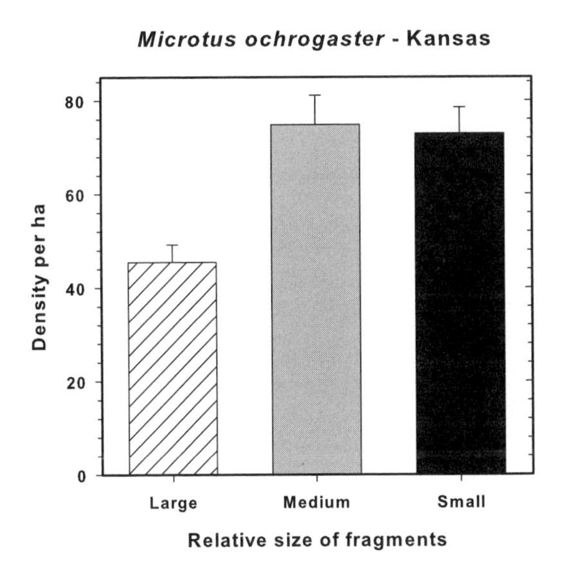

FIGURE 7.4 The effect of landscape patch size on the average density of the prairie vole in Kansas grasslands. The landscape design is shown in figure 6.8. Smaller patches support higher densities of these voles. Error bars are 1 S.E. After Diffendorfer et al. 1995.

the cotton rat (*Sigmodon hispidus*), does not live in small patches, so that fragmenting the habitat removes this competing species (Diffendorfer et al. 1995). The mechanism of the competition is not clear, but it most likely involves interference rather than exploitation. We found the same effect in *Microtus townsendii* along farm fence rows in grassland near Vancouver (Krebs, unpublished). In this case, the small patch effect on vole density cannot be dependent on the absence of a competing species, because there were no other small rodents in this habitat.

Alternative Approaches

These two general approaches to studying small rodent population dynamics represent two world views that are discussed in philosophy of science texts as *deduction* and *induction*. Dennis Chitty and his followers subscribe to deduction, the statement of a clear hypothesis and its predictions, including exactly what the hypothesis forbids, so that rejection or tentative acceptance is the goal. Bill Lidicker and his followers prefer induction, and the amassing of evidence that favors one view over another. Coupled with this distinction is Chitty's judgment that there will be a unifying generality or a set of necessary conditions that will explain rodent population changes, as compared with Lidicker's idea that generality will be elusive, local, and conditional on variables like landscape configuration. These distinctions extend further than those in the story of the blind men exploring the elephant, and they have led to much useful discussion about how to advance the study of rodent populations (Gaines et al. 1991; Lidicker 1991). There is general agreement on a set of principles of investigation, which I list in table 7.2, so the similarities between these two views of how to proceed are important to remember.

One approach to small rodent fluctuations that is not covered by either Chitty or Lidicker is the one taken by mathematical modelers. Both Chitty and Lidicker and their followers support the idea of models, but those models are verbal or graphical rather than rigorously quantitative. There is probably no more contentious issue in ecology than the role of mathematical models. Views range from those at one extreme—that without an explicit mathematical model, no one should do any field studies (Turchin 2003)—to others that question the formulation of current population models (Aber 1997; Oreskes 1998; Ginzburg and Jensen 2004; Ginzburg

TABLE 7.2 **Points of agreement and disagreement in the two alternative approaches to understanding population changes in small mammals**

Points of agreement	Points of disagreement
Experiments must be performed to test hypotheses.	Experimenters should begin with simple models and add complexity only as necessary.
Common factors and general explanations within and between species are a major goal of research.	Laboratory studies of confined rodent populations are not useful for testing models.
The hypothetical-deductive method should be promoted.	The density-dependent paradigm is useful for investigating population fluctuations.
Field studies must consider habitat heterogeneity.	Explanations for population change must be site-, time-, and species-specific.
Dispersal must be considered as a major demographic parameter in addition to reproduction and survival.	Landscape models are useful for understanding the characteristics of population fluctuations.
Both intrinsic and extrinsic factors can be involved in determining the rate of population change.	Specific testable predictions can be made from multifactor models.
More field studies of diverse species and environments will further our scientific understanding of fluctuations.	Holism is a useful perspective in dealing with complex systems.
Operational definitions of terms are essential to progress.	Induction is a desirable approach to scientific questions.

et al. 2007). I cannot summarize here all the discussion about the role of models in population dynamics. The main disagreement between field workers and mathematical modelers has always been that models contain parameters that cannot possibly be measured, that the assumptions of any particular model are rarely stated clearly, and that many models do not lead to any useful proposal for field experiments, so that there is no feedback between the computer modeling and the field work.

A whole host of mathematical models have been useful for population ecologists. The best examples are those outlined in Caswell (2001) and used in an elegant manner by Oli and Dobson (2003). These models use demographic data to test ideas about the relative importance of various parameters of birth and death on the rate of population change. Other good examples come from the work of Stenseth and his colleagues (Stenseth et al. 2001, 2004; Bjørnstad et al. 1999) in which there is a clear feedback loop between field data and model development and prediction. So there are good models to be developed, as Aber (1998) has pointed out, but much modeling of small rodent population fluctuations has been of little use in clarifying the empirical issues we discuss in the next six chapters.

Conclusion

A simmering controversy in studies of small rodent population fluctuations mirrors another ongoing controversy about how to do population studies in plants and animals. The two polar views are illustrated by the important work of Dennis Chitty, who adopted a Popperian, experimental, and deductive framework of testing simple one-factor models first, and the equally important work of Bill Lidicker, who adopted an inductive, descriptive, and holistic view of populations changing under the influence of many different factors. While Chitty hoped to achieve a hypothesis of universal applicability to higher vertebrates, Lidicker emphasized the great variability in time and space both within and between species. Both perspectives are useful, and both strongly support the hypothetical-deductive approach to gaining knowledge.

CHAPTER EIGHT

The Food Hypothesis

Key Points:

- There is considerable disagreement about exactly what the food hypothesis is and how it can be tested.
- All small rodents must eat, and adequate food supplies are a necessary condition for population growth.
- Rodents are selective foragers, and we should expect food supplies to be reduced as density increases.
- The key question is whether quantitative or qualitative food shortages are necessary for population declines.
- Experimental food additions have consistently failed to prevent population declines in cyclic voles and lemmings, and single-factor models of food shortage causing population fluctuations in rodents have been rejected.
- Whether food shortage interacts with other factors, like predation or disease, to determine population growth rates is not yet clear.

One of the first hypotheses suggested for the explanation of rodent population fluctuations was the food hypothesis. In its simplest Malthusian form, it suggested that rodents simply ran out of food when they reached peak numbers, and then declined from starvation. Under the simple assumption that voles and lemmings were mowing machines, this hypothesis could be tested by simple visual observation: Was the vegetation completely destroyed at peak voles and lemming densities? It was obvious to Charles Elton 80 years ago that this was not the case, and consequently the food hypothesis had to be either rejected or altered.

Every ecologist knows the importance of food for life, so for many the food hypothesis must be correct from first principles. Consequently, the response to these early observations went in two directions, as has occurred with every hypothesis we shall discuss in the next four chapters. First, some ecologists rejected the food hypothesis on the basis of visual and quantitative evidence that food shortage was not evident in declining populations. Other ecologists altered the food hypothesis to make it more complicated and thus in need of further study. The important point here is that there is no way to know a priori which of these two paths is correct. A second point to remember is that it is always possible to save a hypothesis by making ad hoc additions. In this chapter I will try to outline the complications that have been introduced to the food hypothesis to rescue it from rejection, and to judge whether these complications have been successful.

What Is the Food Hypothesis?

There is no single food hypothesis that ecologists agree on, so it seems best to state four variants of the hypothesis and the observations needed to test each. Note that I do not consider as a proper hypothesis the statement that food is important, since every ecologist accepted that idea more than 100 years ago.

The Food Quantity Hypothesis

The food quantity hypothesis is the simplest of the four variants. It suggests that the population growth rate should be related to the amount of food available in the habitat. As such it is nonquantitative, because it does not set a lower limit on how much food must be left to warrant rejection of the hypothesis, although perhaps one could start with some figure like 10% to 20%. This hypothesis predicts widespread vegetation destruction for the plant species eaten by the voles or lemmings living in the area, so it could be rejected if a considerable amount of edible forage remained available during a population decline or low phase. It raises the important question of exactly how much of the preferred food plant biomass is harvested at different densities of rodents. It also raises the issue of essential food plants, and whether one or a very few of them are critical for the rodents to reproduce and survive.

The Food Quality Hypothesis

The food quality hypothesis begins with the well established fact that not all that is green is edible for herbivores. To test this hypothesis, we need to determine the diet of specific rodents, their food preferences, and whether they are generalists or specialists in food selection. The main observation involved in this hypothesis is that the world may be green but still quite unsuitable for the population growth of a herbivore (White 1993). The food quality hypothesis raises the important issue of the nutritional value of food plants, and the effects of secondary chemicals on food choice and food use (Coley et al. 1985). This hypothesis is more difficult to test, and it can lead to an infinite regress of hypothesis rejection reversals by repeatedly claiming that yet another nutrient or plant species is the key factor limiting population growth rates.

The Specific Food Resource Hypothesis

This is the strongest food hypothesis, and thus the most preferred approach of those who use the hypothetical-deductive approach to population ecology. It makes very specific predictions about exactly which nutrient (e.g., nitrogen), plant species, or growth form is critical for population growth. It can thus be rejected by showing that the postulated nutrient or plant type is not in short supply during a population decline. This hypothesis can be made even more complex by adding in seasonal factors (e.g., that species *x* is required as food in spring) or interactive nutrients (e.g., protein/carbohydrate ratios) so that again it is possible to fall into an infinite regress of increasingly complex hypotheses.

We should recognize that food is a critical resource and that many investigations of the food supplies of small rodents have produced valuable insights into the nutrition of voles and lemmings. This is important whether the data support or reject the various food hypotheses. Often in ecology, data gathered to test one hypothesis are found to be useful for providing other ideas and understandings of how natural systems work.

The experimental approach to the food hypothesis has been an attempt to avoid the potential infinite regress of the various food hypotheses

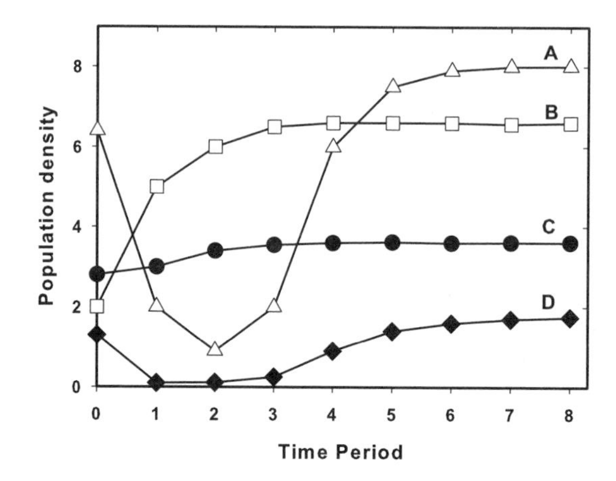

FIGURE 8.1 Hypothetical annual censuses of four populations of the same species occupying different-quality habitats. The food-habitat quality hypothesis asks why there are, on average, more organisms in the good habitats A and B (white symbols) than in the poor habitats C and D (black symbols). In many cases, for small mammals the difference between good and poor habitats will lie in the quantity, quality, and productivity of relevant food plants. Average density in these habitats is sometimes referred to as "carrying capacity," but that concept is more confusing than useful (Dhondt 1988).

discussed above. The idea is simple: to provide adequate food supplies to a population that is declining in abundance, to see if it is possible to halt or reverse the decline. The experiment must be coupled with adequate control populations in the usual manner. The complaints made about these kinds of manipulative experiments are that since the added foods are not natural, they may be nutrient-deficient or not properly digestible. One possible response is to use the same foods on confined animals to show in the laboratory that they are sufficient.

The Food-Habitat Quality Hypothesis

This hypothesis states that habitats with a higher biomass of food plants will support a higher average density for the species under study (figure 8.1). It entails the assumption that one can identify the preferred food items for a particular rodent and then measure the abundance of those plants in a series of study areas. The test of this idea should be relatively straightforward, yet there are fewer data on it than one would like. The best tests involve experimental food additions.

Note that the first three hypotheses above are attempts to relate aspects of the food supply to the *population growth rate* of voles and lemmings. This is not the same as trying to relate food supplies to "carrying capacity" or average density. The abundance and quality of food supplies for voles and lemmings are part of habitat suitability, and most ecologists would agree with the hypothesis that average density can be related to some measure of food availability. Rich habitats support a higher biomass than do poor ones, and this is accepted by virtually all rodent ecologists. We will see some examples later in this chapter.

A simple example illustrates the problem. Brown lemmings (*Lemmus trimucronatus*) on the tundra of the arctic coastal plain near Point Barrow, Alaska, fluctuate in numbers with peak densities of at least 200 to 300 per hectare (Pitelka and Batzli 2007). At Komakuk Beach, Yukon, 650 km to the east but in the same general tundra habitat on the arctic coastal plain, the brown lemming reaches peak densities of four to eight per hectare. Why should brown lemmings be 40 to 50 times more abundant in northern Alaska than they are in the north Yukon? No one knows, but George Batzli (1983, and pers. comm.) suggested that it might be tied to the abundance of one of their preferred food plants: the tundra grass *Dupontia fischeri*. This hypothesis is yet to be tested.

Tests of the Food Hypothesis

Given the variety of ideas that are gathered under the aegis of the food hypothesis, it seems best to illustrate the attempts to test some of them by small rodent ecologists.

One of the first quantitative attempts was carried out by Summerhayes (1941), on the impact of field vole *Microtus agrestis* populations on grassland vegetation at two sites in Scotland and Wales. Summerhayes worked with Charles Elton and Dennis Chitty, who had been investigating vole populations at these two sites during the 1930s. Summerhayes set out fenced exclosures (7.3 × 3.7 m) to protect areas from vole herbivory, and then compared the vegetation inside the exclosures with controls exposed to vole feeding in open grassland nearby. Figure 8.2 illustrates some of his results. Vegetation changes inside and outside the vole exclosures are similar in magnitude and in direction, such that the correlation between plant frequency inside and outside the exclosure was 0.93. Voles declined dramatically over the winter of 1932–33 and in the spring of 1934,

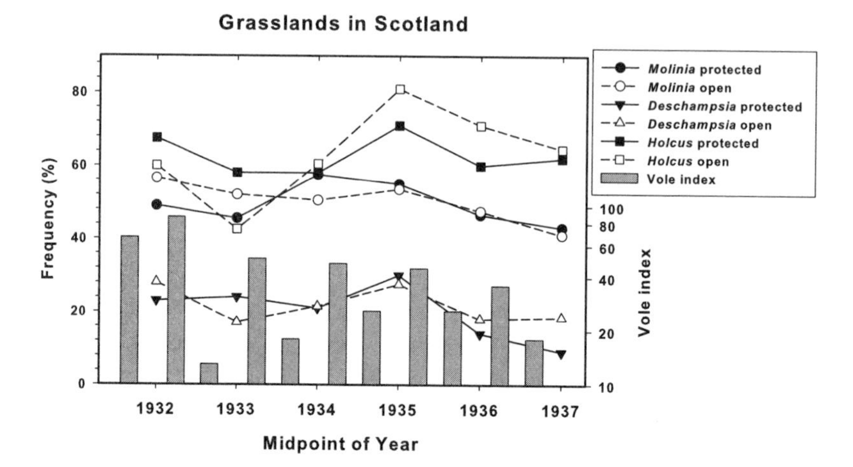

FIGURE 8.2 Changes in the frequency of dominant plants in the vole exclosures and open grasslands of southern Scotland in relation to changes in vole density. The voles were sampled in May and September, the plants in July. Data from Summerhayes 1941.

associated with only slight or no impact on the vegetation. Summerhayes (1941) could measure the impact of vole grazing in *Molinia caerulea* by measuring average tiller length for this grass. He found it to be reduced by 12% when voles were scarce, and by 47% when they were most abundant. Summer rainfall also had an effect on grass growth. Summerhayes's study highlighted two observations on rodent fluctuations: that (1) high vole densities impact vegetation significantly, and (2) even at maximum density of voles, there remains a superabundance of the voles' main food species. Thus, as Chitty (1996, page 39) has pointed out, there is no evidence that starvation is a significant cause of reduced survival in declining populations.

But starvation is too crude a criteria, critics will argue, and not all that is green is good food, so while variants 1 and 2 of the food hypothesis may be rejected with this type of data, other good options are available to save it. An example is the silica hypothesis (Massey et al. 2008). Grasses contain silica, which is an important plant defense chemical that can reduce grazing pressure. Silica in the diet reduces food quality by restricting nitrogen absorption in the rodent gut (Massey and Hartley 2006). Figure 8.3 shows that silica in the diet of field voles in winter can change their body mass growth rate from positive to negative. The silica hypothesis is currently undergoing field tests on vole populations in Britain. As a general hypothesis for food limitation in small rodents it is problematic, because not all cyclic small mammals use grasses as a dominant part of their diet (Rodgers and Lewis 1985), and only grasses respond to clipping by increasing silica levels.

One of the most influential food hypotheses has been the nutrient recovery hypothesis postulated by Pitelka (1964) and Schultz (1964) as an explanation for brown lemming population cycles at Point Barrow, Alaska. Brown lemmings in the Barrow area consume a considerable fraction of the winter green biomass of sedges and grasses, and this overgrazing was the initial observation that suggested the nutrient recovery hypothesis (figure 8.4). The causal chain then led from overgrazing to a deepening of the thaw cycle (melting of permafrost) in summer, so that grasses and sedges would root deeper into the soil in the zone of low nutrients. This in turn would produce low-quality plants, so that breeding would be poor in summer and winter reproduction would be impossible, thus leading to low lemming densities. The lack of overgrazing at low lemming numbers

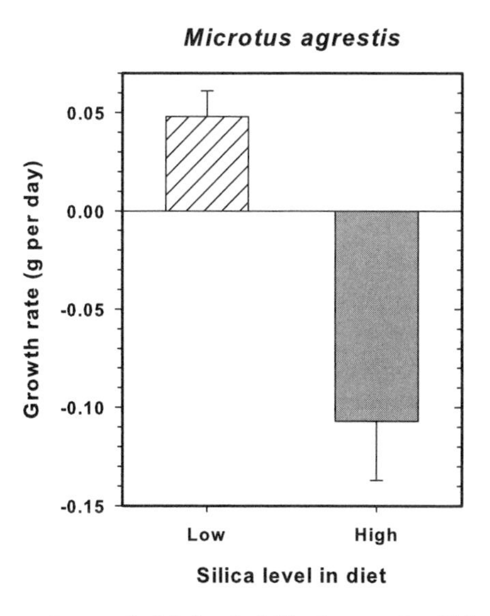

FIGURE 8.3 Winter growth rates of adult female field voles reared on high-silica and low-silica diets of the grass *Deschampsia casepitosa*. Low-silica grass was 1.8% dry mass, high-silica grass was 6.6% dry mass. From Massey et al. 2008.

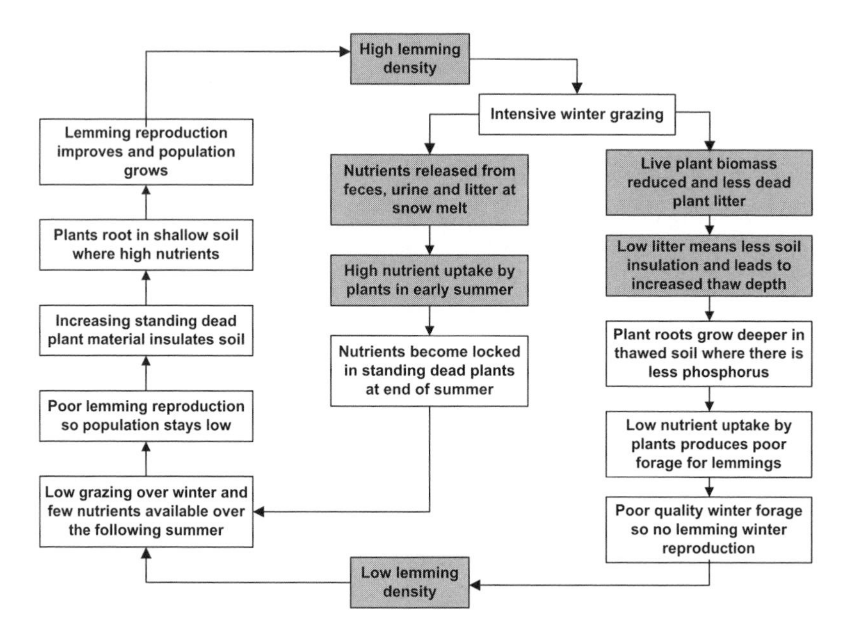

FIGURE 8.4 The nutrient-recovery hypothesis for arctic lemming cycles. Shaded boxes indicate the parts of the hypothesis that were confirmed by later research. After Schultz 1964, Batzli et al. 1980.

would lead to the reversal of these effects, with the permafrost level rising and plants making use of high nutrient soils near the ground surface.

Batzli et al. (1980) evaluated the predictions of the nutrient recovery hypothesis and found that it could not explain lemming fluctuations on its own. The postulated deepening of the thaw cycle was small, and plants did not root deeper in the soil during the phase of declining lemming populations. High-quality plants (measured by phosphorus content) were present when brown lemmings were declining in numbers, so that the links in the causal chain outlined in figure 8.4 were broken. Vegetation quality had a strong effect on lemming reproduction, but there appeared to be only weak feedback from lemmings to vegetation quality. Batzli et al. (1980) compared lemming populations at Prudhoe Bay to those at Barrow. Lemmings are only about one-tenth as abundant at Prudhoe Bay as they are at Barrow, and their impact on the vegetation at Prudhoe Bay was small by comparison.

A fertilizer experiment was carried out on the Barrow tundra by Schultz (1969) to test the nutrient recovery hypothesis, but the detailed results have never been published. According to Batzli et al. (1980), the fertilizer treatment on 2.4 ha of coastal tundra improved the quality of the graminoids, and may have increased winter reproduction of lemmings in one year (evidenced by winter nests), but its effects were then attenuated in subsequent years. No population response was ever indicated. No clear predictions were made from this experiment, and after the fact any results can be explained away. While it was an early attempt in the experimental manipulation of nutrients in graminoids, its effects on brown lemmings are mostly unknown.

The bottom line from the extensive studies of lemming fluctuations in northern Alaska by Frank Pitelka, George Batzli, and many others was that vegetation quality was one significant factor affecting lemming population changes, but that other factors such as predation and winter weather could also be involved, so that a multifactor explanation was essential.

Early studies of the fluctuations of rodents in northern Finland were carried out by Olavi Kalela (1957, 1962) who supported a food quality hypothesis as an explanation. Kalela (1962) was impressed by the flowering cycles of boreal plants in northern Finland, and he looked at the possibility that there might be a link between flowering cycles and the resulting variations in primary production and the rodent fluctuations he was studying. In 1962 he had no specific data to test his ideas, and he stimulated others to test his ideas in Fennoscandia.

The plant production hypothesis of Kalela (1962) was tested by Andersson and Jonasson (1986) during two population cycles of rodents in northern Sweden. The gray-sided vole *Clethrionomys rufocanus* was the most common species; it varied more than hundredfold in density over the cycle. Reduction in the quantity of available food through grazing was rejected as a proximate cause for the rodent cycles. During the second population peak, voles ceased breeding and began to decline while preferred food plants were still abundant. Nor was there any consistent relationship between plant nutrients (nitrogen, phosphorus, calcium, magnesium, potassium, carbon) and the rodent cycle (figure 8.5). Nutrient levels in food plants seemed to depend mainly on processes other than rodent grazing. These results are at variance with the food hypothesis of White (1993): that nitrogen is the key factor limiting population growth in voles and lemmings. Andersson and Jonasson (1986) concluded that factors other than food quantity and nutrient levels seemed to be important in the short-term causation of these two cycles in northern Sweden.

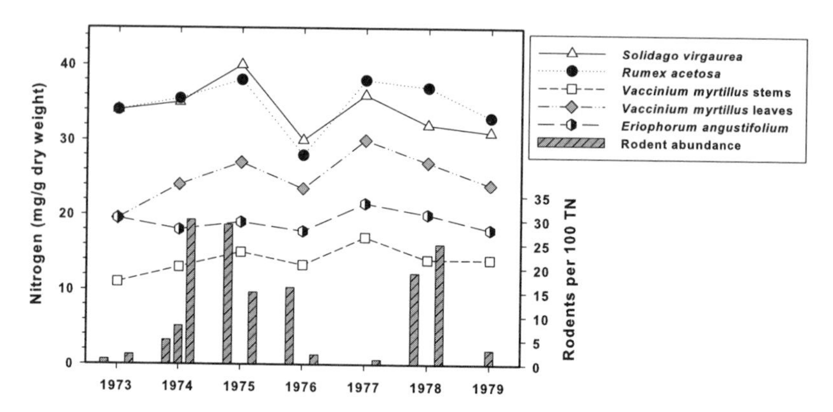

FIGURE 8.5 Changes in the abundance of small rodents over two cycles in northern Sweden measured by snap trapping (histograms), and the mean nutrient content of the dominant plants: *Solidago virgaurea* (open triangles), *Rumex acetosa* (filled circles), *Vaccinium myrtillus* stems (open squares) and leaves (filled diamonds), and *Eriophorum angustifolium* (hexagons). There is no correlation between the rate of population change and the nutrient condition of the plants in this alpine heath. Data from Andersson and Jonasson 1986.

Secondary Compounds and Food Quality

Already by the 1970s, ecologists had begun to explore the role of secondary compounds as a determinant of food quality. Freeland (1974) suggested that the loss of individuals from vole populations at peak density was the direct result of reduced viability induced by the consumption of plants or plant parts containing toxic compounds. Consumption of small amounts of plant toxins by rodents and grazing mammals can inhibit protein synthesis (Rhoades and Cates 1976) and stop growth and reproduction. There is a large literature on the role of secondary compounds in determining food choice in mammals (e.g., Coley et al. 1985; Sinclair and Smith 1984), and that is not the issue here. The question is whether changes in plant secondary compounds affect the rate of population growth in small rodents.

Most studies of the impact of secondary compounds on small rodents have been made at the individual level. Bergeron and Jodoin (1987) analyzed the food plant selection of a *Microtus pennsylvanicus* population in Quebec. Voles selected plant species high in protein, low in phenolics, and low in crude fiber (figure 8.6). Neither the availability of the plant species nor its caloric value seemed to be important in food choice. Bergeron and Jodoin (1987) concluded that, using a very conservative figure of 20 g per day that a vole needs to feed itself, and as much again for nest building

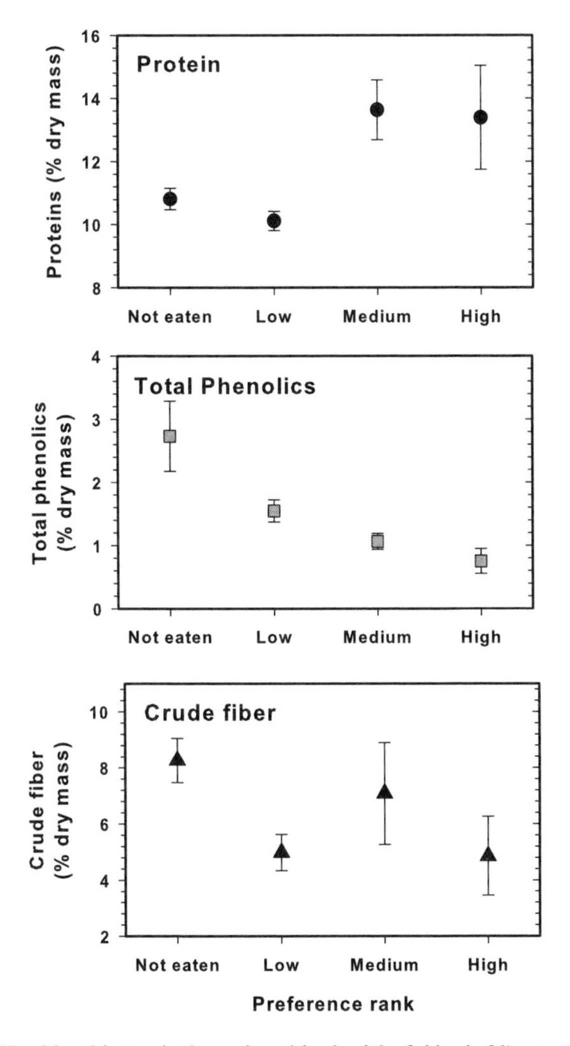

FIGURE 8.6 Nutritional factors in the preferred foods of the field vole *Microtus pennsylvanicus* in Quebec. Means ± 1 S.E. Crude fiber is acid detergent fiber (ADL). Data from Bergeron and Jodoin 1987, figure 4.

and other activities, one animal requires only approximately one kilogram per month of the preferred plants during summer. If this is correct, in their study, in which high-quality food resources comprised between 30 and 52% of available biomass, there was available more than 1,500 kg of high-quality plant mass per hectare. Consequently, at a density of 100 voles per hectare, high-quality plant species were in ample supply to meet more than 10 times the basic feeding requirements of the vole population. Furthermore, 13 other plant species were used on a regular basis by voles on this Quebec site, which added another 2,000 kg perhectare of fresh biomass each month. This suggests that the nutritional requirements of wild voles in this population are easily met during the study period.

Nevertheless, Bergeron and Jodoin (1993) showed that intensive grazing by voles in summer and autumn could reduce habitat quality the following winter and spring. Imposing grazing at an extreme density of 1,000 to 1,300 voles per hectare in autumn reduced green biomass by 15% in late autumn and by 52% the following spring, even though voles were removed from the fenced plots in autumn. But the nutritional content of the forage was not changed by this intensive grazing, and there was no indication that secondary compounds (total phenolics) in the grasses had increased in response to heavy grazing.

Dahlgren et al. (2009) followed the fate of nine plant species on nine islands in northern Finland, and described the impact of gray-sided voles (*Myodes rufocanus*) on plants with more or less secondary chemical defenses. Seven of the nine islands were very small (0.02 to 0.4 ha), and grazing effects were more severe on islands with higher vole densities and fewer predators. While this research was a useful test of the impact of voles on plants with different secondary compounds, it tells us very little about the specific predictions of the food hypothesis. It confirms the observation that more voles consume more plant material and that fewer predators may allow higher vole populations, as long as one assumes that island effects on vole numbers are not significant (cf. Krebs 1996). There are few who doubt these specific conclusions about food and predators as population limitation agents.

While there has been much interest in the possible role of secondary compounds in population dynamics, there has been no evidence to date that these chemicals affect birth and death rates sufficiently to change the population growth rate in voles and lemmings. At best, this is a hypothesis to be tested, but for now it seems unlikely to be a major mechanism of population changes.

Feeding Experiments

The difficulty of testing the food quality hypothesis by measuring forage and diets can be circumvented by adding food to a population. Cole and

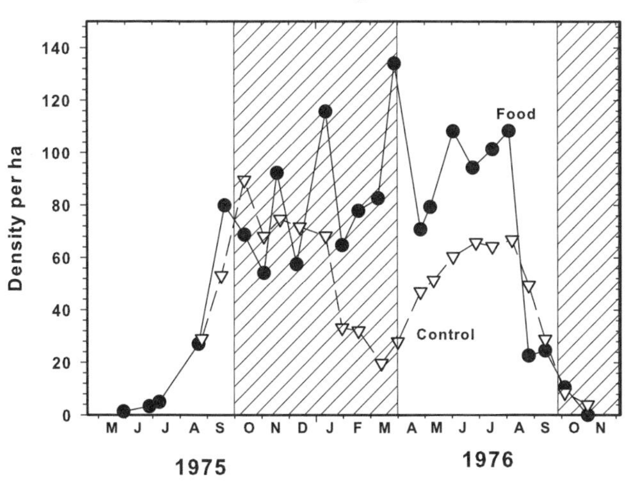

FIGURE 8.7 Density changes in the prairie vole *Microtus ochrogaster* on a food-supplemented site and on a control area in central Illinois. Rabbit pellets were added to the food site throughout the study. Winter months are shaded. After Cole and Batzli 1979.

Batzli (1979) added high-quality food (rabbit chow) to a 0.55 ha abandoned bluegrass pasture in central Illinois, and compared population changes on this site with those on an adjacent 0.8 ha control area. Figure 8.7 shows the resulting population trends. The food-supplemented population reached densities 50% higher than those of the control population (135/ha versus 90/ha), thus validating the idea that habitat quality could be changed by adding food resources. During the winter months the food-supplemented population showed more females breeding, better adult survival, and higher body growth rates than the control population. However, supplemental feeding did not prevent or delay the population decline that occurred in late summer and autumn 1976 in both populations. Cole and Batzli (1979) concluded that the quality of available food influences the amplitude of population fluctuations jn the prairie vole, but that some other factors must have caused the population decline.

Taitt and Krebs (1981) carried out a similar food supplementation experiment on *Microtus townsendii* near Vancouver, Canada. For approximately one year, voles on three areas were provided with different levels of extra food in the form of laboratory chow and oats. On areas with intermediate and high extra food, population density increased to twice the control density. Extra food increased immigration and reproduction, and shortened the winter nonbreeding season. Males and females had smaller home ranges on areas with extra food, and this may have facilitated immigration to these populations. These experiments confirmed the observation that habitat quality or "carrying capacity" was strongly affected by the quality of the food supply.

In a few cases, food supplies for small rodents have been manipulated indirectly by fertilizing the plants. As part of the US International Biological Program Grassland Biome study in Colorado, six one-hectare plots of shortgrass prairie were manipulated for six years (1971–76) through the addition of nitrogen, water, or both (Abramsky and Tracy 1979). Two additional one-hectare plots served as controls. *Microtus ochrogaster* invaded the two nitrogen-water combined treatment plots in 1971, and they increased in annual peak density to 100/ha until the study's completion in 1976. Plots with only additional water, or with only nitrogen added, showed neither increased grass growth nor increased vole densities. Since the fertilizer improved grass quality as well as increased grass cover (Birney et al. 1976), those two variables are confounded as to which is more important to increasing population density.

These experiments, along with several others, have supported two generalizations for small rodents: (1) that habitat quality, or "carrying capacity," can be increased by increasing the food supply and reduced by reducing it; and (2) that population changes (particularly cyclic declines) cannot be prevented by food manipulations (Batzli 1983). The power of field experiments to answer questions about rodent population dynamics has been shown most clearly, in contrast with the many inconclusive attempts to evaluate the food hypothesis since Lack (1954) enunciated it.

Indirect Methods for Testing the Food Hypothesis

One way to test for the role of food in driving population fluctuations is to measure the usage of food plants over winter in northern environments, and then to relate potential food shortage to individually measured physiological traits of body condition. The presumption is that food shortage will result in reduced measures of individuals'condition in declining populations. Huitu et al. (2007) studied a population of *Microtus agrestis* in western Finland during two years of a population fluctuation, and measured the standing crop of grasses at the end of winter and the condition of the voles. They sampled in winter (February), and also in spring

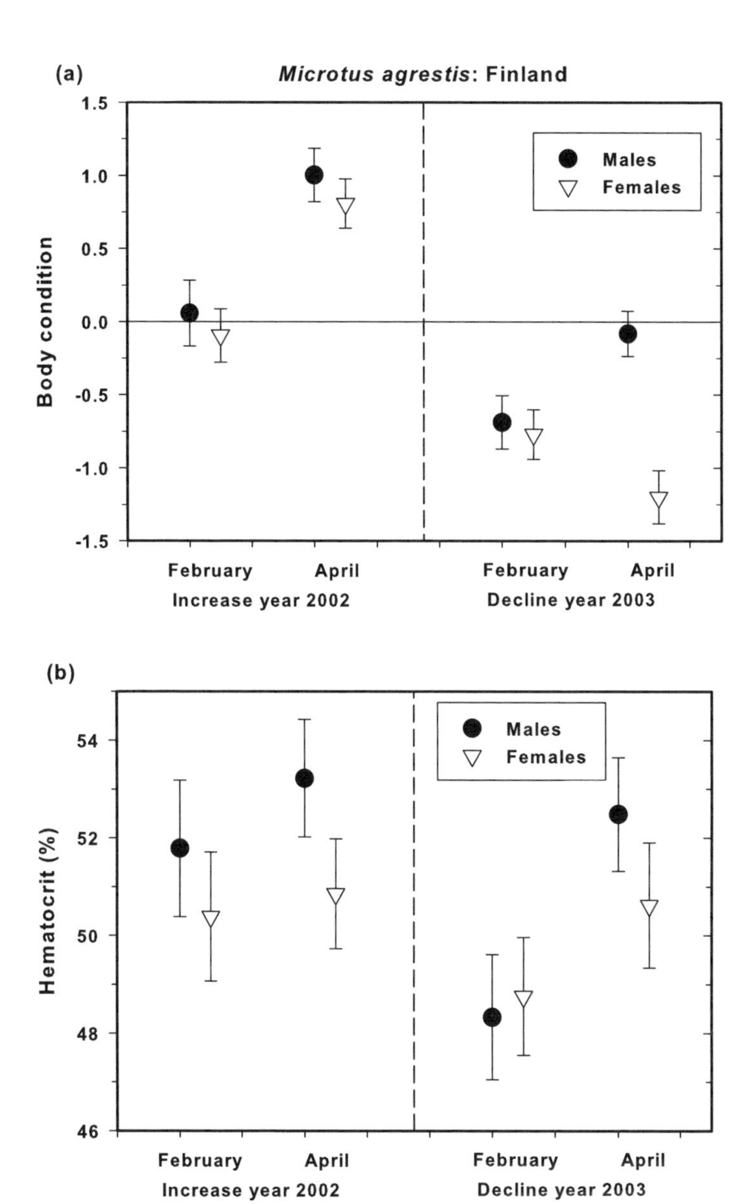

FIGURE 8.8 Mean (± SE) values of body condition index (a), and haematocrit (b), of field voles (*Microtus agrestis*) sampled in Finland in February and April of 2002 and 2003, during increase and decrease phases of their population cycle. Data from Huitu et al. 2007.

(April) just as the snow melted. Figure 8.8 illustrates their results for a body condition index and for hematocrit levels in the blood. In the spring of the decline year, the physiological condition of both males and females was relatively low, which the researchers interpreted as a result of over-winter food shortage. One problem with their results was that year and cyclic phase were confounded, and we do not know if the winter weather in 2003 might have caused the physiological condition of the voles to deteriorate. An improved study design would be to have a replicate from a second cycle or an out-of-phase population for comparison. Nevertheless, the Huitu et al. (2007) study is a good example of how hypotheses about food shortage could be integrated into studies of individual physiological measures of condition.

To test the food quantity hypothesis convincingly, we must move to experimental tests. Indirect methods leave too much room for potential arguments about variables that have not been directly measured.

Conclusion

The extensive research on the relation of food supply to changes in small rodent populations has led to major increases in our understanding of how food supplies affect numbers. Food quality is of major concern to all animals, and we need to understand exactly which aspects of food chemistry are important for food selection and food limitation. There is widespread agreement that food quantity and quality are critical variables defining habitat quality, so that on a landscape scale, good and poor habitats for many species of small rodents can be defined on the basis of their food plants. From a management perspective this is important information, and a conservation biologist wishing to increase the abundance of a threatened species should consider the idea that lack of food abundance could be limiting average density.

The major hypothesis that food shortage alone can explain changes in small rodent populations has now been rejected, and replaced by including food effects within multiple-factor hypotheses, which we will discuss in chapter 12. In this way we have moved away from a simple, readily tested idea toward more comprehensive models of what causes populations to rise and fall. We can all agree that food is a critically important variable for all rodent populations; but it is not the only variable that may drive dynamics.

CHAPTER NINE

Predation as the Explanation for Fluctuations

Key Points:

- Predators kill many small rodents, but knowing only how many animals they kill does not tell you if they determine the growth rate of the prey population.
- Predation theory states that the impact of predation can be measured if you know the numerical and functional responses of predators to prey density, and whether these responses are directly density-dependent or delayed density-dependent.
- If prey populations can compensate for predation mortality, predator kill rates may not dictate rodent population responses to predation.
- Experimental tests of predation limitation can be done by removing predators or by reducing their efficiency. Such experiments have led to much controversy.
- Most researchers now link predation with other limiting factors to provide a multifactor explanation for rodent population changes. The simple predator-prey cycles of ecological theory have become outdated.

Predators are a key component of any ecosystem, and the graphic television documentaries that show lions attacking zebras or grizzly bears fishing for salmon give their viewers the impression that the ecological world is run by predators. Ecologists have dissected the process of predation for more than 60 years after the key papers by Solomon (1949) and Holling

(1959) laid the groundwork for developing a precise theory of predation. Insect ecologists and large mammal ecologists have carried out much of the fundamental research on predation as a process that can determine the rate of population change. In this chapter we review the basic principles of predation and attempt to apply these ideas to small rodents. We begin with the simple, one-factor hypothesis that predation is both necessary and sufficient to determine population growth rates in small rodents.

Components of Predation

Predation theory begins with the assumption that the key interactions take place between a number of predators and a number of prey animals. Underlying this simple model is the assumption that all predators are identical in their preferences and all prey items are identical and equally vulnerable. As these simplifications are made more realistic, the models become more complicated, but let us start with a simple model. Figure 9.1 illustrates the three types of *functional responses* that might be expected in a predator-prey system. The expectation is that for most systems, predators may become satiated at very high prey densities, so the type 2 and

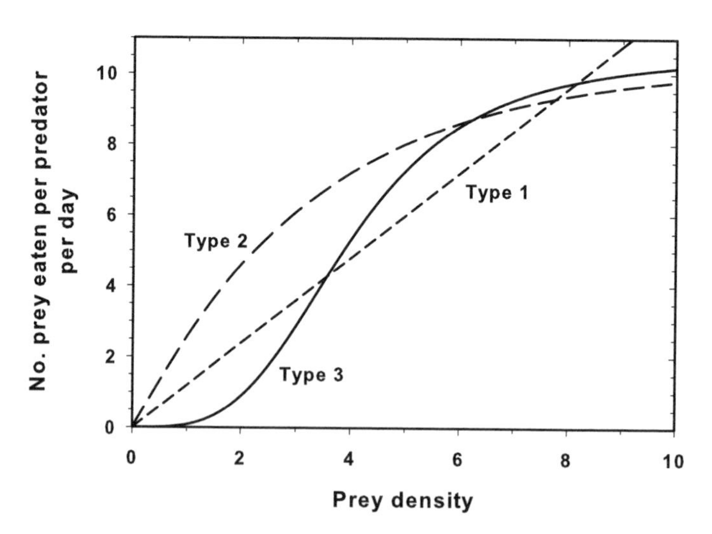

FIGURE 9.1 Three types of possible functional responses by predators to changes in the abundance of their prey. Type 1 or linear responses show a constant consumption of prey with no satiation. Type 2 and type 3 responses reach satiation at high prey densities.

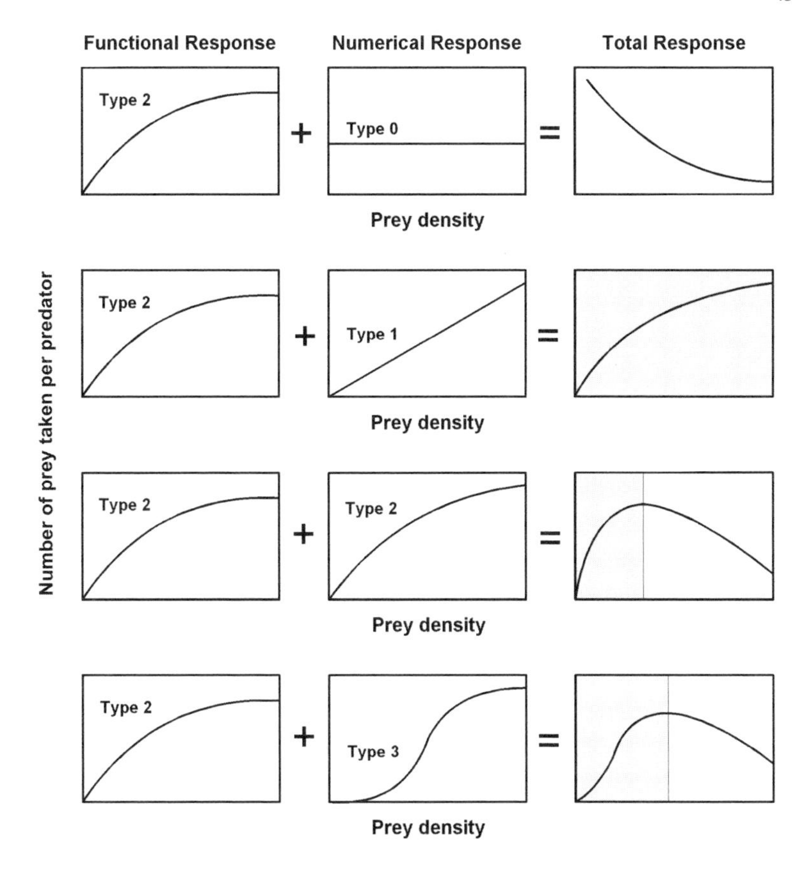

FIGURE 9.2 Relative changes in the percentage of the prey population killed by predators, as a function of prey density. Type 2 functional responses are assumed (since they are most common), and four types of numerical responses are illustrated. The gray zones on the total response curves at the right illustrate the density range of the prey for which predation will be density-dependent. These curves illustrate simple predation models and what they predict. After Messier 1992.

type 3 curves reach a plateau. In real predator-prey systems, the prey may never attain high density and the functional response may be more like a type 1 linear response.

The second component of predation is the *numerical response*, which is the density of predators in relation to that of prey. Most authors use the same type 1, 2, and 3 response curves for the numerical response, with an added type 0 curve for predators whose density does not change as prey density changes (Messier and Joly 2000). Figure 9.2 illustrates schematically how the functional and numerical responses can then be combined to give *total response*, the total predation rate or percentage of the prey population killed by the predator.

Complications of these simple models are not difficult to find. Predators do not respond immediately to changes in prey abundance because often they are seasonal breeders, so there is a time lag in the predator responses. Predator-prey systems often have several prey species and several predators, so that complexity soon overwhelms these simple models. While they are useful in understanding the impact of predators on prey numbers, they are only a starting point (Messier 1994).

Predation Hypotheses

For populations that fluctuate periodically, predation can be isolated as a cause of the fluctuations only if it is delayed density-dependent. Consequently, the exact hypothesis that is to be tested in any study must be stated precisely. There are at least five predation hypotheses that can be starting points for precise predictions that can be tested with field data.

1. The predation-as-limitation hypothesis, which states that a particular predator (or suite of predators) limits population density of the prey species so that its density averaged over a few years will be lower if predators are present. This hypothesis implies that predation mortality is directly density-dependent. It can be tested by predator removal studies or by observations of habitats where the particular predator is absent.
2. The hypothesis that predation is necessary and sufficient to produce prey population fluctuations. This hypothesis states both that predation mortality destabilizes prey populations because it is delayed density-dependent, and that the predation mortality is so high that the prey cannot compensate for the losses. This is the strongest of the predation hypotheses that have been postulated as single-factor explanations for population cycles or fluctuations since the days of Lotka and Volterra.
3. The hypothesis that predation is necessary but not sufficient to produce prey population fluctuations. This can be viewed as a decomposition of hypothesis 2. It states that predation mortality is one component of a multifactor explanation for population changes in small mammals. It is restrictive in stating that if predation mortality is very low or absent, population fluctuations will not occur.

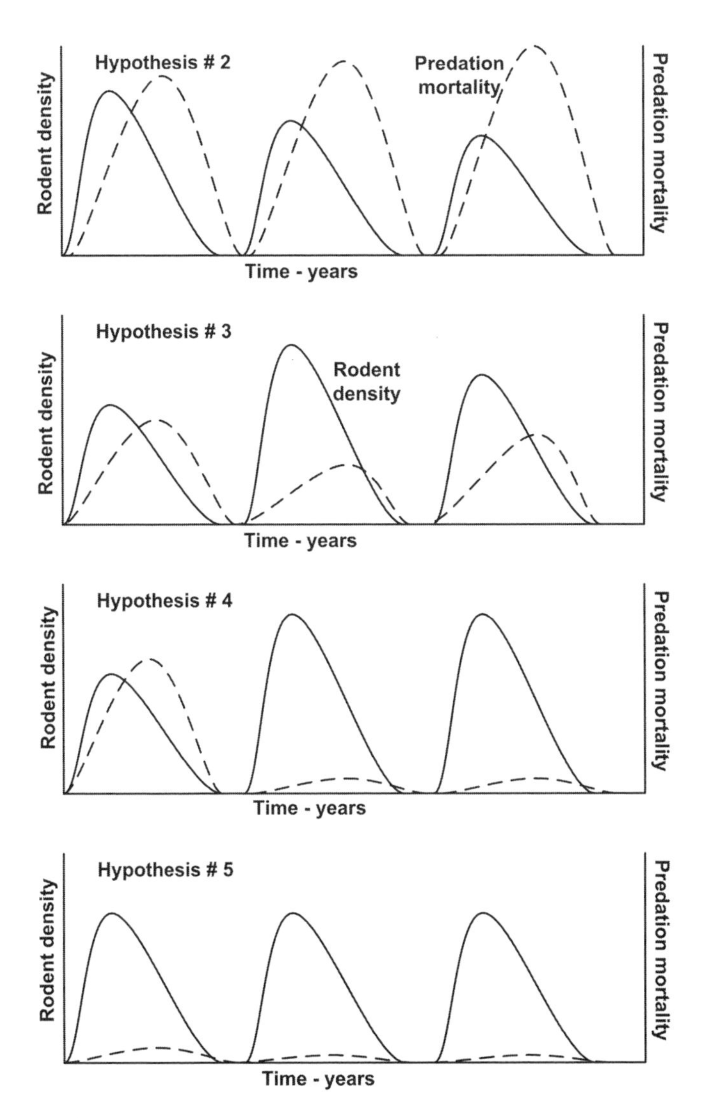

FIGURE 9.3 Schematic illustration of the types of observation that can be used to test the predation hypotheses described in the text. The scale of predation mortality (dashed lines) is meant to reflect nearly 100% mortality at the top of the scale and nearly 0% at the bottom of the scale. The time scale would be approximately 10 to 12 years.

4. The hypothesis that predation is sufficient but not necessary to produce prey population fluctuations. This is the opposite of hypothesis 3 in stating that if predators are present, they will generate population fluctuations on their own, but that if for some reason they are at low numbers or absent, fluctuations will still occur without them. In this sense it implies that there are two or more classes of explanations for rodent fluctuations, so that there is a loss of generality. If correct, it reflects interest in defining the classes of explanation.
5. The hypothesis that predation is neither necessary nor sufficient to produce prey population fluctuations. For the sake of completness, we can state this hypothesis: that predation mortality is both sporadic and too weak to generate population changes. To test it we need a quantitative evaluation of the impact of predation and a long-term set of observations to distinguish it from hypothesis 3. Under this hypothesis, predation can be a major factor in one cycle but nearly absent in the next.

Figure 9.3 illustrates schematically the kinds of observations that are consistent with hypotheses 2 through 5. The key point is that to test these simple models, one needs a series of observations on several population fluctuations, and good quantitative data on predator densities and food intake. The alternative method of manipulating predator density gives a faster answer to these questions, if such experiments are logistically feasible. We turn now to analyze some empirical studies of predation in small rodents.

Observational Studies

One way to measure the impact of predators on small rodents is to measure their offtake and relate it to the prey's ability to replace themselves. One of the first to use this technique was Dennis Chitty, who attempted to estimate predator offtake of field voles (Chitty 1938). Short-eared owls, the principal predator on the field site at Newcastleton in Scotland, were censused through a cyclic decline in the field vole *Microtus agrestis*. Extensive field work by Goddard (1935) showed that there were at least three but not more than four pairs of short-eared owls at Newcastleton in 1934 on an area of 817 ha. There were few other predators in the area (Chitty 1996). Chitty (1938) calculated that a single short-eared owl would consume a maximum of 1,100 voles per year, so that on this study area a total of 8,800 voles would be killed by these avian predators. Yet the density

TABLE 9.1 **Total predator impact on a high brown lemming population at Barrow, Alaska. From Maher 1970, table 8.**

Predator	Age class	Density per km²	Daily food consumption (grams per individual)	per individual	Lemming consumption per hectare 25 May–15 July	16 July–31 August	Total
Pomarine jaeger	Adult	15	250	338	25	52	77
	Young	15	200	167	—	—	
Snowy owl	Adult	0.8	250	350	3	3.8	7
	Young	2.7	150	160			
Least weasel		25	50	100	12	12	25
Glaucous gull		8	250	125	2		2
Surplus kills					10		10
Totals					52	69	121

of voles in April 1934 was between 120 and 300 per hectare, or a total of 100,000 to 250,000 voles on the entire site. The vole population declined throughout 1934, and by the end of that year their density was extremely low, so virtually all these individuals and their offspring were gone. The simplest calculation is that the owls consumed about 3.5% to 9% of the voles. There was no influx of predators into the area, and Chitty concluded that these observations were consistent with hypothesis 5: that predation losses were small, and not a cause of the population decline.

Birds of prey are significant predators of brown and collared lemmings in tundra areas of Canada and Alaska. Pitelka et al. (1955) showed that the pomarine jaeger (*Stercorarius pomarinus*) is a major lemming predator in northern Alaska, along with the snowy owl (*Nyctea scandiaca*), the least weasel (*Mustela nivalis*), and the arctic fox (*Vulpes lagopus*). Bird predators in the arctic are particularly mobile, and they move from place to place in search of high densities of lemmings. Their impact is typically concentrated in the peak summer of the lemming cycle, producing mortality that is density-dependent but not delayed-density-dependent. A careful analysis of these predators' impact on a lemming peak in northern Alaska was made by Maher (1970). He calculated the offtake of brown lemmings by these predators for high lemming populations, with the results given in table 9.1. Maher (1970) pointed out that the effects of predation on lemming density in the peak year depended critically on the spring density of lemmings (table 9.2). In the lemming peak of 1956, which reached a spring

TABLE 9.2 **Hypothetical impacts of predation on high brown lemming populations of varying spring densities at Barrow, Alaska. Predation can cause a severe decline only if lemmings begin the summer breeding period at low abundance in this ecosystem. From Maher 1970, table 9.**

	Low abundance	Moderate abundance	High abundance
Spring lemmings (per ha)	50–75	75–100	100–125
Number of females (per ha)	31	43	56
Females lost before breeding	26	26	26
Females left to breed	5	17	30
Average number of embryos	6	6	6
Young produced	30	104	178
Total adults remaining	10	34	60
Total midsummer population	40	138	238
Predation loss after breeding	69	69	69
Population density after predation losses	–30[1]	69	168

[1] Clearly the population density could not be negative, so the calculation suggests that for this case predators could limit density close to zero.

density of 100 to 125 per hectare, predation was strong enough to prevent further lemming population growth, while during the 1960 peak, with a spring density of 175 to 200 hectares, the predation impact on summer population growth was very small.

The role of mammalian predation has not been well studied in the Barrow ecosystem, and Maher (1970) argued from work on Banks Island that least weasels, which persist over winter and hunt under the snow, could have a strong effect on lemming populations in winter, driving their numbers down to low levels. The Barrow lemming work has been some of the best in quantifying the avian predation offtake for this ecosystem. Maher (1970) concluded that the role of avian predators in northern Alaska's lemming cycle has been to exploit the peak population and truncate the top of the cycle by their predation mortality. These results are most consistent with hypotheses 1 and 3 above. Integration of the impact of all the predators of lemmings requires information on all elements of the food web from single sites, and this has not yet been accomplished.

Birds of prey typically have a type 2 numerical response to small rodent density, which implies a plateau in bird density at high rodent densities (figure 9.4). At Barrow, pomarine jaegers reach their own maximum density at a spring brown lemming density of about 100/ha. In France, Montague's harrier (*Circus pygargus*) shows a type 2 numerical response and reaches maximum density above about 10 voles per 100 trap nights (Millon and Bretagnolle 2008). In most cases there is density dependence within the avian species based on territoriality, which sets an upper limit to

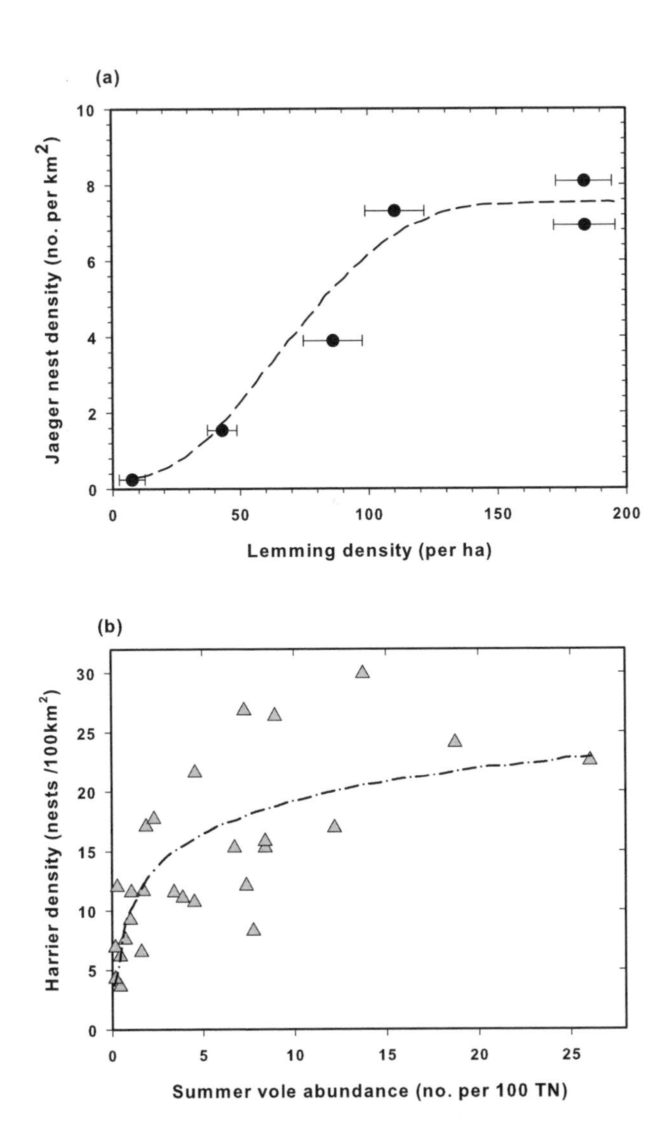

FIGURE 9.4 Type 2 and 3 numerical responses of migratory avian predators on small rodents: (a) type 3 numerical response of the pomarine jaeger at Barrow, Alaska, in relation to density of the brown lemming in spring; (b) type 2 numerical response of Montague's harrier in France to abundance of the common vole in summer. It is often very difficult to distinguish type 2 and type 3 responses from field data. Data from Maher 1970, figure 2, and from Dupuy et al. 2009, figure 3a.

breeding density. This restricts the vole or lemming density within which population regulation might occur (cf. figure 9.2); and with migratory raptors, predation operates on the local population of rodents during only part of the year.

Many studies have shown heavy predation losses from small rodent populations (e.g., Maher 1967; Dupuy et al. 2009; Bernard et al. 2010). Predators have been classified into *generalist predators*, which eat a great variety of prey, and *specialist predators*, which feed mainly on one or a few kinds of prey (Erlinge et al. 1984). Predator-prey theory dictates that generalist predators stabilize populations because their predation mortality is directly density-dependent, whereas specialist predators destabilize populations because their predation mortality is both directly density-dependent and delayed density-dependent. This simple classification is obscured by some species that operate as generalists in some habitats and time periods and as specialists in others (Dupuy et al. 2009).

Two approaches can help us sort out the role of predation in small rodent population fluctuations. We can construct models of predator-prey systems, an approach we will discuss in chapter 12; or we can carry out experiments with predator removal or predator addition, an approach we discuss next.

Experimental Studies

Experimental studies of predation can be observational or manipulative. Observational studies ask the question of whether there are small rodent populations that fluctuate in environments with no predators. Since very few if any small rodent populations lack predators, there are no detailed studies of this type to report. Wrangel Island off Siberia, for example, has well-documented cycles of collared and brown lemmings, but no least weasels or ermine (Travina 2002). Consequently, if a particular predation hypothesis depends on the presence of least weasels, this island provides a Popperian counterexample. The island does, however, have arctic fox, snowy owls, and pomarine jaegers as lemming predators, so conclusions about the role of predation cannot be reached until they have been studied in detail.

Erkki Korpimäki and his colleagues in Finland have been testing the predation hypothesis with a series of removal experiments. The first such experiment was carried out on avian predators. For three years Norrdahl and Korpimäki (1995) removed breeding kestrels and Tengmalm's owls from five sites of three square kilometers each. They removed the birds' nest sites before the breeding season each year, effectively removing all breeding pairs. They subsequently found no difference in abundance of *Microtus* and *Myodes* (*Clethrionomys*) voles between control and experimental plots. They therefore concluded that nomadic avian predators in this ecosystem in western Finland tracked vole abundances rapidly without any time lags (Korpimäki 1994), and thus did not significantly help to produce vole population fluctuations. They speculated that increased least weasel predation on voles might compensate for decreased avian predation losses when avian predators were removed.

If nomadic avian predators did not affect vole abundance, the next likely hypothesis is that mammalian predators are the key. In particular, specialist vole predators like the least weasel (*Mustela nivalis*) have the potential for producing both density-dependent and delayed density-dependent mortality on vole or lemming populations (Korpimäki 1993). In two decline phases of the rodent cycle in western Finland, Korpimäki and Norrdahl (1998) carried out experiments to reduce the abundance of vole predators. In the 1992 vole decline, about 70% of the least weasels were removed from six paired areas (control, removals) with no effect on the rate of decline of *Microtus* voles (figure 9.5a). In the 1995 decline they were able to remove least weasels as well as ermines (stoat) and nomadic birds of prey, thus effecting a complete reduction of all the main vole predators. Figure 9.5b shows that the complete predator removals of 1995 were successful in stopping the summer decline in vole numbers. The researchers concluded that specialist predators drive the summer decline of cyclic rodent populations in western Finland.

Because the Korpimäki and Norrdahl (1998) removal experiments could be carried out only during the summer months, the question remained concerning what the consequences of a long-term predator reduction experiment would be. A short pulse experiment can produce immediate results that are lost by compensation over the annual cycle, while press experiments, where feasible, are more informative about long-term dynamics. Korpimäki et al. (2002) followed the experiments described above by a three-year experiment in which all major predators were reduced on four areas that were matched with four controls. The predator removals could be conducted only in summer, so there was some recolonization of predators over the winter period. Figure 9.6 shows the results of these manipulations. The reduction of predator densities

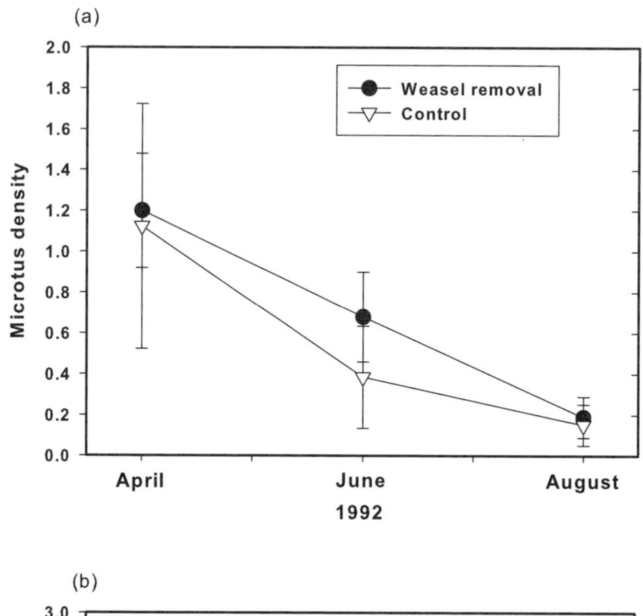

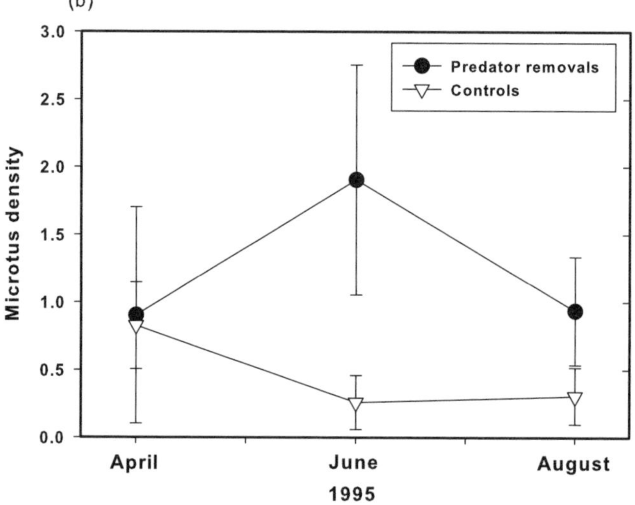

FIGURE 9.5 The experimental predator reduction experiments carried out by Korpimäki and Norrdahl in western Finland: (a) the 1992 collapse of the *Microtus* spp. population on control (unmanipulated) plots and on areas from which about 70% of least weasels were removed; (b) the 1995 collapse, in which least weasels, ermines, and nomadic birds of prey were all removed. Six paired plots, each two to three square kilometers in size, were used in these highly replicated experiments. Means ± 1 S.E. Removal of predators could be carried out only during the summer months. Data from Korpimäki and Norrdahl 1998.

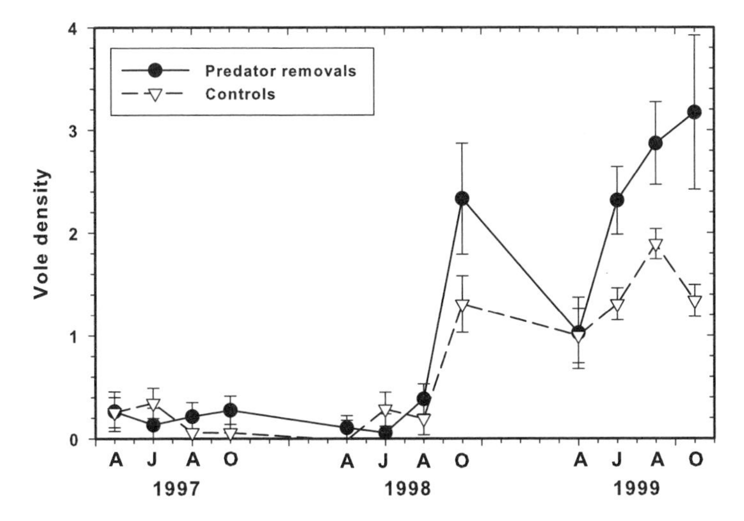

FIGURE 9.6 Density changes in *Microtus* voles in predator reduction areas and in control areas of western Finland, 1997–99. All major mammal and bird predators were removed during April to October of each year. Means of four areas ± 1 S.E. Data from Korpimäki et al. 2002, figure 3.

increased the autumn density of voles fourfold in the low phase during 1997, accelerated that increase twofold in 1998, and increased the autumn density of voles twofold in the peak phase of 1999. They concluded that specialist predators may generate cyclic population fluctuations of voles observed in northern Europe.

The Finnish experimenters could not continue their predator removals during winter because of the logistical problems of working in landscapes with deep snow, but other areas with vole population cycles have less severe winters and offer an opportunity to do continuous predator removal experiments. In Scotland, Graham and Lambin (2002) tested the specialist predator hypothesis for population cycles in a five-year removal experiment. They removed least weasels continuously and measured the response of *Microtus agrestis* in three populations with removals and in three control populations where weasels were present. Because of continual immigration, least weasels could never be completely removed from experimental areas, but with continuous trapping they could be kept to a low level. Graham (2001) could detect no differences in density changes between plots with and without least weasels (figure 9.7a). Survival rates showed a negative effect of weasel density in adult voles but a positive effect in juveniles (figure 9.7b). The positive effect of weasels on juvenile

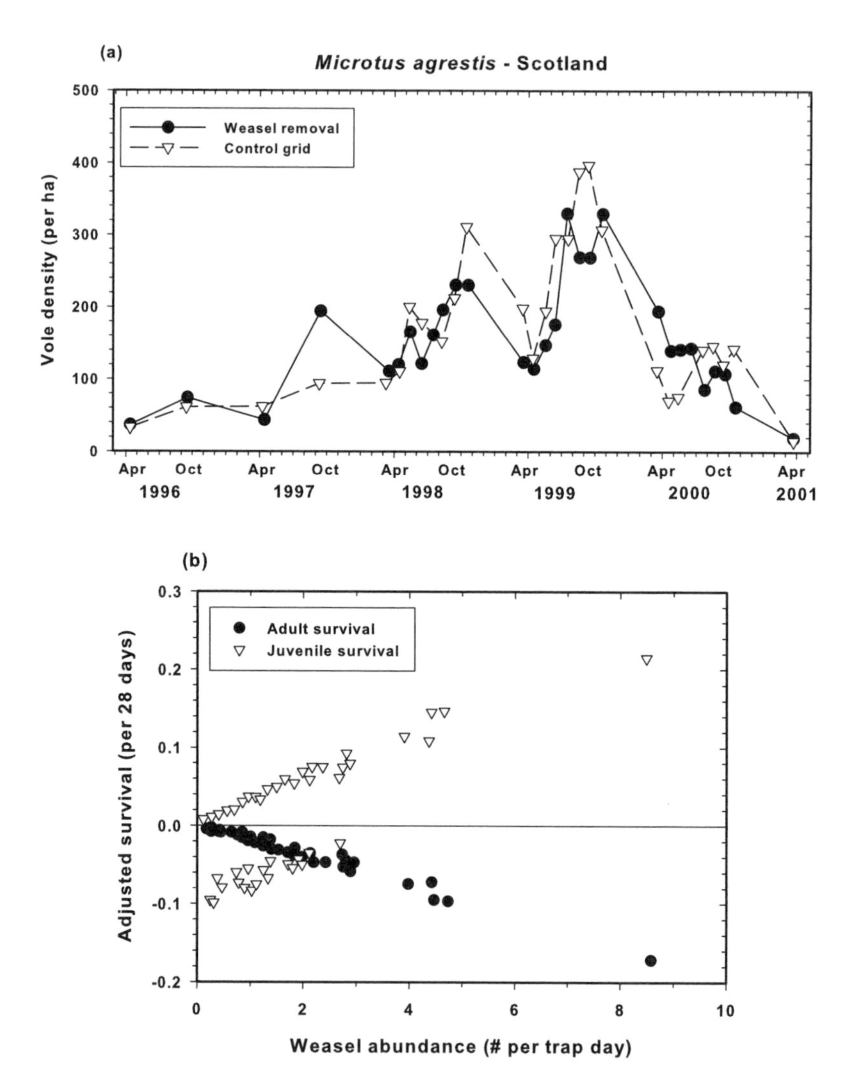

FIGURE 9.7 (a) Density changes in the field vole *Microtus agrestis* on a pair of sites in Scotland from which least weasels were removed (circles) and from control areas (triangles). (b) Survival changes in adult and juvenile voles in relation to least weasel density. Paradoxically, juvenile voles survive better when there are more weasels. From Graham 2001, figure 5, and Graham and Lambin 2002, figure 2.

survival could have arisen from reduced infanticide when adults were more abundant, or as a result of increased emigration from areas with higher adult densities. From this evidence, Graham and Lambin (2002) rejected the specialist predator hypothesis as a necessary explanation for vole cycles, thus supporting hypothesis 4 or 5 above.

The Graham and Lambin weasel removal experiment set off a minor row in the scientific literature. In a review of the study, Oli (2003) accepted the view that the specialist predator hypothesis was now rejected by these experiments. Korpimäki et al. (2003) responded with a vigorous attack on the Graham and Lambin experiments. They suggested that the Scottish population fluctuations studied by Graham and Lambin were fundamentally different from the fluctuations found in voles in Fennoscandia and the rest of Europe. The density of the low phase in Scotland was much higher than the equivalent phase in Fennoscandia (25/ha versus <1 ha), and the amplitude of the Scottish fluctuations was only about tenfold compared to fifty- to five hundredfold in Europe. They also criticized the small spatial scale of the Scottish experiments and the fact that only least weasels were removed, so that all other predators that were not being studied could have been contributing to vole mortality rates. In reply to these criticisms, Lambin and Graham (2003) noted that they had been testing only the specialist predator–least weasel hypothesis, not a more complex model with multifactor interactions.

From 2002 to 2009 Maron et al. (2010) carried out predator exclusion experiments for the vole *Microtus montanus* in Montana grasslands. The experiment was done in two phases. From 2002 to 2005 all avian and mammalian predators, except weasels, were excluded from four replicate sites of one hectare each. Beginning in September 2005, weasels were also excluded, so that there was no access for any of the predator species. Figure 9.8 shows the results of the experiment. The exclusion of all predators except weasels had no effect on vole abundance, but once weasels were also excluded, vole density increased about fourfold in the cyclic peaks of 2006 and 2009 as well as in the low phase between these peaks. This experiment confirms hypothesis 1 of predator limitation; but it is clear that in the absence of predation, the vole population went through a cyclic increase, decline, and low phase. Hypothesis 5 is thus confirmed for this particular ecosystem.

A replicated experiment on *Microtus agrestis* in western Finland was carried out to test whether predation or winter food shortage limited vole numbers over winter (Huitu et al. 2003). Twelve areas were treated with a full two-factor experiment: (1) fencing to exclude all predators, plus winter food supplementation ($n = 4$ areas); (2) fencing, plus no food supplementation ($n = 4$ areas); (3) no fencing, plus winter food supplementation ($n = 2$ areas); and (4) no fencing, plus no food supplementation ($n = 2$ areas). The problem of a fence effect was alleviated by having dispersal sinks inside each enclosure. Figure 9.9 shows the experimental results. In general, adding food in winter by itself roughly doubled vole density, while fenced populations with no predation increased about eightfold over unfenced populations. The largest effect was obtained by fencing out predators and adding winter food: density increased about twelvefold with this combined treatment.

There were some anomalies in the Huitu et al. (2003) experiments. Two fenced populations declined to extinction during the first winter and had to be restarted. Both of these populations increased during the winter of 1999–2000, yet no reproduction was occurring in winter, so the demographic changes in the first year are anomalous. Nevertheless, these experiments show that both predation and winter food supplies can have strong effects on average population density in this habitat.

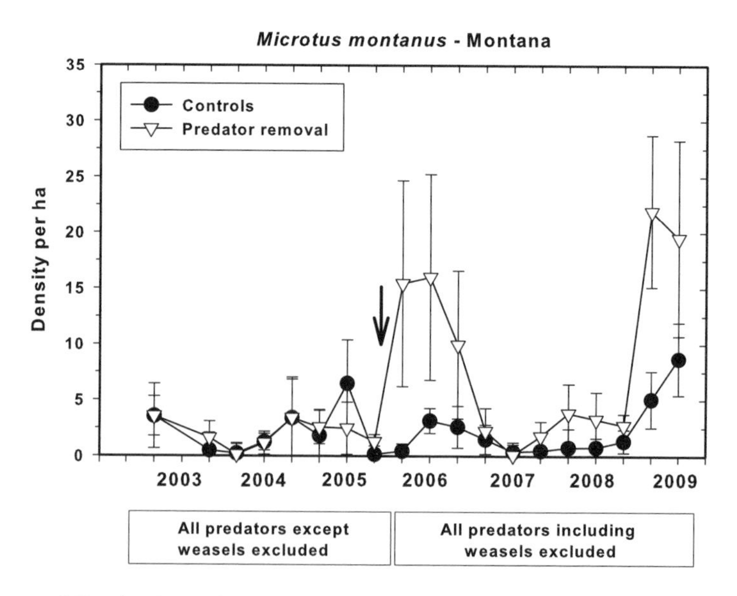

FIGURE 9.8 Density changes in the field vole *Microtus montanus* on sites in western Montana from which all avian and mammalian predators were excluded from 2002 to 2005, and all weasels were excluded from 2005 to 2009. Excluding weasels increased average density about fourfold, but did not prevent the cyclic fluctuation. Modified from Maron et al. 2010.

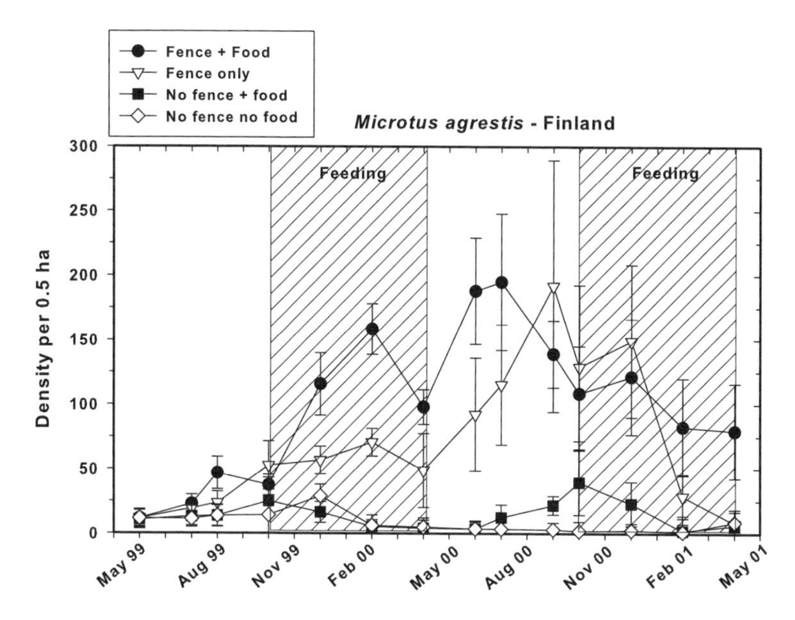

FIGURE 9.9 Density changes in the field vole *Microtus agrestis* on sites in western Finland from which all avian and mammalian predators were excluded by fences from 1999 to 2001, and where food was added during the winter months. Excluding predators increased average density, as did adding food in winter. Modified from Huitu et al. 2003.

A Possible Synthesis of Contrasting Views

Much of the conflict over the predator hypothesis rests on the failure to state clearly and precisely the predictions of any particular model that involves predation. Every ecologist will accept the general view that predators are an important source of mortality for small rodents. Consequently, there is no need for further testing of this idea by compiling lists of all the rodents killed by predators.

Korpimäki's predator removal experiments show convincingly that reducing all the major vole predators can alter the summer pattern of population change and thereby increase autumn density, but these experiments do not show that predation by itself can generate population fluctuations in small rodents. Graham and Lambin's weasel removal experiment rejected the idea that least weasels on their own can cause rodent population fluctuations. We are left to conclude that predators can modify the shape of the population density curve, but cannot on their own generate population fluctuations. To achieve that goal, it is necessary but not sufficient that the predator or suite of predators produce a mortality rate in the prey that is delayed density-dependent (Moss and Watson 2001). Graham (2001) showed that in Scotland, least weasels produced losses in voles that were directly density-dependent but not delayed density-dependent.

The early demonstrations from mathematical models that predator-prey systems would produce cyclic population fluctuations have biased many ecologists into thinking that the predation hypothesis (hypothesis 2 above) must be the explanation for small rodent population fluctuations (e.g., May 1981: Sherratt 2001). While this may be true in some systems, there are many other mechanisms that can also produce population fluctuations. All that is necessary is a mechanism that has a strong effect on births, deaths, or movements and has a delay with respect to population density. Many models of these kinds of effects are now available, as we shall see in chapter 12, and consequently the early bias toward predator-prey dynamics needs to be tempered. But we do need more long-term, large-scale experimental manipulations of vole and lemming predator-prey systems to quantify exactly how much of the change we see in a population can be attributed to predation mortality. The demonstration that predation risk can affect reproductive rates in snowshoe hares (Sheriff et al. 2009) raises the question of whether this indirect effect of predation could be operating in cyclic vole and lemming populations.

Predation may interact with other mechanisms of population change, such as disease and food limitation, and these kinds of multifactor models are becoming increasingly popular. We will address them in chapter 12.

Conclusion

The predation hypothesis is not a unitary concept, and there are many different hypotheses about the role of predation in population dynamics. The first conclusion we need to state is that is it essential to specify the exact hypothesis under evaluation, and then to consider how it can be rejected. Predation is one of the easiest of the major hypotheses to be accepted after the fact by ad hoc arguments. The demonstration that predators kill many individuals is not sufficient for understanding the role of predators in any system. The common vertebrate system, in which many predators feed on a particular prey species, is far more complex than any of the existing predation models allow, yet it must be evaluated with future experiments. For the present, I summarize what I think the evidence shows with respect to the five hypotheses presented earlier in this chapter.

1. The predation-as-limitation hypothesis generally predicts that where there are more predators, there are fewer prey on average. It is generally accepted for many animals, but hard evidence for voles and lemmings is scarce. The experiments on *Microtus montanus* support this hypothesis, which is tentatively confirmed.
2. The hypothesis that predation is necessary and sufficient to produce prey population fluctuations is the strongest of the predation hypotheses and is now not accepted by most small mammal ecologists. It is strongly rejected.
3. The hypothesis that predation is necessary but not sufficient to produce prey population fluctuations states that predation mortality is one necessary component of a multifactor explanation for population changes in small mammals. This hypothesis is tentatively confirmed but future research may show that it is a special case of hypothesis 5. It will be discussed further in chapter 12.
4. The hypothesis that predation is sufficient but not necessary to produce prey population fluctuations is the opposite of hypothesis 3, and seems to have little support in the small mammal literature. It is tentatively rejected.
5. The hypothesis that predation is neither necessary nor sufficient to produce prey population fluctuations suggests that predation mortality is both too sporadic and too weak to generate population changes. For some systems this seems correct and the issue that is currently unresolved is whether to accept hypothesis 3, 4, or 5. We need a quantitative evaluation of the impact of predation and a long-term set of observations to distinguish hypothesis 5 from hypothesis 3.

I conclude that predation can be a major factor affecting the growth rates of vole and lemming populations, but there seem to be several cases in which predation mortality seems to play a minor role in rodent population dynamics. The task is to sort out when predation is a necessary component of population changes, and when it is a contingent component that is present but not necessary.

CHAPTER TEN

Disease as a Potential Factor in Population Changes

Key Points:

- Small rodents, like all mammals, are host to a variety of diseases, and by 1920 disease epidemics were suggested as the most likely cause for rodent population fluctuations.
- Diseases have been shown to cause dramatic changes in rodent population size, but only in sporadic events.
- Disease incidence is typically density-dependent and could limit average density, but cannot be responsible for population fluctuations.
- Disease agents could reduce the body condition of individuals and make them more susceptible to predators.
- The impact of disease on small rodent population dynamics is most likely to be evident as part of a multifactorial model.

Disease is a common cause of death in individual mammals, and the incidence of disease could thus play a role in small mammal population fluctuations. Dennis Chitty (1996) has briefly recounted the history of the disease hypothesis in vole and lemming research. In the 1920s, when Charles Elton at Oxford was beginning research on small rodent populations, most ecologists believed that the fluctuations in numbers of lemmings and voles were due to recurrent epidemics of infectious disease. This belief inspired a 20-year program of research at Oxford on the role of disease in vole population changes. The results of all this research are sobering to recount. Disease organisms like the protozoan *Toxoplasma* were found in declining populations in some areas but not others, and in some cycles they showed high prevalence while in the following cycles they were not found at all. How should one proceed in this kind of situation?

Methods of Testing Causal Hypotheses

The classical methods of testing for a disease association were developed by Robert Koch in 1884 and have been discussed by Chitty (1996, p. 52). Begin with a simple hypothesis: that a particular epidemic disease is a necessary and sufficient cause of population declines in a small rodent. The method of inductive elimination proceeds as follows. First, construct the decision table, which includes the causal agent (C) and the suggested effect (E). In our case, the cause is the disease and the effect is the population decline. There are four possible kinds of observations:

		Supposed cause	
		Present	Absent
Effect to be explained	Present	$C^+ E^+$	$C^+ E^-$
	Absent	$C^- E^+$	$C^- E^-$

The observations that support the hypothesis are in the unshaded boxes: C^+E^+, or cases in which the cause (e.g., epidemic disease) and the effect (e.g., population decline) are both present; or C^-E^-, in which both cause and effect are absent. The key observations are in the shaded boxes, where the cause is absent and the effect is present (so that the postulated cause is not necessary to produce the effect), and C^+E^+, in which the cause is present but the effect does not occur (so the cause is not sufficient to produce the effect).

The problem with this classic approach to elimination is that some of the observations may be faulty because of technique problems, and the effect may have several different causes, so that no single cause (e.g., one particular disease) can be linked to the effect. But the main point is that we have to be careful in every case to look for evidence in all of the four boxes in this decision table. Having many observations of C^+E^+ is not convincing unless you have made an effort to look for situations that fit the other three possible relationships. Hence, the general admonition is to look at many diverse ecological situations to determine the generality of your particular hypothesis.

Observational Studies

The classic case of a test of the disease hypothesis was the study carried out by Dennis Chitty and A. Q. Wells from 1936 to 1939 on fluctuating vole populations and disease. The results of this study, interrupted by World War II, were described by Chitty (1954), but the capsular summary was that Dr. Wells discovered vole tuberculosis and Dennis Chitty rejected the disease hypothesis. Tuberculosis was unknown in wild mammals in the 1930s, and it was a new discovery that it could occur in field voles.

By defining the cause (C^+) as a high prevalence of vole tuberculosis and the effect (E^+) as a rapid population decline, Chitty (1954) produced the following decision table for populations studied in 1938 and 1939 (one observation per study area per year)

		Vole tuberculosis	
		Present	Absent
Rapid population decline	Present	3	1
	Absent	2	2

Chitty concluded that vole tuberculosis was neither necessary nor sufficient to produce the observed population declines in field voles in Wales. He suggested that disease was only one in a complex group of factors that could kill individuals, and that attention should shift to finding out why individuals might vary in sensitivity or susceptibility to a wide variety of factors.

A central problem with studies of disease effects is that there are many diseases one can study, so that if you take a single-factor or single-disease approach you will face almost an infinite regress. This problem, coupled with an ever-changing ability of molecular techniques to detect disease, means that disease issues should always be revisited as methods improve, to see if further fieldwork is required.

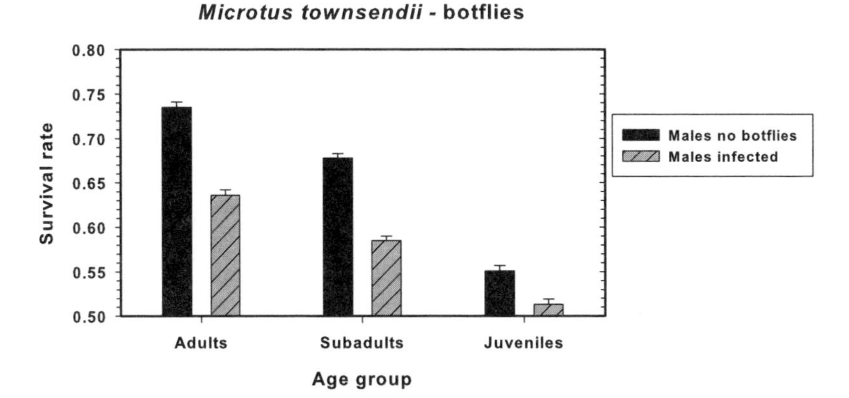

FIGURE 10.1 The difference in survival rates per 14 days between noninfested and infested male Townsend's vole during the botfly season at Vancouver, British Columbia, August through October. Sample sizes were 7,076 adults, 2,759 subadults, and 1,272 juveniles. Females showed the same pattern. Confidence limits of 95% are shown. Data from Boonstra et al. 1980.

There are many papers about rodents and disease in the literature, but they are largely irrelevant to us at this point because the question is typically about human deaths caused by a particular rodent-borne disease. Plague in gerbils of central Asia is a classic example (Davis et al. 2004). Much valuable work and modeling has be devoted to trying to predict outbreaks of plague with regard to human transmission, but we know next to nothing about what happens to the rodent host. Kallio et al. (2009) have shown that human epidemics of the PUUV hantavirus in central Finland can be predicted if one knows the season of the year and the population density and cyclic phase of the host (bank vole). This type of study shows that if one knows the disease ecology of the hosts, it may be possible to make useful predictions about human epidemics without necessarily studying the pathogen incidence directly.

Many papers show the impact of parasitism and disease on the survival rates of small rodents. One simple example comes from botflies. Many different rodent and lagomorph species are annually parasitized by the larvae of cuterebrid flies (botflies). We studied botfly parasitism in *Microtus townsendii* near Vancouver (Boonstra et al. 1980). Botflies can potentially kill their host through secondary skin infections, and we measured survival of voles by live trapping during the botfly season (August through October). Figure 10.1 shows that botfly parasitism reduced survival rates in Townsend's voles between 7% and 16% per 14 days.

The red-backed vole (*Myodes gapperi*) in the boreal forest is also parasitized by botflies. LeMaitre et al. (2009) showed that botfly parasitism reduced survival of these voles and also reduced the rate of midsummer population growth (figure 10.2). There was no indication that female reproductive rates were affected by botfly parasitism in midsummer, and the reduction in population growth rate was presumably driven by increased mortality among infected voles. Since botflies occur only in midsummer, they have only a transient effect on annual rates of population change, so they are unlikely to be a major mechanism of such change. Because botflies in one year produce pupae that mature only in the following year, if other things are equal botfly parasitism rates should oscillate in a two-year pattern. This delay in the density-dependent effect due to the life cycle of botflies should produce instability in red-backed vole populations. Botfly parasitism can thus be a limiting factor, and it might contribute to the

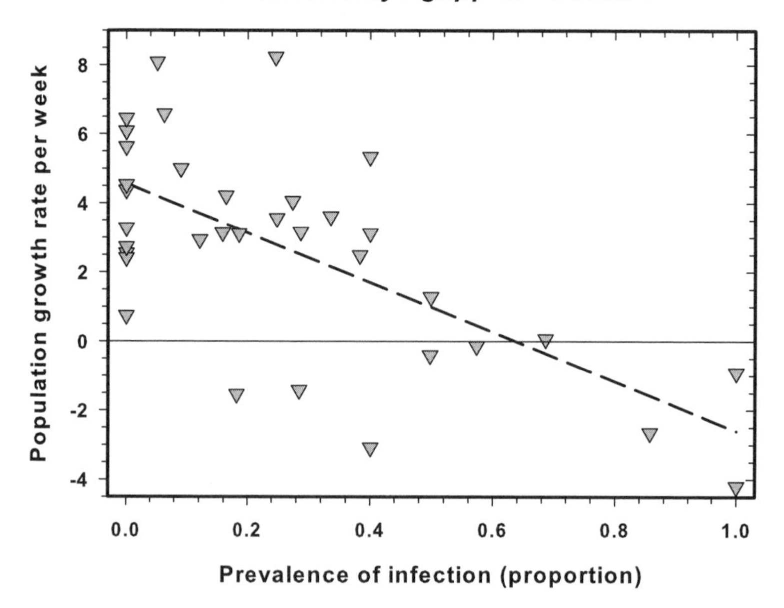

FIGURE 10.2 Prevalence of botflies in relation to the midsummer population growth rates of red-backed voles (*Myodes* [*Clethrionomys*] *gapperi*) in the boreal forest of Quebec. Population growth was measured as (August abundance – July abundance)/four weeks. Data from LeMaitre et al. 2009.

TABLE 10.1 **Survival rates of bank voles over winter in Finland, in relation to their hantavirus infection status in autumn. Data from Kallio et al. 2007.**

	Not infected with hantavirus		Infected with hantavirus	
Sex	No. in autumn	% surviving	No. in autumn	% surviving
Female	308	22.4	89	4.5
Male	285	11.2	64	3.1

instability of red-backed vole populations in the boreal forests of eastern Canada.

Hantaviruses are widespread in rodents and are another example of a virus disease that causes considerable human mortality. Much literature has been published on these viruses, but only a few papers provide data on what they do to their host rodent species. The prevailing view is that the rodent hosts are well adapted to their hantaviruses, and that hantavirus infections will have very little effect on rodent population change. Prevetali et al. (2010) supported this view because after an extensive analysis they could find no effect of the Sin Nombre hantavirus on survival rates in deer mice (*Peromyscus maniculatus*), the main vehicle for transmitting this lethal virus to humans in the United States. By contrast, Kallio et al. (2007) analyzed the overwinter survival of the bank vole *Myodes* (*Clethrionomys*) *glareolus* on islands in Finland, and found that hantavirus-infected males and females survived less well than uninfected individuals (table 10.1). But only 20% of the individuals were infected, and this level of infection may not greatly reduce overall winter population survival. This is the first step in showing that diseases can affect population dynamics: showing that they can influence reproduction or survival. But the next important step is to show that these effects on the rodent host are related to rodent density, and are large enough to affect demographic change.

Experimental Tests

The weakest hypothesis for disease is that prevalence is related to population density or rate of population growth. This is a necessary condition for the overall hypothesis that epidemic disease can generate population fluctuations, but it is weak because it is a correlation. Many aspects of organisms and their environment can rise and fall with population density and not be a necessary part of the machinery that causes population change. Every statistical textbook tells us that correlation does not imply causation, yet many ecologists seem not to have listened. The key experiment will involve increasing disease prevalence, or reducing it in a wild population, and observing inverse changes in the population growth rate. Useful evidence can also be obtained by following parasite and disease prevalence through several phases of a population fluctuation.

The water vole (*Arvicola terrestris*) has been the subject of two major studies that have tried to relate parasite prevalence to the phase of the population cycle. Deter et al. (2007) studied the prevalence of the gastrointestinal nematode *Trichuris arvicolae* in water voles in 13 populations that were out of phase with one another. Deter et al. (2007) were able to show that the prevalence of this parasite varied with phases of the population cycle (figure 10.3). Prevalence increased toward the water voles' peak of density, and then collapsed in their late peak and declining populations. Juvenile voles had a higher prevalence than adult animals, which may explain why late peak and declining populations showed low prevalence. Gastrointestinal nematodes are present in most species of mammals and are commonly thought to be relatively benign to their hosts. There is no

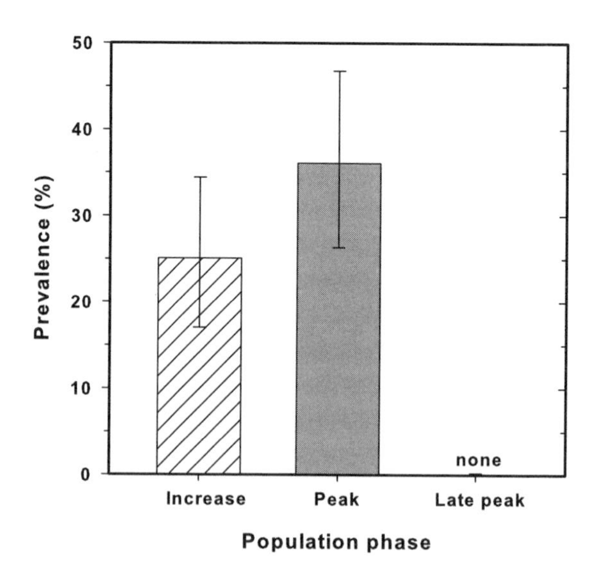

FIGURE 10.3 Prevalence of the nematode *Trichuris arvicolae*, found in water voles in the central part of France, in relation to the stage of population fluctuation. Sample sizes for the three population phases were 105, 92, and 38 individuals respectively. Data from Deter et al. 2007.

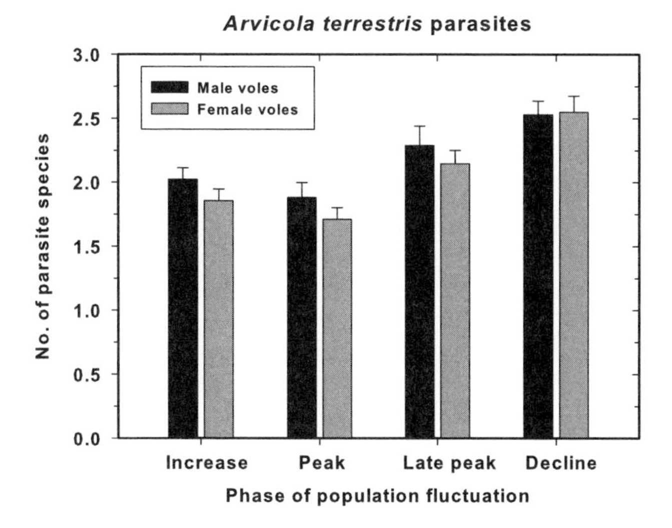

FIGURE 10.4 The average number of trematode and cestode species found in male and female water voles in eastern France over a population fluctuation. Water voles in this region have six-year cycles. Five nematodes and seven cestodes were sampled. Data from Cerqueira et al. 2007.

evidence that this nematode affects the survival or reproduction of the water vole, and it is possible that it might contribute to population changes as part of a community of several species of parasites interacting together.

Cerqueira et al. (2007) investigated how a helminth community (five nematode and seven cestode species) changed over the course of a full six-year water vole (*Arvicola terrestris*) population cycle. They measured the parasites' intensity by how many of these 12 species were carried per individual vole during different phases of the cycle. Figure 10.4 shows that voles in the late peak and decline phases of the cycle had more parasite species present. This delayed density-dependent effect thus may produce cyclic dynamics. The next step in research must be to find out if the effects of these parasites on vole survival and reproduction are sufficiently large to cause the decline phase. The parasite community may produce effects that are synergistic, and this important idea needs investigation.

Cowpox virus is found throughout much of Europe and western Asia. In Great Britain, cowpox antibody has been found occasionally in house mice (*Mus musculus*), but the highest seroprevalence is in bank voles (*Myodes glareolus*), wood mice (*Apodemus sylvaticus*), and field voles (*Microtus agrestis* L.); these three species are believed to be the reservoir

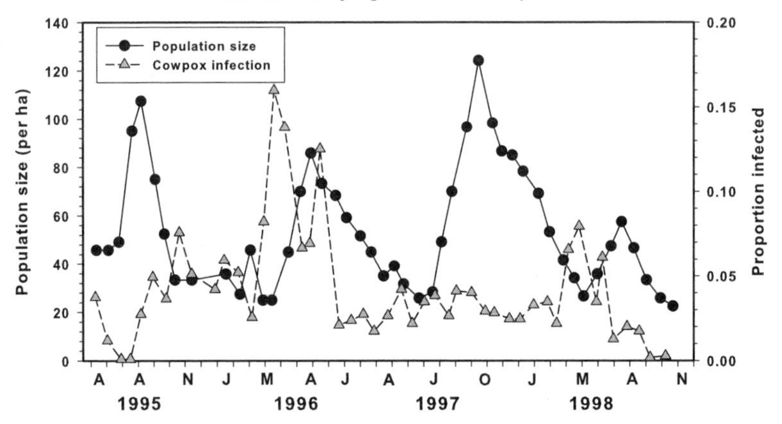

FIGURE 10.5 Changes in population size (per hectare) and prevalence of cowpox infection for bank voles in Manor Wood, northwest England, from 1995 to 1998. The prevalence of cowpox peaked in 1996 but was low in 1997. Data from Telfer et al. 2002.

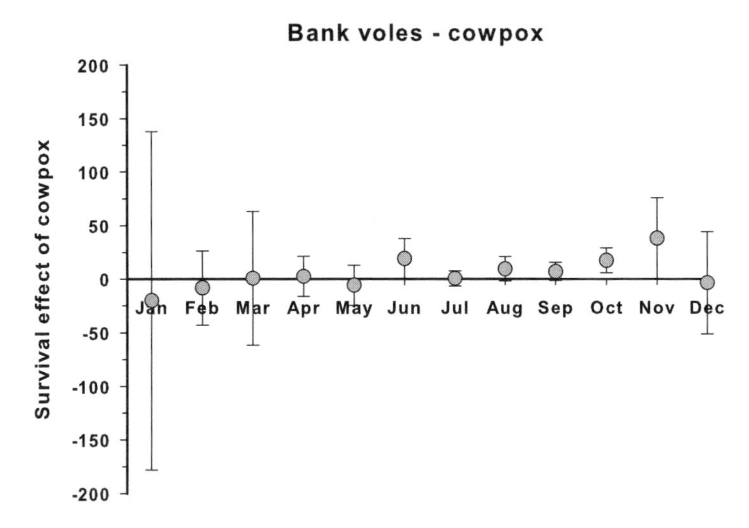

FIGURE 10.6 The estimated effect of cowpox prevalence on survival of bank voles over the annual cycle in northwest England, from 1995 to 1998. A zero value indicates no effect of cowpox infection on survival, and a positive value indicates that infected voles survived better than uninfected voles. Bars give 95% confidence limits. From Telfer et al. 2002, figure 4.

hosts (Telfer et al. 2002). Although it does not cause obvious clinical signs or increase mortality in voles or mice in the laboratory, experimental studies have demonstrated that cowpox can affect fecundity by delaying the onset of reproduction. Telfer et al. (2002) followed the effect of cowpox virus on survival in bank voles and wood mice using four years of data from two sites in northwest England. Figure 10.5 shows the temporal change in the size of bank vole populations and in the incidence of cowpox infection, which peaked at 17%. Cowpox had variable effects on the survival rates of bank voles (figure 10.6). In winter there was slightly lower survival rate in infected voles; but in autumn, infected voles survived better than uninfected ones. Parasite infection could increase apparent survival in small rodents by reducing their tendency to disperse or move about and thus expose themselves to predators, or by compensatory reduction of energy devoted to reproduction in voles that are infected.

Could cowpox infection influence the reproductive rate of bank voles? Telfer et al. (2005) measured the age at sexual maturity of female bank voles that became infected with cowpox, and of those that did not. Figure 10.7 shows that the uninfected females matured at a median age of

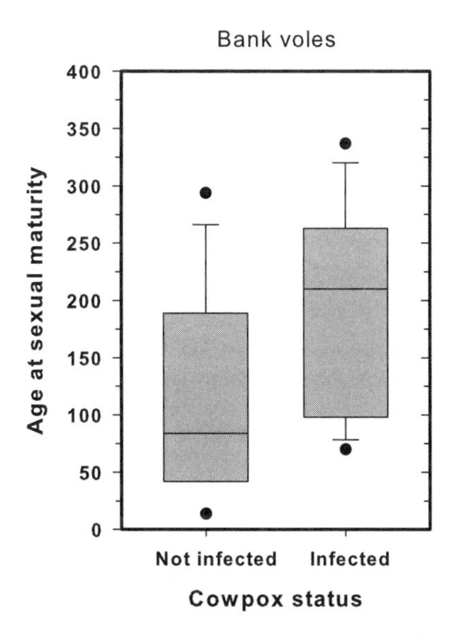

FIGURE 10.7 The age at sexual maturity of female bank voles that were and were not infected with cowpox before reaching sexual maturity. The box plots give the median ages, the grey boxes includes the 25% and 75% percentiles, and the error bars delimit the 10% and 90% confidence limits for the median. Dots indicate outliers. Data from Telfer et al. 2005, figure 1.

84 days, while the infected females did not reach sexual maturity until a median age of 210 days. Juvenile voles can mature in their first summer of life or in the following summer, since there is almost no winter breeding in this species. Cowpox infections delay the sexual maturity of many females into their second summer of life.

But can a delay in sexual maturity affect population growth rates in small mammals? As we discussed in chapter 3, reproductive rates change dramatically in fluctuating rodent populations. Oli and Dobson (2003) showed that for small rodents, age at sexual maturity had the largest effect on population growth rate of all the demographic variables. Smith et al. (2008) produced a mathematical model of the effect of delayed sexual maturity and reduced reproductive output on population growth rates in voles. Their model predicts disease-induced multiyear population cycles for rodent populations that recover relatively slowly following a disease-induced population collapse. Diseases for which the period of infection is brief but full recovery of reproductive function is relatively slow could generate high-amplitude multiyear population cycles. Chronically reduced fecundity following recovery can also induce multiyear cycles. For the Kielder Forest *Microtus agrestis* populations illustrated in figure 9.7, the model predicts that a disease must chronically reduce host fecundity by more than 70% following recovery from infection for it to induce such cycles. That is a very large effect on reproductive rate, and it remains to be seen if a disease like cowpox, or a combination of parasitic diseases, could produce an effect this large.

These studies of disease in small rodents have opened the door to the possibility that some of their population fluctuations could be driven by chronic infections of one or several parasites. There is a model for trying to determine the role of parasites and disease in rodent population dynamics, from the earlier investigations of the red grouse cycle in Scotland and England. The role of parasites was thought to be critical in that system (Hudson et al. 1998) until a large-scale field experiment showed that parasite burdens affected some red grouse population changes but did not determine the birds' cyclic dynamics (Redpath et al. 2006).

Ostfeld (2008) has suggested two models for how disease might interact with other factors in causing population declines (figure 10.8). For small rodents he suggests a model based on food limitation, which affects both levels of physiological stress and can lead directly to mortality. Physiological stress is known to reduce immunocompetence, which in turn can aggravate parasite loads (Lochmiller and Dabbert 1993). The second model

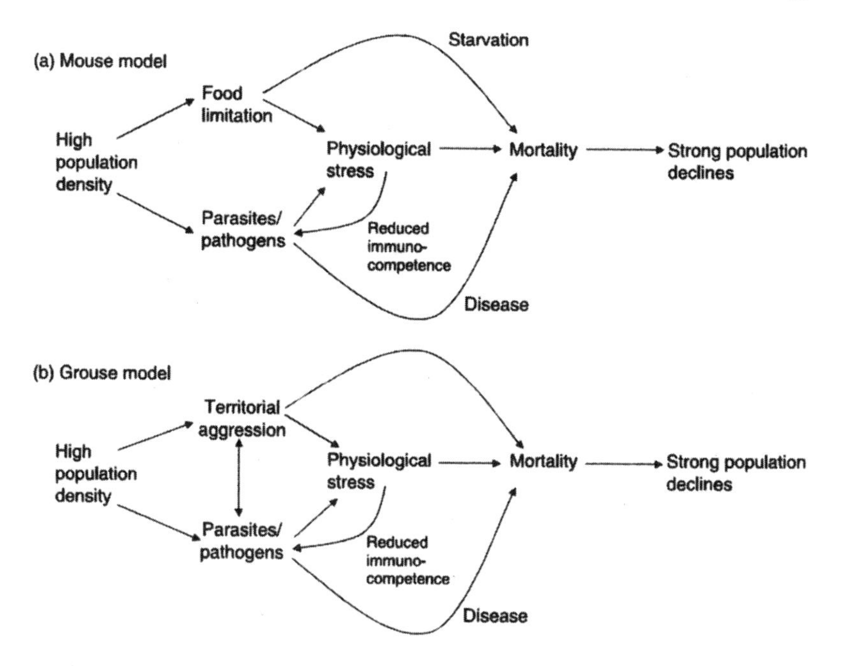

FIGURE 10.8 Two conceptual models for how interacting factors could cause strong population declines in fluctuating populations of small rodents. The mouse model uses food limitation as a key variable that causes physiological stress, the grouse model uses territorial behavior as a direct influence on parasite loads. Both variables contribute to stress. The key point is that much more experimental work is needed to test these models. From Ostfeld 2008.

Ostfeld (2008) suggested is the grouse model, in which territoriality and aggression can directly affect parasites and pathogens. These two models have much in common and provide an outline of how future research on disease in small rodents might proceed.

Conclusion

The role of disease in population dynamics has been an item of great interest since the classic papers of Anderson and May (1979) and May and Anderson (1979). The first step has been to show that disease agents affect reproduction, mortality, or movements in the host species. This step is necessary to demonstrating that they affect population growth rates (cf. figure 7.1); but since populations can compensate for changes in demographic rates, it is not sufficient proof. The next step is to show a correlation between population growth rates and disease incidence. These correlations may be density-dependent or delayed density-dependent, and only those that are delayed can contribute to fluctuations in numbers. This raises the important question of how long infections persist, and how long their effects on individuals persist. There is a clear potential for synergy between infection and body condition: poor condition predisposes individuals to infections, which in turn further reduce their body condition (Beldomenico et al. 2008).

Interactive suites of parasitic infections further complicate testing of the role of disease in rodent population dynamics. Much more basic research on disease agents must be carried out before we will be in a position to do classic field experimentation on these questions by removing infections or increasing them, as has been done with red grouse (Redpath et al. 2006) and deer mice (Pedersen and Greives 2008). For the present we can conclude only that disease agents are possible causes of instability in rodent populations, and that they must be considered in any investigation of the population dynamics of rodents. The classic view that rodents are well adapted to their parasite and disease agents may be correct, but new evidence is now available to put a question mark after this hypothesis.

CHAPTER ELEVEN

Self-Regulation Hypotheses for Fluctuations

Key Points:

- Self-regulation hypotheses all assume that interactions between individuals in a population cause demographic changes that produce population fluctuations.
- Self-regulation hypotheses suggest that interactions between individuals produce changes in physiology and behavior that may be phenotypic or genotypic, and the resulting time lags generate population fluctuations.
- The two general pathways for possible self-regulation are physiological stress involving the adrenal-corticosteroid axis or genetic selection of aggressive or spacing behavior.
- Early tests of the physiological stress hypothesis were negative, and this mechanism was rejected prematurely because of the crude measures of stress available during the 1950s and 1960s. Newer methods are more promising and now provide ways of testing for maternally inherited stress effects on reproduction and survival.
- Genetic selection for aggressive or spacing behavior has proven difficult to test, and results to date are negative. Genetic changes are difficult to uncouple from maternal effects in wild populations, and maternal effects are now a focus of study.

Extensive research by Hans Selye and his associates during the 1930s and 1940s led to an increasing understanding of the role of stress in the

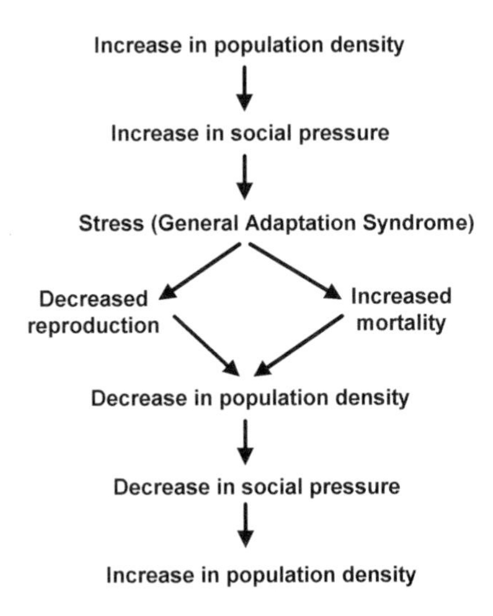

FIGURE 11.1 The stress hypothesis as a suggested mechanism for population fluctuations in rodents, first suggested by John Christian in the 1950s. After Krebs 1964.

lives of mammals. The stress response involves the hypothalamic-pituitary-adrenal axis and is a generalized response to any type of environmental disturbance that disrupts homeostasis. In his early studies Selye showed that stress in laboratory rodents could reduce reproductive rates, hinder growth and development, and cause an increase in disease (Selye 1936, 1955). Selye called this set of physiological responses to stress the General Adaptation Syndrome, and postulated that it was a universal vertebrate response to stress.

John Christian (1950, 1956) recognized in 1950 that population density could itself be a stressor, and activation of the General Adaptation Syndrome by rising population size could be a factor in limiting population growth. Christian and his colleagues did extensive research on laboratory colonies of house mice and rats, and concluded (Christian and Davis 1964) that the endocrine responses associated with stress could be a method of population control of cyclic rodents as well as pest mammals. Christian's stress hypothesis (figure 11.1) was important because it focused attention on the interactions between individuals, the effects they had on physiology, and the consequences of stress for reproduction and survival. While Christian and his coworkers focused on population density as a stressor, Selye had shown clearly that any number of physical or chemical agents could cause stress as well.

Stimulated by the early work of Christian, Dennis Chitty developed a general hypothesis for self-regulation of population size. The central hypothesis of the self-regulation school was stated first by Chitty (1960): "All species are capable of limiting their own population densities without either destroying the food resources to which they are adapted, or depending on enemies or climatic accidents to prevent them from doing so. . . . Under appropriate circumstances, indefinite increase in population density is prevented through a deterioration in the quality of the population" (p. 99). Populations stop growing, according to Chitty, because they decay in "quality" as density goes up. But what is "quality?" On this point Chitty (1960) was not explicit. Quality may be defined as any individual morphological, physiological, or behavioural attribute that influences population fitness.

Two types of self-regulation mechanisms are possible because changes in individual quality may be genotypic or phenotypic. The stress hypothesis (figure 11.1) suggests that mutual interactions lead to physiological changes, phenotypic in origin, that reduce births and increase deaths. These effects could be density-dependent or delayed density-dependent. In order to generate population cycles, the physiological changes generated by stress must have a time lag (May et al. 1974).

The second mechanism of self-regulation has been labeled the Chitty hypothesis or the polymorphic behaviour hypothesis (figure 11.2; Krebs 1978). This hypothesis suggests that mutual interactions involving spacing behavior have genetic consequences and produce changes in birth, death, and dispersal rates that are subject to natural selection. As Chitty (1967) stated: "All species of animals have a form of behaviour that can prevent unlimited increase in population density. . . . Mechanisms for the self-regulation of animal numbers are thought to be a consequence of selection, under conditions of mutual interference, in favour of genotypes that have a worse effect on their neighbours than vice versa."

Hypotheses about self-regulation have been fruitful in developing cross-linkages between biological disciplines. The stress hypothesis has excited endocrinologists, ecologists, and sociologists and has led to an exploration of the area of social stress (Selye 1955) and reproductive physiology (Clarke 1977). The polymorphic behavior hypothesis has been particularly fruitful in forging a linkage between social behavior and population regulation. It has helped to bring the discipline of population genetics into the ecological area and has fitted into the wider movement to infuse ecology with natural selection and to ask the important question of how rapidly natural selection could operate in wild populations (Sinervo et al. 2000).

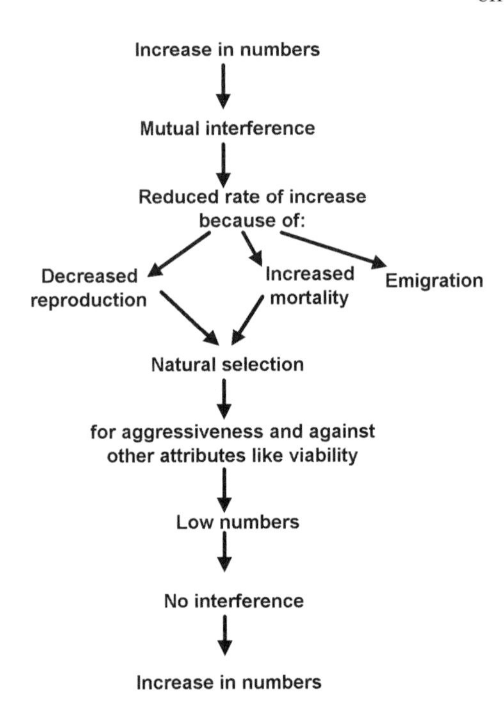

FIGURE 11.2 The polymorphic behavior or Chitty hypothesis, a suggested mechanism for generating population fluctuations in rodents. After Krebs 1964.

Testing of Christian's Stress Hypothesis

Christian's stress hypothesis was characterized in stressed animals by an enlargement of the adrenal glands, and during the 1950s and 1960s this became its defining prediction. In laboratory colonies of house mice and Norway rats, adrenal enlargement was a characteristic feature associated with population density (Christian and Davis 1964). But in field populations existing at population densities much lower than those of laboratory colonies, adrenal size changes did not map onto population size changes as Christian's hypothesis predicted they would. (Chitty 1961; Krebs 1964). One example of this, from *Microtus agrestis* in Wales, is illustrated in figure 11.3 (Chitty 1961). There are large seasonal changes in adrenal weight, as well as year-to-year shifts, but none of those changes correlates with population rates of change.

The result of these early studies was a rejection of Christian's stress hypothesis and the conclusion that it was an artifact of confinement in

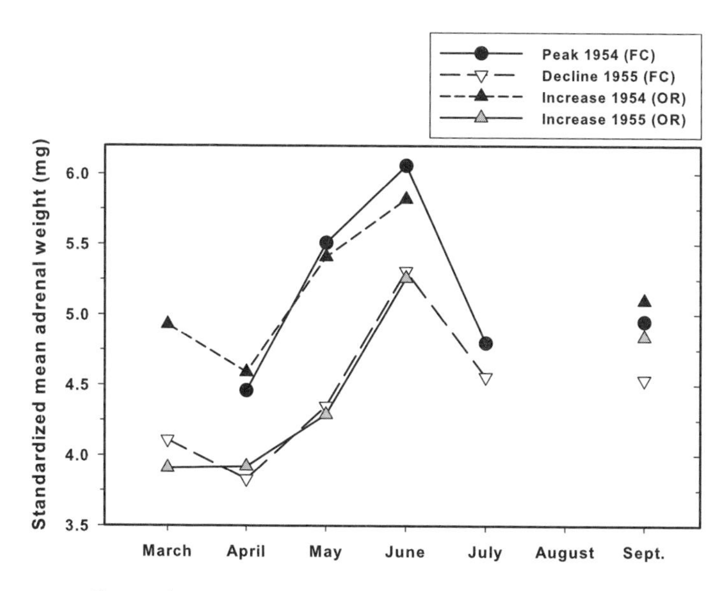

FIGURE 11.3 Changes in mean adrenal weights in two field populations of *Microtus agrestis* in Wales that were out of phase. Adrenal weights were relatively high in 1954, regardless of the phase of the population fluctuation, and were relatively low in 1955. Data from Chitty 1961, table 2.

laboratory colonies whose population densities were much higher than those of field populations. As we shall see below, this conclusion was based on physiological measurements now considered too crude and obsolete, and new methods have resurrected the stress hypothesis in a modern form (Boonstra et al. 1998; Sheriff et al. 2011), as we will see below.

Testing of the Chitty Hypothesis

Attention turned in the 1960s and 1970s to attempts to test the Chitty hypothesis. Two aspects of the polymorphic behavior or Chitty hypothesis needed investigation:whether there were changes in spacing behavior as populations fluctuated, and whether specific genotypes could be identified that were subject to fluctuating natural selection. Increasingly sophisticated studies of small mammal behavior brought forth a series of ideas on how behavioral interactions, via territoriality and possibly infanticide, might affect rates of population change in field populations (e.g., Wolff et al. 2002: Wolff 2003a: Wolff and Sherman 2007). But showing that spacing behavior could vary with population density was only a start to providing support for the polymorphic behavior hypothesis (Watson and Moss 1970).

Testing behavioral hypotheses on small rodents is made difficult by their use of heavy cover and their nocturnal activity. The earliest attempts in enclosed areas were useful for describing the kinds of interactions they had with one another (Clarke 1955), but the high densities of enclosed populations made it difficult to translate any effects to natural populations. A series of removal experiments on field populations of *Microtus* were the next attempts to manipulate behavior indirectly by changing density, and we have described the results of some of the many studies in described in chapter 5 (cf. figures 5.3 and 5.8).

The key experiments used to test for behavioral limitation of numbers manipulate behavior directly with implants such as testosterone. The elegant experiments on red grouse by Watson (1964, 1967) and by Watson and Jenkins (1968) showed how this could be done in diurnal, highly visible birds, but it was much more difficult to conduct these experiments in small rodents. One example was an attempt to increase the rate of population decline in the vole *Microtus townsendii* by implanting males with testosterone at the start of breeding in spring (Krebs et al. 1977). No increase in aggression could be measured, population changes were similar on control and experimental fields, and we concluded that in this species testosterone did not control male aggressive behavior.

Similar problems have plagued attempts to determine whether infanticide could be a significant factor in reducing population recruitment in small mammals (Labov et al. 1985). In laboratory populations, infanticide can be severe. Figure 11.4 illustrates one study on collared lemmings in the laboratory, and shows the significance of male relatedness to the likelihood of neonatal deaths. In spite of these laboratory studies, very few data are available on the magnitude of infanticide in field populations, principally because it is so difficult to pinpoint (Agrell et al. 1998; Ylönen et al. 1997). In bank voles in the laboratory, female infanticidal behavior has a heritability of 0.56 (Poikonen et al. 2008), and infanticide caused a 20% decrease in offspring survival. At present we can only speculate about how these laboratory results might apply to fluctuating field populations of voles and lemmings.

The original formulation of the polymorphic behavior hypothesis envisaged that we could find two divergent ecotypes of voles and lemmings:

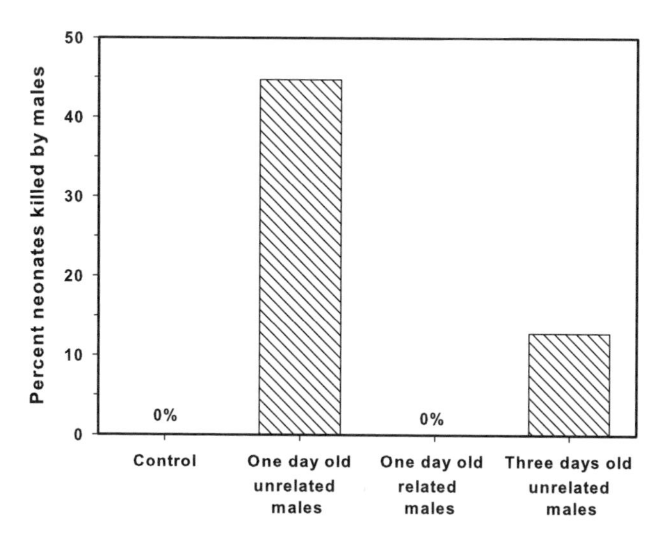

FIGURE 11.4 Differences in neonate mortality induced by unrelated and related male collared lemmings (*Dicrostonyx groenlandicus*) in laboratory tests. Each control cage had a mated pair throughout the test period. In the other treatments related males (fathers) or unrelated males were introduced to the cages at the specified age of the neonates. Data from Mallory and Brooks 1978, table 1.

one an *r*-selected ecotype, at an advantage when population density was low and crowding was not important, and the other a *K*-selected ecotype, at an advantage when density was high and spacing behavior was the key to success. Unfortunately, no one has been able to find these ecotypes in any small mammal population. This could mean either that not enough research has been carried out, or that the hypothesis is not correct.

If individual "quality" is fixed, one should be able to transfer a population of individuals from a declining population into a new area and see them continue to decline. Conversely, if individuals from a declining population are transferred to an area of increasing density and take on the characteristics of the increasing animals in that area, it would suggest that there nothing in "quality" is intrinsic or fixed, contrary to the Chitty hypothesis (Krebs 1978). Ergon et al. (2001) did this experiment on *Microtus agrestis* populations in Scotland and found that transplanted individuals took up the characteristics of animals at the new site, so there was no evidence of the fixed, intrinsic "quality" of individuals predicted by Chitty (1967).

There are many reasons why transfer experiments in the field may fail. Many transferred animals simply disappear, so it is possible that the transfer process selects for certain phenotypes or genotypes. No one has shown this to be the case. But "quality" is a social construct, and immigrants to a new area must fit into the social milieu, it is possible that "quality" would shift automatically. The key, however, is that we have no evidence in voles or lemmings that different behavioral ecotypes exist and play an important role in population changes. Additional experiments of the Ergon et al. (2001) type are desirable so that we can be more strongly convinced that absence of evidence for polymorphic behavior types can be interpreted as evidence of absence.

The suggestion by Chitty (1967) that genetic changes might underlie population fluctuations has stimulated a search for genetic shifts that might accompany those fluctuations. Genetic methods have changed so much in the last 50 years that the methods of electrophoresis for single loci we began using in the 1960s now seem so primitive that the early work can be dismissed as irrelevant to testing the genetic aspects of Chitty's views.

By the 1970s, attention had changed to a heritability approach toward testing the Chitty hypothesis. If life history characteristics such as age at sexual maturity were subject to natural selection, they should show relatively high heritability. The first attempt I know to try to quantify the value of high heritability needed to validate Chitty's hypothesis was a model developed by Judith Anderson (1975) in her PhD thesis. Anderson estimated that heritability would need to be around 80% to be able to drive population changes through genetic shifts in life history characteristics like age at maturity. This heritability was much higher than Anderson (1975) found in life history traits of *Microtus townsendii* in British Columbia, and she was sceptical that the Chitty hypothesis could operate in field populations of this species.

The best test for genetic changes via heritability analysis has been that of Boonstra and Boag (1987) on *Microtus pennsylvanicus* in Ontario. Implicit in the Chitty hypothesis is the assumption that a suite of life history traits is simultaneously undergoing selection, and that the traits are strongly heritable. Boonstra and Boag (1987) tested this assumption in two ways. First, they determined whether the year-to-year differences in phenotypes of fluctuating meadow vole populations in the field were maintained in samples of young animals raised in the laboratory. Second, they measured whether the variation seen in the field was inherited as de-

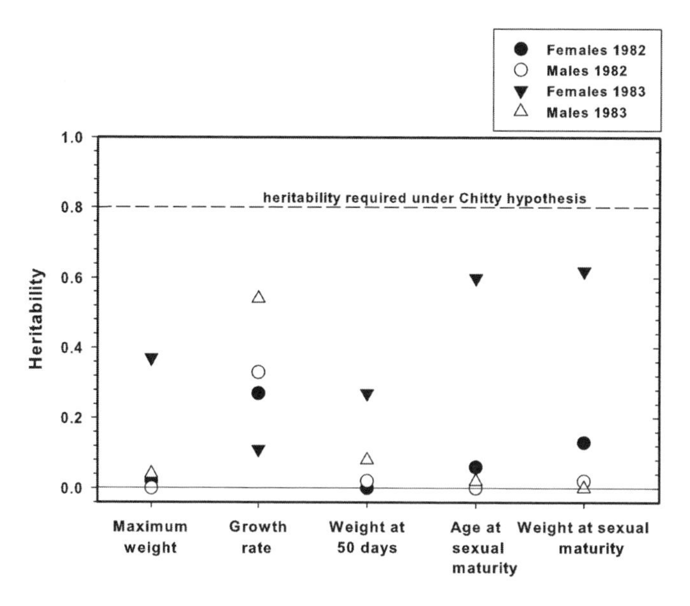

FIGURE 11.5 The heritabilities for life history characteristics observed in progeny of *Microtus pennsylvanicus* over two years in southern Ontario. Almost all the observed heritabilities are low, and do not reach the theoretical level of 0.8 that would be required for rapid genetic change in cyclic vole populations. Data from Boonstra and Boag 1987, tables 6 and 7, and on theoretical heritability from Anderson 1975.

termined by half-sibling analysis. In the field, young from a year in which population size was increasing grew more rapidly than those from the year of peak population. Boonstra and Boag (1987) calculated the heritability of body weight, growth rate, and age and weight at sexual maturity (figure 11.5). Virtually all these traits showed significant dam effects but small or nonexistent sire effects. Thus, most of the variation was nongenetic in origin; which suggests that maternal and other environmental effects were of overriding importance. They concluded that these life history traits have low heritability and thus rejected the Chitty hypothesis. Their calculations backed up those of Anderson (1975), and they suggested that heritabilities of 0.82 would be needed over the five to seven generations that constitute an average vole or lemming fluctuation. Such high heritabilities have never been observed in voles or lemmings. They suggested as an alternative hypothesis that maternal effects could be an important mechanism by which changes in life history traits are produced in fluctuating populations.

TABLE 11.1 **Characteristics of two lines of female bank voles selected for high and low reproductive effort. Reproductive effort is defined as (litter size × mean body mass of offspring$^{0.75}$) / mother's postpartum mass. These two lines were used in the field experiments of Mappes et al. (2008) on density- and frequency-dependent selection described in the text. (Means ± S.E.)**

Measure	Low-reproductive-effort line	High-reproductive-effort line
Reproductive effort	0.61 ± 0.02	0.88 ± 0.02
Litter size	3.82 ± 0.14	6.15 ± 0.18
Mean body mass of offspring (g)	1.97 ± 0.03	1.84 ± 0.02
Mean head width of offspring (mm)	8.28 ± 0.05	8.09 ± 0.042
Body mass of mother postpartum (g)	22.4 ± 0.4	24.6 ± 0.4

Mappes et al. (2008) showed that if one selected lines of the bank vole (*Myodes glareolus*) in the laboratory for high and low reproductive output (table 11.1) and put these females in a large pen in the field, there resulted strong selection that was both frequency-dependent and density-dependent. Females with low reproductive effort were at an advantage in low-density populations, while females with high reproductive effort were most successful when they were rare in high-density populations. These experiments show that selection is possible in fluctuating populations, but it is not clear if the observed selection acts as a cause of density change or is a consequence of it. Until females with high and low reproductive effort can be recognized and studied in natural populations in the field, we will not know whether these studies are just another example of laboratory selection or are consistent with Chitty's genetic mechanism for helping to generate population cycles.

A Revised Stress Hypothesis

By the 1990s, most population ecologists had rejected intrinsic or self-regulation hypotheses as a possible explanation for population fluctuations, since both the original stress hypothesis and the polymorphic behavior hypothesis had been rejected. By this time, however, physiologists had developed new and more sophisticated methods for measuring stress levels (Sheriff et al. 2011), and had explored in much more detail the physiological machinery behind stress effects in mammals.

A revised stress hypothesis (figure 11.6) was developed by Boonstra et al. (1998) and Sheriff et al. (2009) to explain population cycles in snowshoe hares (*Lepus americanus*). Although here we are concentrating on small rodents, the ideas suggested by Boonstra et al. (1998) are relevant to the need for future testing of the revised stress hypothesis in small rodents. The central idea is that stress can arise from a variety of sources. In snowshoe hares it seems to arise mainly from predation risk, since nearly all hares are killed by predators. But in voles and lemmings, stress could result from a multitude of causes, from predation risk to food shortages to mutual interference via spacing behavior. Thus, this hypothesis has the potential to integrate what have in the past seemed to be quite separate schools of thought on population regulation. The manifold effects of stress now recognized are diagrammed in figure 11.7.

For stress to drive a population fluctuation, its consequences must be delayed density-dependent, as shown in figure 11.6. Chitty (1967) recognized very early that adult voles did not seem greatly affected by high densities, but that their offspring were impaired. For this to occur, there must be a memory in the system that traditionally has been thought of as genetic but now is increasingly considered to be maternal. Can stress effects in natural populations be inherited in offspring via maternal effects? We do not know this for voles or lemmings, but it is clear that it occurs in snowshoe hares (Sheriff et al. 2010b).

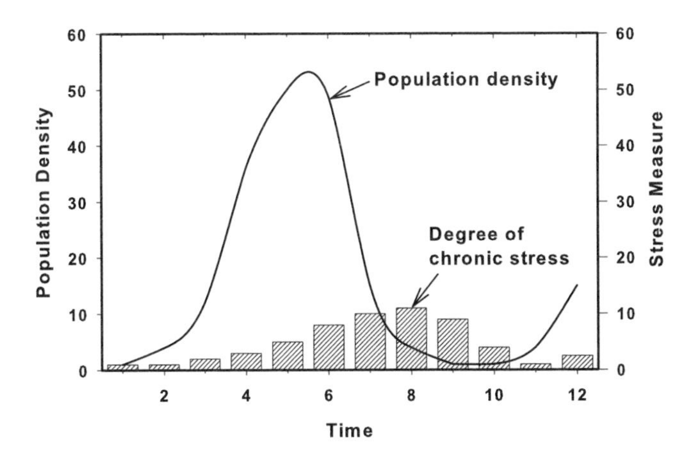

FIGURE 11.6 A schematic representation of the revised stress hypothesis of Boonstra et al. (1998). To explain population fluctuations, stress must have a delayed density-dependent effect on individuals. This hypothetical model has been applied to snowshoe hares but not yet to small rodents.

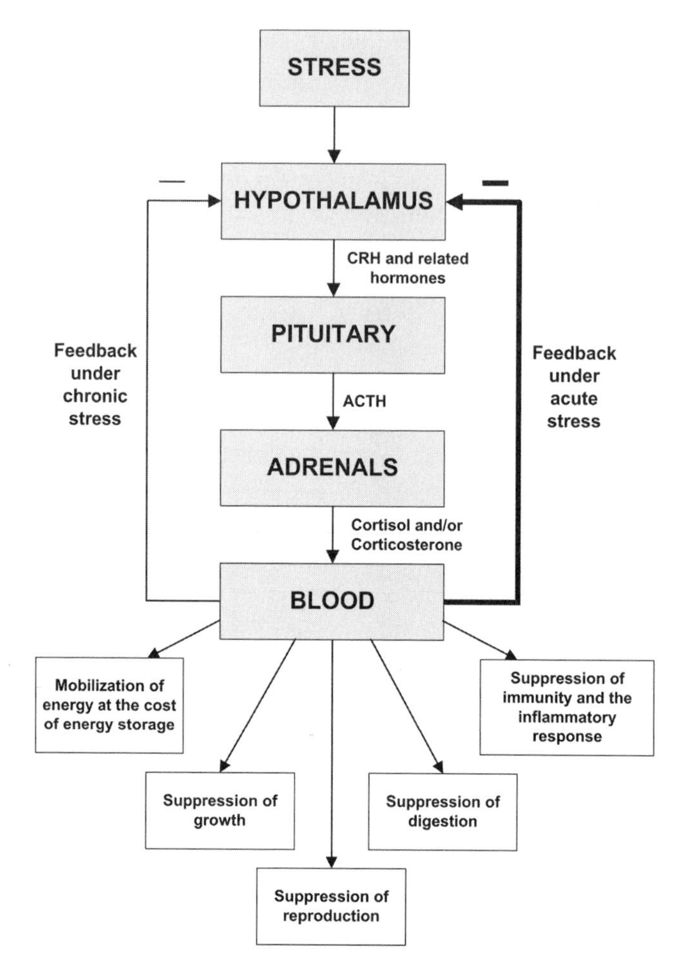

FIGURE 11.7 The current pathways and influences of stress on mammals. These effects remain to be studied in fluctuating populations of small rodents. From Sheriff et al. 2011.

Territoriality and Population Fluctuations

One puzzling feature of population fluctuation in rodents is that many of the species that fluctuate strongly are territorial, exhibiting male territoriality, female territoriality, or both (Wolff and Sherman 2007). Territoriality in its simple form is supposed to limit population growth at high density (figure 11.8a), and if this were so, territorial species would never fluctuate above the territorial threshold. Eccard et al. (2011) suggested that there could be a tipping point at which dominant territorial owners can no longer defend their space and suppress female maturation. The population would be released from density control at this point and the social system, breaking down, would release all females to breed (figure 11.8c).

Eccard et al. (2011) tested this model in the bank vole in Finland (figure 11.9) by the use of field pens of 0.25 ha. At low and moderate densities (<20 females per ha) the dominant breeders suppressed breeding by juvenile females, as others have found in other species of *Myodes* (e.g., Gilbert et al. 1986). But above a density threshold of about five females per enclosure (>20 per ha), the dominance of breeders could not be sustained (a condition of incomplete control). This threshold was reached when, after territories could shrink no further, the number of intruders continued to

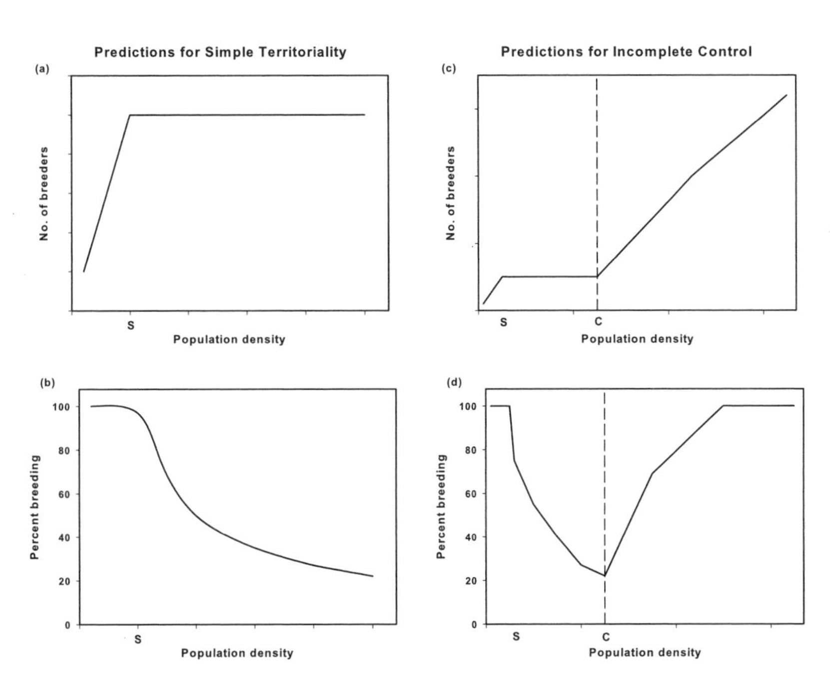

FIGURE 11.8 The incomplete control model for explaining density outbreaks in territorial rodents. (a) Under a simple model of territoriality, the number of breeders is limited by territorial occupancy (S = saturation density). (b) The percentage of females breeding falls with density in the simple model. (c) Under the incomplete control model of Eccard et al. (2011), once density reaches a tipping point (density C), the number of breeders and the percentage of females breeding (d) increases again. This is because the dominant animals cannot defend their space and the social system shifts, thus allowing outbreaks of density. Modified from Eccard et al. 2011.

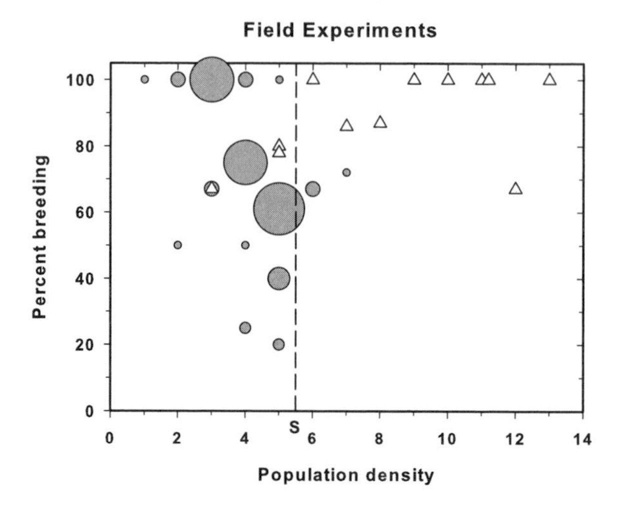

FIGURE 11.9 A field test of the incomplete control model in the bank vole (*Myodes glareolus*) in Finland. Earlier tests (bubble plot data) on 38 populations in 0.25 ha enclosures showed the traditional pattern of reproductive suppression in females as population density increased from one to five per enclosure. When further experiments were carried out at double and triple the density used in the early experiments, reproductive suppression could not be sustained and the number of females breeding increased to 100%. S marks the tipping point of about five females per enclosure (20 per hectare). Data from Eccard et al. 2011.

increase. Suppression probably becomes too costly for the dominant females. In wild rodent populations, crossing this threshold would allow for rapid density increases or population outbreaks, enabling territorial species to escape from territorial control of density. Stress hormone levels also increased at high densities, perhaps feeding back via the mechanisms described by Sheriff et al. (2009) for snowshoe hares.

Conclusion

Self-regulation hypotheses arose first from observations on confined populations of house mice and rats in situations where there were no predators, diseases, or food shortages to hold down population growth. They were then applied by John Christian and Dennis Chitty, among others, as possible explanations for population fluctuations of voles and lemmings in the wild. At least some wild populations that seemed to have few predators, few signs of disease, and no explicit food shortages, nonetheless fluctuated strongly in abundance from year to year. The question then became:

Was it possible that some mechanism of self- regulation was involved in population regulation in small rodents? Because the self-regulation model was applied to cyclic rodents, it was implicit that there had to be some kind of memory in the system, and this could only be either a genetic memory or a nongenetic physiological memory that might involve maternal inheritance.

The view now is that it is most unlikely that a genetic memory is involved, and the best model for self-regulation must be a maternal effects model. If this model operates through the hypothalamus-adrenal axis, as is suggested for snowshoe hares, it is possible that a wide variety of stressors could generate the required memory. At present we do not know whether small rodents could show these stress effects in field populations, nor do we know how much of a role could be played by mutual interference or aggressive behavior over social space (Krebs et al. 2007; Wolff 2003a). How much self-regulation mechanisms might interact with traditional external factors like predators, disease, and food shortage is at present an open question, which we shall address in the following chapters. Since both density-dependent and delayed density-dependent processes must be present to generate cycles, self-regulation mechanisms could contribute to either or both of these processes.

CHAPTER TWELVE

Multifactor Explanations of Fluctuations

Key Points:

- More explanations of population changes in rodents are relying on multiple factor hypotheses.
- Multifactor models are useful only insofar as they specify clear predictions. Vague multifactor models cannot be tested, and should be considered fairy tales.
- Most multifactor explanations invoke an intrinsic component and an extrinsic factor, which is usually food supplies, predation effects, or both.
- Multifactor explanations are needed to explain why the average abundance of a particular species of small rodent varies from place to place.
- Multifactor explanations encapsulate the quintessence of truth, but need to be carefully evaluated by long-term and large-scale observations and experimental manipulations.

One continuing discussion in the analysis of population dynamics has concerned the utility of multifactor explanations of population trends. The tension in this discussion has been between those who believe that to achieve understanding of population processes one should proceed from simple hypotheses to more complex ones, and others who appreciate the complexities of natural systems and wish to adopt a holistic approach from the start. In the small world of rodent science these two schools have been typified in the approaches of Dennis Chitty (1996) and Bill Lidicker (1988). These two schools have led to an interesting and important discussion of the role of logic in scientific explanation, and how one should approach problems that may have multiple causation. Hilborn and Stearns (1982) have discussed this problem with respect to the problem of population cycles in rodents, and we have made a preliminary exam of this issue in chapter 7.

Single- and Multiple-Factor Hypotheses

The key point in resolving this issue of apparent disagreement is to specify the ecological process to be explained and then clarify the meaning of necessary and sufficient conditions. If we cast a wide net in population dynamics, we can say that we wish to explain all population fluctuations, whether in insects, bacteria, fish, or mammals. Theoreticians have solved this general problem by defining density-dependent reproduction, mortality, and movement and by adding delays in density dependence (Turchin 2003). Knowing which processes are density-dependent or delayed density-dependent is a useful start to understanding. But every biologist knows that within this skeleton framework, many processes differ between insects and mammals, bacteria and birds, so that the details of the biological mechanisms that produce density dependence must be examined to achieve a mechanistic understanding.

If we consider mammals alone as a class, there are many different social systems, feeding habits, and life histories to consider. Again, some generalities can be suggested (Wolff 1997). Consider the rodents alone, and again we have a great variety of social systems and behaviors (Wolff and Sherman 2007). So the question is where to start—that is, how should we define a class of events that might have a common explanation? Some small rodent ecologists have chosen cyclic populations as a class of events for which a common explanation should be sought. This could well be a poor choice, and we could, for example, subdivide these rodent populations into those that live in snowy environments and those that live in temperate or tropical environments, since they face such different conditions of life. Again, there is no way to know which subdivision will turn out to be most useful. But this is no cause for despair, because in reality one studies a particular system of one or a few species for a number of years. The background assumption to all of this scientific exploration is a general belief in the uniformity of nature: that if we understand a natural process operating in the 20th century, the explanation will apply in the 21st century as well, or that if we understand lemming population dynamics in northern Alaska, we will also understand them in the tundra of eastern Canada. Our background assumptions may turn out to be wrong, but we can find that out only by doing the necessary research.

Events within a class, like population cycles, can be considered similar if they share a set of defining characteristics, so we can gain some insight into the definition problem by treating cycles as a taxonomist would treat insect or bird specimens, to see whether they are all the same species. The long history of such attempts is outlined in chapters 1 through 4 of this book. We begin with the assumption that *rodent population cycles with a common structure also have a common causal explanation*. This could be completely in error, but we can find out for certain by testing the assumption and trying to falsify it. To avoid confusion with the mathematical use of the term *cycle*, we can substitute the word *fluctuation*—a more inclusive term.

But what do we mean by a common causal explanation? For a population ecologist, causes are factors or mechanisms of population change: weather, food supply, predators, parasites, social mortality, and disease. Causes can be broken down into three broad types: (1) contingent causes, which may be present or absent in any particular population fluctuation; (2) necessary causes, which must be present in every cycle or fluctuation or else the cause is not necessary; and (3) sufficient causes, which by themselves can cause a population fluctuation or cycle. This classification leads us to several possibilities, outlined in table 12.1. In this hypothetical table only reduced reproduction can be considered a necessary cause, and if this were the case, the next investigation would be to determine if it was quantitatively sufficient to explain the measured changes in abundance. In every case some items are not studied, and because other factors might be present that we do not even know about, this logical approach has problems when it is applied to the real world.

In particular cases it is important to determine if the conclusions of a study are robust. For example, in table 12.1 heavy predation was not present in populations B and G; otherwise it would have been a promising candidate for a necessary condition. We would need to check these particular studies B and G to make sure their methods of studying predators were sufficiently reliable to reach a strong conclusion about predation.

TABLE 12.1 **Schematic illustration of some general mechanisms that could cause population fluctuations in rodents. Nine general factors are identified here, but they could be made more specific. Note that these items are used for illustration only; they are not based on any data analysis. Modified, after Chitty 1996.**

Population or species	Poor weather	Food shortage	Heavy predation	Parasitic disease	Reduced breeding	Plant nutrients	Plant secondary chemicals	Social mortality	Emigration	Other factors
A	+	–	+	–	+	–	–	0	+	0
B	–	–	–	+	+	+	–	0	–	0
C	–	+	+	–	+	+	+	+	0	0
D	+	+	+	–	+	+	–	0	0	0
E	–	+	+	+	+	–	+	0	–	0
F	+	+	+	+	+	–	+	+	+	0
G	–	+	–	+	+	–	+	0	–	0

+ present in this particular fluctuation
– absent in this fluctuation
0 unknown or not studied

So in all these cases, methodology is critical to good science. Because so many factors impinge on small rodent populations, it is very difficult to define sufficient conditions for a population fluctuation, and at this stage we should probably be trying to determine only the necessary conditions for generating population fluctuations.

Multifactor explanations for population changes are often, but not always, based on holistic philosophy (Lidicker 1978). *Holism* is loosely defined as "systems thinking," and has as its antonym *reductionism*, or the tendency to analyze the parts of a system as a path to understanding it. The central idea of holism is that the whole is more than the sum of its parts, and that the behavior of a system cannot be predicted from a study of its parts. Its scientific expression is in systems thinking and complexity science. Clearly, extreme forms of holism are not accepted by many ecologists, who continue to build models of population dynamics (see chapter 13). But a moderate form of holism could be fitted into current ecological modeling frameworks.

Specific Multiple-Factor Hypotheses

I will try to specify some multiple-factor hypotheses that bridge the gap between Chitty and Lidicker (see chapter 7), in order to move this discussion forward. Gaines et al. (1991) has provided a useful critique of the multiple-factor approach to explaining population changes in rodents, and Lidicker (1991) responded by clarifying his point of view on multiple-factor explanations. Batzli (1992) has presented an approach similar to Lidicker's on these issues, and has helped to clarify many apparent arguments in the literature that showed more items and approaches to be in agreement than in disagreement.

Multiple-Factor Hypothesis 1: Many Factors Affect Population Changes

This is the most mild of the multiple factor hypotheses and states a belief that every ecologist can accept. Thus, we all accept that predation, food, disease, weather, social behavior, and parasites affect population rates of change. This hypothesis is sometimes stated in shorthand as, for example, "Predation is important." This hypothesis is not in question and no one should be trying to base a thesis program or a research program on it, since it is universally accepted.

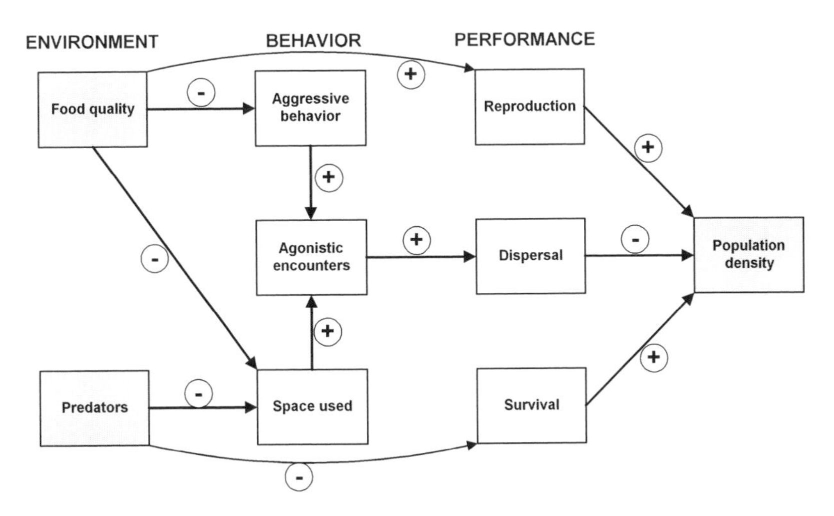

FIGURE 12.1 A multifactor hypothesis developed by Batzli (1992) to investigate the role of predators and changes in the availability of high-quality food on prairie vole populations in enclosures. After Batzli 1992, figure 3.

Multiple-Factor Hypothesis 2: Food Quality and Predation Are Key Parameters Causing Density Changes in Small Rodents

This multiple-factor hypothesis picks out two key variables as the drivers of density change, and it could be stated in two ways. First, the effects of food and predators could be independent and additive. Alternatively, the effects of food and predators could be synergistic and produce an interaction term, which is another way of saying that the whole result would not be simply the sum of the two parts. In the form proposed by Batzli (1992), this hypothesis is stated in an interactive form. It also does not ignore intrinsic changes in individuals' behavior, but assumes that those changes are driven by changes in food supply and predators. Figure 12.1 diagrams this hypothesis as presented by Batzli (1992). The important point is that the hypothesis became the basis of a set of experiments carried out by Desy and Batzli (1989).

The hypothesis outlined in figure 12.1 could be tested only in enclosed populations. Desy and Batzli (1989) constructed eight enclosures (each 39 × 33 m, or 0.13 ha) and introduced prairie voles *Microtus ochrogaster* into them in each of three years. The experiments ran for 14 weeks in the first year and 24 weeks in each of the next two years. Figure 12.2 shows the

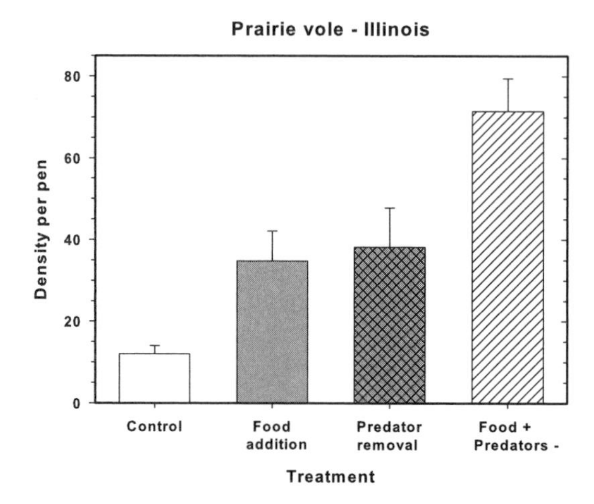

FIGURE 12.2 The density of prairie voles in 0.13-hectare enclosures after 24 weeks during 1985 and 1986 in central Illinois grassland. Predators were excluded from the predation treatment areas by netting and fencing. "Food +" means added food; "predators –" means predators excluded. Two replicates for each treatment were used in each of two years. Data from Desy and Batzli 1989.

resulting population density changes averaged for the two replicates. Both food addition and predator removal treatments increased population density at the end of each season. Reproductive rates of females were higher in food addition areas, but no survival differences could be detected between treatments. Desy et al. (1990) investigated changes in behavior in these same experiments. They postulated that adding food would reduce aggressive behavior and territory size. They did find that voles raised with supplemental food displayed less aggression toward one another, but they detected no effect of food on home-range size when comparing treatments for a given density. Food appeared to act indirectly on home-range size via its effect on population density.

These important experiments had several limitations. To secure replication Desy and Batzli (1989) used small enclosures of 0.13 ha, which were only the size of three to four male home ranges or five to seven female home ranges for this species (Madison 1985). This must produce crowding at a more intense level than is normally encountered in field populations. Secondly, they could run these studies for only about half of the year; the winter period had to be omitted. These kinds of limitations are present for many experimental studies, and the use of enclosures can be particularly controversial. Dispersal was mimicked by having dispersal sinks in two corners of each enclosure, but the effectiveness of this approach to allowing dispersal is not clear. These problems are no reason to ignore these findings, but they do warrant reservations that can be alleviated only by additional studies that try to circumvent these limitations.

One assumption of these kinds of experiments is that the behavior of the individuals can be changed only by extrinsic factors, particularly food supply and predation risk. Thus, processes like infanticide and, to a lesser extent, dispersal have to be left unstudied or overcome by dispersal sinks. The social behavior of voles may be strongly affected in enclosures, and the resulting stress could have effects independent of food supply and predator presence.

Multiple-Factor Hypothesis 3: Food Quality, Predation, and Social Behavior Interact to Cause Density Changes in Small Rodents

This now becomes a three-factor hypothesis and, as in the analysis of variance, it is thus more complicated to analyze. Because these experiments are difficult to carry out, they have been too few in number and have rarely been long-term. Taitt and Krebs (1983) carried out a short-term manipulation of food, cover, and predators on Townsend's vole, *Microtus townsendii*, near Vancouver. The focus of these manipulations was the period in spring between February, when breeding was just beginning, to May, when it was in full swing. The experiments were carried out in open fields but were not replicated, so they can be considered only tentative in their conclusions. One control trapping area and five manipulations were carried out:

- reduction of cover by mowing
- addition of net to prevent bird predation
- addition of net and reduction of cover by mowing
- addition of net and cover in the form of straw
- addition of net, cover, and food

There are missing combinations in this three-factor design, so it remains a far from perfect experiment. In the study area, which was on an island, there were no mammalian predators and the netting prevented avian predators (mostly great blue herons) from gaining access to the grassland.

Straw was chosen as cover because it has no food value for the voles. During the experiment the grasses began growing, so by the beginning of April natural cover was increasing.

The results of these manipulations were that all three factors contributed to population change. Populations protected from predation by a net had higher density, survival, and immigration than unprotected populations in the six-week period at the onset of breeding in February and March, while populations with extra cover and food increased in density and the control population declined (figure 12.3). Voles with added food were heavier and reproduced earlier than voles on all other grids. All three factors helped to increase population size in the early part of the breeding period by early reproduction or immigration. Once breeding was fully underway in mid-March, all populations, irrespective of treatment, exhibited population decline as the breeding season developed into April. In this period, males moved more and had more wounds, while females were lactating. Taitt and Krebs (1983) concluded that the early part of the spring decline in the year of study was the result of avian predation, and that the latter part of the decline, when all voles were breeding, was the result of density-dependent dispersal, possibly induced by aggressive interactions with mature females. Figure 12.4 shows that the voles' rate of disappearance during the main breeding season was density-dependent. Disappearance is a mixture of emigration and deaths, and circumstantial evidence suggested that this was mostly emigration rather than deaths.

There were many deficiencies in this experiment of Taitt and Krebs (1983), in spite of all the effort that went into it. Replication was missing, and some treatments were missing in a full three-factor design because of a lack of grassland space and a shortage of person-power. The use of radio collars would have been desirable for finding out where the disappearing animals went. Finally, the whole experiment should have been repeated in several different years, to secure the generality of its findings.

Unfortunately, there have been almost no attempts to replicate the experiments conducted by Desy and Batzli (1989) and by Taitt and Krebs (1983). The reasons for this are not hard to find: such experiments are expensive, require much person power, and are typically beyond the resources of a single PhD student. But until we can mount experiments over several years on a sufficiently large scale, we will not be able to test in a rigorous quantitative way the hypotheses that flow from multifactor ideas. I would emphasize that these experiments should be done in an open habitat rather than in small enclosures. The spacing behavior that is an integral

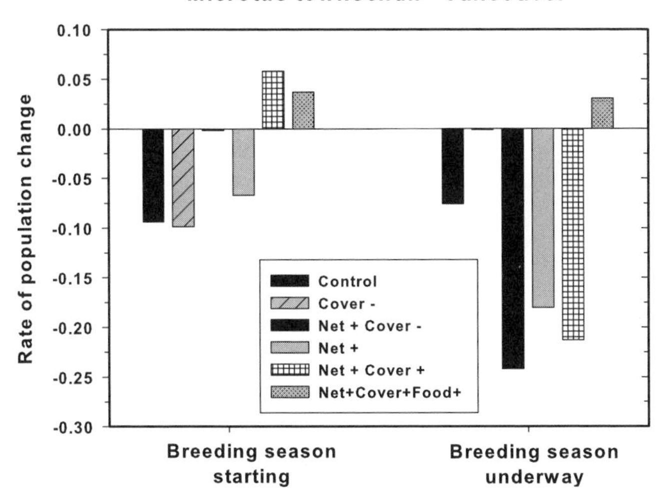

FIGURE 12.3 The rate of population change in male *Microtus townsendii* during the six weeks when breeding was just beginning and the following six weeks when breeding was well established. Instantaneous rate of change per two weeks: + means added; – means removed. In these open-grassland experiments, immigration and emigration were confounded with births and deaths as causes of population change. Data from Taitt and Krebs 1983, figure 3.

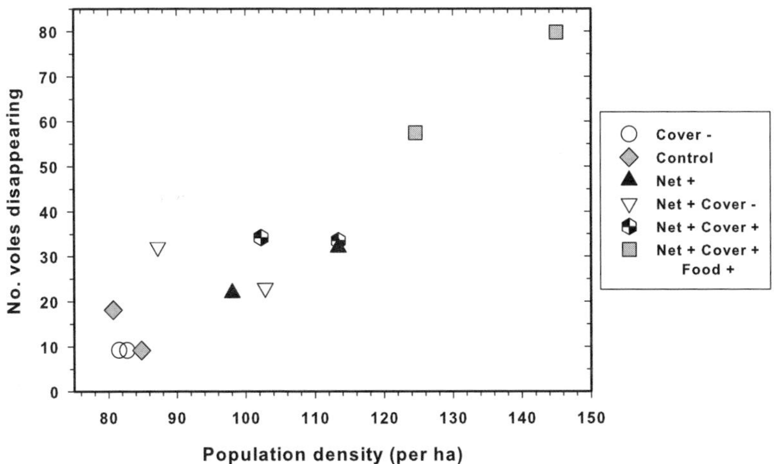

FIGURE 12.4 Disappearance rate of voles in relation to population density during the start of breeding (March and April) in the multifactor experiments of Taitt and Krebs (1983). Disappearance could be due to death or emigration but was thought to be mostly emigration and was directly density dependent. (Data from Taitt and Krebs 1983.)

part of all intrinsic social hypotheses of regulation (cf. Wolff 1997; Krebs et al. 2007) require dispersal as a critical element, and we do not know if we can mimic these social processes in small enclosures. Wolff et al. (1996) suggest that we cannot.

Additional Multiple-Factor Hypotheses

I have not attempted to specify additional two-, three-, or four-factor hypotheses to explain population fluctuations in rodents. At the moment we are challenged to test two- and three-factor hypotheses adequately, and if we use ANOVA as a model, designs with four or more factors become extremely difficult to decipher, not to mention the effort needed for sufficient replication.

Progress in the analysis of multiple factors and their interactions or lack of interactions will come from a variety of observational studies and experimental manipulations on small rodents. I suggest that when dealing with multiple-factor models we should strive to be as specific as possible in our predictions of what they predict and do not predict. Natural history observations can be useful for suggesting ideas, but rigorous testing is needed to temper wishful thinking, to which we are all prone.

Population Limitation by Multiple Factors

The previous discussion has assumed that the critical independent variable is the rate of change in population density. Another important question in population dynamics is the question of which variables set average density, often loosely called "carrying capacity." This question is separate from and independent of the question of explaining rates of change. It is a question that theoreticians have answered very simply with density-dependent models (e.g., Krebs 2009a, figure 14.2). But in the real world things are not so simple. Explanations of population limitation run the gamut of single-factor to multiple-factor types, and some of the experiments carried out to determine the factors that affect rates of increase also provide insight into the factors that limit density.

The simplest situation in which population limitation is important is the case of a rare species of conservation interest. If a species is rare, what factors would you change to make it more abundant? The typical starting hypothesis for this type of investigation would be that the species' food

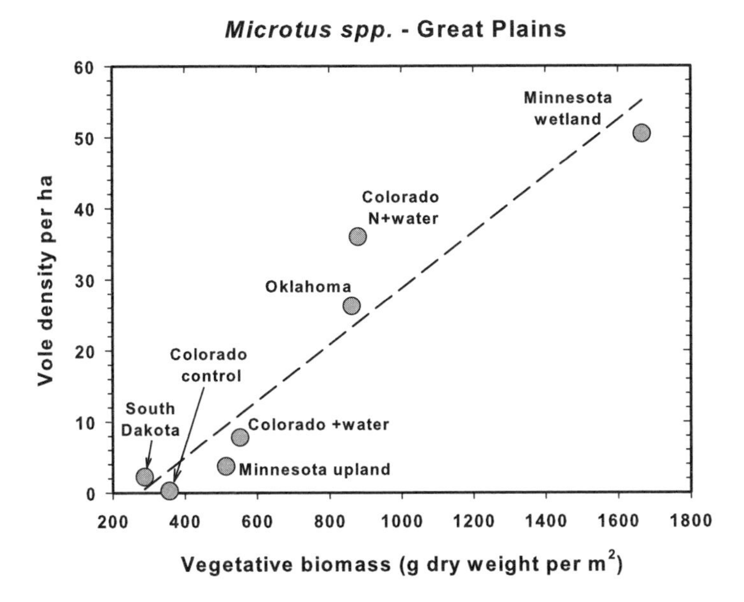

FIGURE 12.5 The relationship of average *Microtus* density in the Great Plains grasslands of North America to vegetative biomass. Below about 300 g dry weight/m^2, there are virtually no voles present. The Colorado sites were experimentally manipulated with nitrogen and water to increase grass growth. Vole density was averaged over three to four years. Data from Birney et al. 1976, figure 1.

supply is uncommon in the habitat where it lives, and that some manipulation of its food supply might increase its average density. In the experiments by Desy and Batzli (1989) shown in figure 12.2, changing both food supply and predation pressure increased density of the species by about the same amount. Wildlife managers often conduct experiments in cover manipulation to restore or increase wildlife at risk. Such cover experiments can do two things to a population: increase food plant abundance and provide more possibilities for escape from predators. Cover can also change microclimates, and consequently cover manipulation is automatically a multiple-factor manipulation. The role of cover has already been discussed in chapter 8, from the work of Birney et al. (1976). Figure 12.5 summarizes the data they presented on the role of vegetative cover in setting the average *Microtus* density in the grasslands of the Great Plains. The available data suggest a linear response to plant biomass with a cutoff around 300 g dry weight per square meter. This linear response may be simplified by the grassland community, and in forest or tundra communities one might expect that the relevant plant biomass would be restricted to a few plant species. Birney et al. (1976) pointed out that cover was a complex variable combining food, predator avoidance, competitive interactions with conspecifics or closely related species, and protection from severe weather. They offered a hypothesis for the role of cover in permitting population cycles to occur (figure 12.6). They suggest two tipping points: one for colonization and persistence, and a second for the ability of the vegetation to support high vole densities and thereby generate population fluctuations.

A more elaborate trophic-landscape model of small rodent populations was outlined by Lidicker (2000): the trophic/ROMPA interaction model. This model combines the intensity of predation by generalist and specialist predators with the landscape model Lidicker designates as ROMPA: the ratio of optimal to marginal patch areas of habitat. This is among the most complex of the multifactor models presented so far. It has been usefully applied to landscapes altered by agriculture and human settlement, which affect the types and abundance of generalist and specialist predators (Delattre et al. 1999, 2009). Figure 12.7 illustrates one example of Lidicker's model, applied to the common vole in eastern France. The landscape configuration of patches can modify the predation schedule for any particular habitat, and the critical point is that average population density within one habitat patch can depend critically on the surrounding areas.

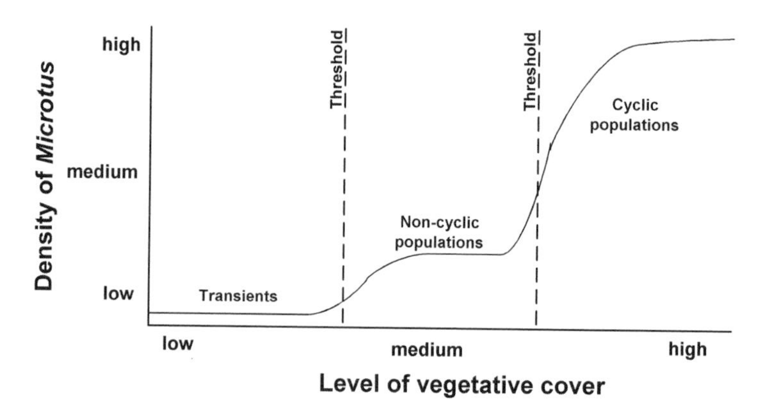

FIGURE 12.6 The cover-threshold hypothesis of Birney et al. (1976). At some low level of cover, *Microtus* populations can persist but cannot increase even to a moderate density. At an upper cover threshold, high densities can be supported and cyclic fluctuations are possible. From Birney et al. 1976, figure 4.

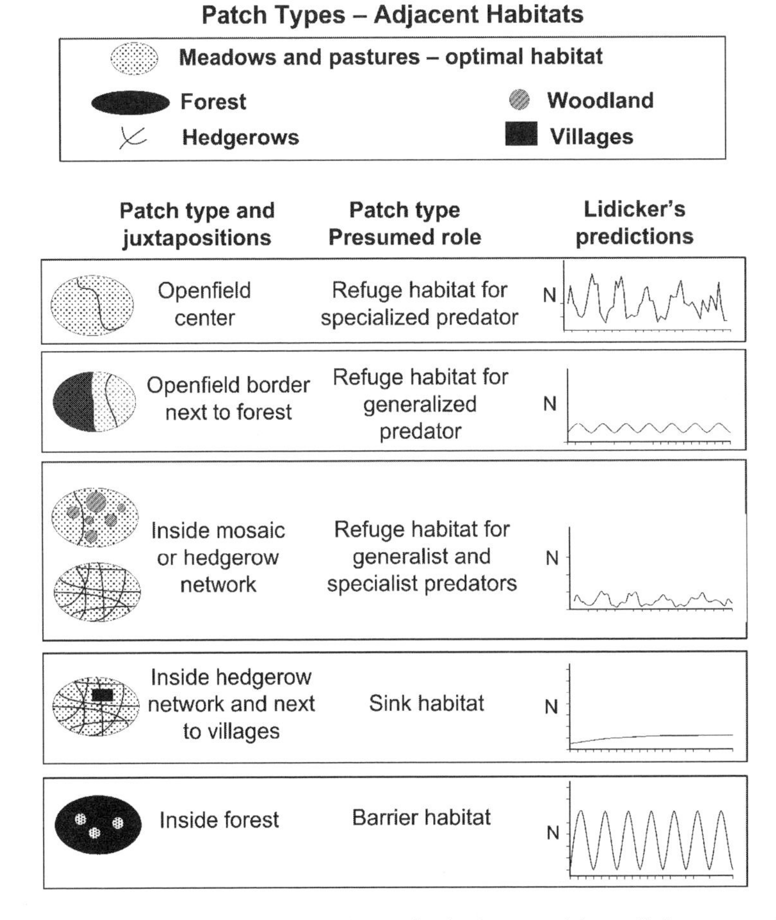

FIGURE 12.7 Spatial arrangements of patch types in a landscape, and the predictions suggested by Lidicker (1995) of how population density of a small rodent in a high-quality habitat might vary, depending on the nature of the adjacent habitats and the predators they contain. Meadows and pastures are the example used here, as they are optimal habitat for the common vole (*Microtus arvalis*). *N* = abundance index; X-axes indicate time in years on all graphs. After Lidicker 1995; modified by Delattre et al. 1999.

The Exploitation Ecosystems Hypothesis

The exploitation ecosystems hypothesis (EEH) is a particularly inclusive but sequential multifactor hypothesis proposed by Oksanen et al. (1981) and further elaborated by Oksanen and Oksanen (2000). As such, it is a particular example of multiple-factor hypothesis 2, stated above. Its

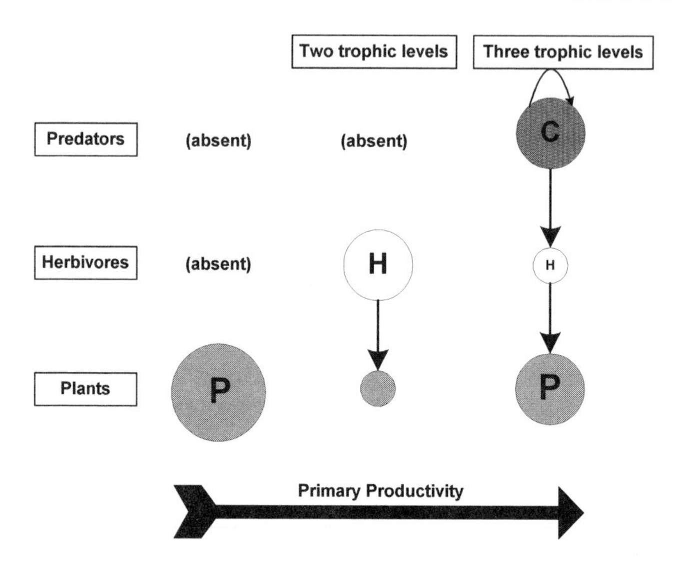

FIGURE 12.8 Schematic representation of the exploitation ecosystems hypothesis (EEH). Along a gradient of primary productivity, trophic dynamics are predicted to split from one to two to three trophic levels, with three levels showing a trophic cascade (Aunapuu et al. 2008). Herbivore numbers are food-limited in areas of moderate plant productivity, and predator-limited in areas of high productivity. C = carnivores, H = herbivores, P = plants. Sizes of the circles indicate relative estimates of biomass.

predictions are based on the potential primary productivity of a site. According to the EEH, productive terrestrial ecosystems are characterized by community-level trophic cascades, whereas unproductive ecosystems harbour food-limited grazers, which regulate community-level plant biomass (figure 12.8). At extremely low levels of primary production, no herbivores can be supported and only plants occur. Figure 12.9 diagrams the predictions of this hypothesis, which has the advantage of incorporating the food hypothesis for systems of low productivity and the predator hypothesis for systems of higher productivity. In low productivity, ecosystems herbivores limit plant biomass (figure 12.9a) and are themselves limited directly by food, responding in a linear manner to plant productivity (figure 12.9b). At higher levels of plant productivity, carnivores can be supported and they begin to limit herbivore numbers (figure 12.9b and c). The clear prediction is that, depending on the site being studied (its primary productivity), investigators will reach different conclusions about food limitation or predator limitation. It is important to note that the EEH is based on an equilibrium model of community dynamics and consequently

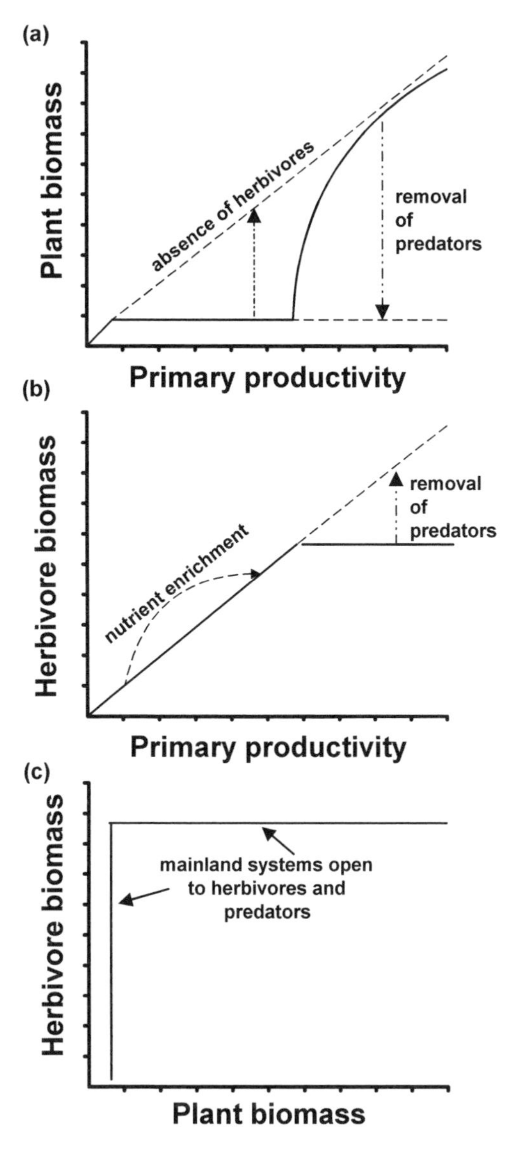

FIGURE 12.9 Predictions from the exploitation ecosystems hypothesis (EEH) of Oksanen et al. (2000). Plant (a) and herbivore (b) biomass are plotted along gradients of potential primary productivity, and the predicted relation between plant and herbivore biomass (c) is inferred from (a) and (b). The solid lines indicate mainland systems open to herbivores and predators. The straight dashed lines refer to herbivore-free areas (a) or to predator-free areas in (b). The dashed lines with arrowheads in (a) and (b) indicate the predicted outcomes of predator removal and herbivore removal experiments and the predicted effects of experimental enrichment of plants with nutrients. Modified from Aunapuu et al. 2008.

has an uncertain but implied application to the demographic causes for fluctuating populations.

Aunapuu et al. (2008) attempted to test the EEH on a local scale in the tundra of northern Norway at the intersection of a higher-elevation plateau and a lower-elevation plateau, where primary production varied with altitude. These studies were broadly in conformity with the predictions given in figure 12.8, but severe limitation of scale (small plots, small manipulations) renders the conclusions only tentative. While Anapuu et al. (2008) argue for the EEH as a good description of their tundra food webs, the model seems to fit the landscapes of northern Scandinavia better than the tundra areas of North America (Gauthier et al. 2004, 2009). I do not know of any North American small-rodent studies that report data consistent with the EEH.

The EEH has the advantage of integrating food-limitation and predator-limitation hypotheses under the overarching control of primary productivity. As such, it is a useful template into which we might fit small rodent dynamics into ecosystems of differing primary productivity. It also raises the important question of how much herbivore biomass can be predicted by primary productivity, as illustrated in figure 12.9, and represents a challenge to ecologists to gather these data to test it in a variety of different ecosystems. The issues raised by Oksanen and Oksanen (2000) are far from settled.

The Role of Weather

While most studies of fluctuating populations of rodents have concentrated on predation, food supplies, disease, and social factors, few have attempted to integrate weather variables, partly because they have commonly been considered density-independent effects. Since fluctuations in numbers could be driven only by a combination of direct and delayed-density dependence, weather could be ignored or viewed as a disturbing influence on population growth. Goswami et al. (2011) have used the long-term data on *Microtus ochrogaster* collected by Lowell Getz in Illinois to investigate the role of weather in a cyclic population. By breaking the time series into phases of the cycle, they obtained insight into the effects that weather might have on population change.

In the increase phase, Goswami et al. (2011) found that population growth rates were high ($\lambda = 1.45$) but declined with increasing density in

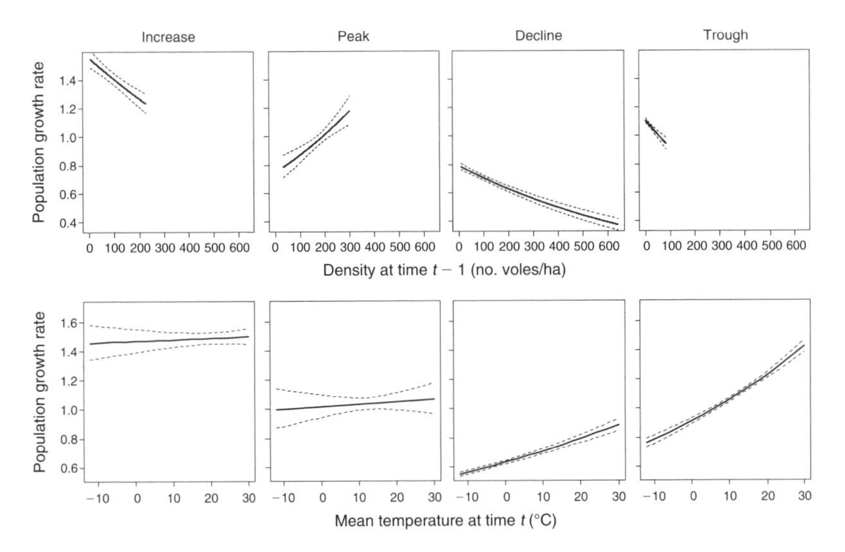

FIGURE 12.10 The relationship of population growth rate to lagged density and mean temperature for *Microtus ochrogaster* in Illinois. Temperature had a significant effect on population growth rate, but only in the decline and low phases. Dotted lines show 95% confidence limits. From Goswami et al. 2011.

the previous month (figure 12.10a). During the peak phase, the population continued to grow (λ = 1.13), but only when lagged density was high (192 voles/ha or more). Mean monthly temperatures did not influence growth rates in the peak and increase phases (figure 12.10b). Population growth was reversed in the decline phase (λ = 0.71), and lagged density contributed negatively to the growth rate. However, the growth rate improved significantly with increasing mean temperature. Finally, the trough phase witnessed a modest population growth (λ = 1.02), but only at low densities and when mean temperatures exceeded 4.4° C. The key point is that both survival and reproduction, and thus population growth rate, were affected by weather variables that had a phase-specific effect. Climatic variables exerted an influence, but not the same influence in all stages of the population fluctuation.

Common vole (*Microtus arvalis*) populations in Europe can increase to hundreds of individuals per hectare during outbreaks and damage crops. In central Europe such outbreaks usually extend across large areas, but there are significant regional differences in outbreak intensity, outbreak risk, and crop damage. Imholt et al. (2011) used climatic variables to determine how much of the regional variation in weather might be associated with regional variation in population outbreaks. With data covering 25 years and 50 monitoring stations in east central Germany, they found that weather parameters especially in winter and early spring are related to the risk of common vole outbreaks in the following autumn. They could explain 51% of the variation in vole density by citing 11 weather parameters, mostly involving temperature and precipitation in winter and early spring. By adding topography and soil properties, this level of precision in prediction could be improved (Blank et al. 2011). The results could be used by farmers in these regions as a simple model to predict the probability of vole outbreaks. They show clearly that weather variables influence rodent densities and cannot be ignored in multifactor models.

These multiple-factor models that include weather are of particular importance with the onset of climatic warming (Berteaux et al. 2006). The impact of climate change is such that some areas will be more affected than others. These differences could be used in a clever experimental design to analyze how climate change may affect rodent population fluctuations.

Conclusion

Multiple-factor hypotheses are the quintessence of truth—since everyone agrees that many factors impinge on small rodent populations—but are also an invitation to multiply explanations for changes after the fact. The way out of this impasse is to specify clear multifactor models that are testable in field populations. The first step, again, is to specify clearly the questions being asked and how specific types of data will be used to determine the validity of the proposed explanation. The second step is more difficult because all multifactor hypotheses require large-scale, long-term research that is ill-suited to most PhD-level programs. The result is that at present the empirical database is too small; we have too few examples of attempts to test comprehensive multifactor models. The temptation to conduct experiments in very small enclosures needs to be resisted unless it is possible to design the enclosures so that the restrictions on dispersal are removed. All of the dominant hypotheses used to explain small rodent population changes invoke two or more ecological factors and may involve climatic effects; and the key work that remains is to tease out the mechanisms that cause population change with clever observations and manipulations.

CHAPTER THIRTEEN

Models for Fluctuating Rodent Populations

Key Points:

- Many different mathematical models can produce rodent population fluctuations that resemble actual field data. Not all of them can be correct.
- Numerical models of food-supply-generated cycles, predator-generated cycles, and intrinsically generated cycles are reviewed and evaluated.
- Verification of numerical models of population cycles or fluctuations is impossible. Models can be confirmed by showing agreement between observation and prediction.
- Much controversy over the form and function of models in population dynamics makes comprehensive confirmation unlikely. All the available models involve multiple-factor explanations.
- Models have great heuristic value and can help us explore assumptions. The best models suggest experiments or observations that are key to model assumptions and model integrity.

The real world is a complex mess of multiple factors impinging on populations that vary in both space and time, and one way to bring order to this chaos is to set up a model. We must begin by distinguishing between several types of models. A model is a hypothesis designed to answer a particular question or set of questions. The simplest models are verbal or "boxes-and-arrows" models, such as we saw in figure 11.1. For a population, a simple model states that population changes are the net result of births, deaths, immigration, and emigration. These simple models are useful in ecology, as we have seen many times in previous chapters. More complex verbal models evolve as researchers probe more and more deeply into population and community dynamics. Does food shortage stop lemming population increase? Do diseases cause population crashes? Is there selection for aggressive behavior as population density grows? For many of these issues, ecologists have resorted to mathematical modeling to clarify in precise terms the forms of relationships and the quantitative value associated with them. The value of mathematical models in science has been demonstrated over and over again, and it is unlikely that ecology will be an exception to this rule.

Nevertheless, many ecologists question the types of mathematical modeling that are currently used in their discipline (Pielou 1981; Loehle 1987; Gilbert 1989; Aber 1997). Many suggestions have been made to improve model presentation in ecology; I outline here only a few major recommendations for models meant to be tested with field data:

- Define the boundaries of the system to be modeled.
- Ask questions that reduce complexity to a manageable level.
- Use as many parameters in the model as necessary, but not more.
- Do not use parameters that cannot be measured in some way if you desire a field test of the model.

A Brief History of Mathematical Models for Fluctuating Populations

One of the first mathematical models in ecology was the predator-prey model of Lotka and Volterra, published independently by Lotka (1925) and Volterra (1926). These models convinced many ecologists that all fluctuations in populations could be ascribed to predation or equivalent processes like parasitism or disease epidemics. Turchin (2003) pointed out that Lotka-Volterra models are basically consumer-resource models, and that changes in food supplies could generate cycles as well as changes in predation. For small rodent ecologists like Dennis Chitty who had decided that predation and food shortage were not necessary causes of rodent fluctuations, attention turned from mathematical models to simpler verbal models. Population ecologists of the 1940s to 1960s were usually concerned with equilibrium dynamics, and cyclic fluctuations were considered atypical and a rare occurrence in nature. But by the 1950s, modelers rediscovered that the key to fluctuations lay in delayed density dependence that destabilized populations (Ricker 1954; May et al. 1974).

Since most small rodent populations fluctuate greatly in abundance, whether in regular cycles or not, attention was directed very early to factors in the environment that could produce delayed density dependence. The problem immediately became that in any mammal that has a generation time, delays could be expected in any factor you could list that affected populations. So no factor could be eliminated a priori, and it became necessary to determine whether any particular factor *x* showed a delayed density-dependent relationship with population density. But almost every factor that one could measure showed delayed density dependence, so the search was no farther ahead. The solution had to move forward in two directions: detailed analyses of factor *x* and how it affects births, deaths, and movements, and quantitative models of the potential effect of factor *x*. We have described the findings of detailed biological analyses of particular factors in the previous chapters. We turn here to consider the quantitative models that have been put forward to explain population fluctuations in small rodents.

In 1996 a group of small rodent ecologists and modelers met to try to find a way out of the impasse that had developed in trying to define the causes of cycles (Batzli 1996). The decision of the group was to construct mathematical models of the three dominant hypotheses: the food hypothesis, the predation hypothesis, and the social behavior hypothesis. A general agreement that multiple factors were involved in population changes capped the discussion, and it led to the decision to try to formulate three types of models that would allow evaluation of competing hypotheses and stimulate further experimental work. I will now describe the models that evolved partly from this meeting. As is typical in ecology, the devil of all models is in the details.

Models of the Food Hypothesis

Turchin and Batzli (2001) provided a clear, detailed model of the properties of the food supply that would generate population fluctuations in small rodents. They asked: What quantitative characteristics of rodents and their food plants are likely to result in vegetation-herbivore cycles?

The key interactions to specify are the dynamics of plant growth and rodent consumption. To make the analysis more precise, they broke down the possible roles that vegetation can play in rodent dynamics. First, the availability of high-quality food plants can be a slow dynamic variable, so that food shortages persist for more than one herbivore generation after the vole or lemming population peak. This generates a delayed density-dependent effect on the rodent population, and is thus capable of generating a population cycle. However, Turchin and Batzli (2001) pointed out that the influence of food supplies is not limited to their potential to generate population cycles. Food shortage, for example, may be a fast dynamic variable, imposing direct density dependence when population density reaches the level at which the amount of food per individual is insufficient for survival or reproduction. In this particular case, there are no delayed effects of food shortage persisting after population density declines, and food will help to stabilize rodent populations. In addition to being a fast or slow dynamic variable, the availability of food may have no influence on rodent population changes because other density-dependent factors stop rodent population growth before significant depletion of the food supply occurs.

The first question Turchin and Batzli (2001) faced was how to model vegetation dynamics, an issue treated by Caughley (1976). The most important factor influencing the food supply for voles and lemmings is the primary productivity of the plant community. Productivity has two very different quantitative aspects: it determines the maximum biomass achieved in the absence of herbivory, and it also determines the rate at which plant biomass is replenished after being depleted by herbivores. Two models have been used to calculate the growth rate of the vegetation: the logistic model and the regrowth model, both illustrated in Figure 13.1. These two models make very different assumptions about the dynamics of plant growth, and they have very different dynamic consequences. The logistic implies that when vegetation biomass V is near zero, its growth rate is an accelerating function of V that reaches its maximum at one-half of carrying capacity, and then slows to zero as V approaches carrying capacity or maximum biomass (figure 13.1). The logic underlying the logistic model is that the more plant biomass is present, the more solar energy it can fix, and the faster it will grow (until it approaches the carrying capacity). By contrast, the regrowth model implies no acceleration period; instead, when V is low, regrowth increases at the maximum rate, and gradually slows to zero as V approaches maximum biomass K. Note that the term

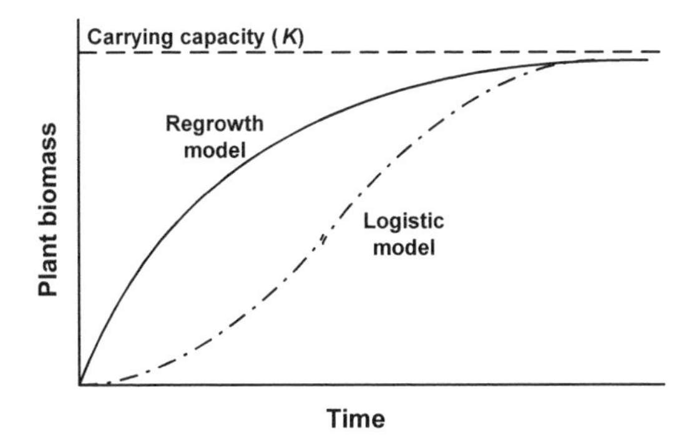

FIGURE 13.1 Two models of the temporal dynamics of vegetation: the regrowth model (solid curve) and the logistic model (broken curve). K is the "carrying capacity," the maximum biomass of vegetation in the absence of herbivory. In the terminology of predator-prey interactions, the regrowth model is a type 2 functional response and the logistic model is a type 3 functional response. After Turchin and Batzli 2001.

"vegetation" refers not to the total plant biomass, but only to the part easily accessible to herbivores (i.e., the stems and leaves on which rodents feed). The regrowth model assumes that initial growth is fueled by energy stored in the roots and rhizomes, and there is no period of accelerating growth.

Starting from this dichotomy of vegetation growth patterns, Turchin and Batzli (2001) constructed six possible models of plant-herbivore interactions and explored their dynamic consequences. Table 13.1 lists the differential equation models they specified and the parameters they contain, and table 13.2 gives the median values of the 11 parameters contained in these models. Details of these values are given in Turchin and Batzli (2001); here we explore only the two important conclusions they reached from their analysis.

First, the dichotomy between logistic growth and linear regrowth of vegetation is very important (cf. figure 13.1). It is extremely easy to obtain population cycles in model I (the Rosenzweig-MacArthur 1963 model). Almost any biologically reasonable combination of parameters produces oscillations, except possibly when maximum vegetation biomass, K, is very low (e.g., in an arctic desert). By contrast, the regrowth model cannot produce population oscillations. According to the modeling results, we should not expect delayed density-dependent rodent cycles driven by the interaction with food when plants regrow rapidly, which seems likely for *Microtus* in temperate grasslands. Support for this prediction comes

TABLE 13.1 **Six models for vegetation-herbivore population interactions. The dynamics of each model are illustrated for the median values of the parameters given in table 13.2. From Turchin and Batzli, 2001, table 1.**

Equations	Dynamics		Variables
	Type	Quantitative characteristics (median parameters)	
Model I. Rosenzweig-MacArthur model (logistic vegetation)			
$\frac{dV}{dt} = uV\left(1 - \frac{V}{K}\right) - \frac{AVH}{V+B}$ $\frac{dH}{dt} = RH\left(\frac{AV}{V+B} - G\right)$	Cycles	Period: 16 years Amplitude: <.001 to 800 ind./ha	V = vegetation biomass H = herbivore density
Model II. Bazykin model (rodent self-limitation)			
$\frac{dV}{dt} = uV\left(1 - \frac{V}{K}\right) - \frac{AVH}{V+B}$ $\frac{dH}{dt} = RH\left(\frac{AV}{V+B} - G\right) - EH^2$	Cycles	Period: 12 years Amplitude: <.001 to 500 ind./ha	V = vegetation biomass H = herbivore density
Model III. Variable territory model			
$\frac{dV}{dt} = uV\left(1 - \frac{V}{K}\right) - \frac{AVH}{V+B}$ $\frac{dH}{dt} = RH\left(\frac{AV}{V+B} - G\right) - \frac{eH^2}{V}$	Stable	Equilibrium density 70 ind./ha	V = vegetation biomass H = herbivore density
Model IV. Regrowth-herbivory model			
$\frac{dV}{dt} = U\left(1 - \frac{V}{K}\right) - \frac{AVH}{V+B}$ $\frac{dH}{dt} = RH\left(\frac{AV}{V+B} - G\right)$	Stable	Equilibrium density 420 ind./ha	V = vegetation biomass H = herbivore density

Model V. Regrowth-herbivory model with seasonality			
$\frac{dV}{dt}=U(\tau)\left(1-\frac{V}{K}\right)-\frac{AVH}{V+B}$ $\frac{dH}{dt}=RH\left(\frac{AV}{V+B}-G\right)$	Cycles	Period: 2 years Amplitude: 1 to 1,400 ind./ha	V = vegetation biomass H = herbivore density π = season ($0 \le \tau \le 1$)
Model VI. Hervivore-logistic / regrowth-vegetation model			
$\frac{dV}{dt}=U\left(1-\frac{V}{K_V}\right)-\frac{AVH}{V+M+B}$ $\frac{dM}{dt}=uM\left(1-\frac{M}{K_M}\right)-\frac{AMH}{V+M+B}$ $\frac{dH}{dt}=RH\left(\frac{A(V+M)}{V+M+B}-G\right)$	Stable	Equilibrium density 40 ind./ha	Two kinds of vegetation V = "logistic" vegetation M = "regrowth" vegetation H = herbivore density

A = maximum rate of vegetation consumption by a herbivore
B = herbivore half-saturation constant
E = herbivore density-dependence parameter (for model II)
e = vegetation/herbivore ratio at equilibrium
G = herbivore consumption rate at zero population growth
H = herbivore density
K = vegetation carrying capacity (maximum biomass per unit area)
K_H, K_V, K_M = carrying capacities of the herbivore population, vascular plants (regrowth vegetation), and mosses (logistic vegetation) respectively
M = moss biomass density
R = conversion rate of vegetation biomass into herbivore biomass
U = maximum rate of regrowth
V = vegetation biomass
α = discounting parameter for consumption of mosses relative to that of vascular plants
μ = maximum per capita rate of logistic vegetation growth
τ = seasonality variable
Ampl. = amplitude of oscillations, minimum and maximum values
Eq. density = equilibrium density in a stable equilibrium

TABLE 13.2 **Parameter values used in simulations of the food models given in table 13.1. From Turchin and Batzli 2001.**

Parameter	Median value	Range
u	2 yr^{-1}	1–10
K	2,000 kg/ha	500–5,000
K_M	2,000 kg/ha	n.a.
K_V	300 kg/ha	n.a.
U	10,000 $kg \cdot ha^{-1} \cdot yr^{-1}$	1,000–10,000
A	15 $kg \cdot yr^{-1} \cdot ind.^{-1}$	10–20
B	70 kg/ha	50–200
G	0.6 A	0.4–0.8 A
r_{max}	6 yr^{-1}	n.a.
R	r_{max} $[AK/(K+B)-G]^{-1}$	n.a.
K_H	1,000 ind./ha	100–1,000
E	r_{max} K_H^{-1}	n.a.
q	10 kg/ind.	1–10
e	r_{max} q	n.a.

r_{max} = maximum per capita rate of herbivore population growth
q = vegetation/herbivore ratio at equilibrium
n.a. = not applicable
All other parameters as defined in table 13.1.

from the study by Ostfeld et al. (1993), who observed strong, immediate, but not one-year delayed effects of high population densities of *Microtus pennsylvanicus* on their food supplies. The take-home message is that better measurements of vegetation recovery after a herbivore outbreak are needed to determine how closely vegetation dynamics conform to either of the two vegetation growth functions illustrated in figure 13.1.

A second conclusion from the Turchin-Batzli analysis is that populations of voles and lemmings should regularly destroy their own food supplies if herbivore-food interactions alone are to produce cyclic population fluctuations. However, data based upon total herbaceous production suggest that *Myodes* (*Clethrionomys*) and *Microtus* voles living in temperate habitats may consume no more than about 5% of available plant material (Krebs and Myers 1974, table XIII). The difficulty with such consumption figures is that all herbaceous plant material is assumed to be food, even though we know that lemmings and voles have distinct food preferences (Batzli 1985). Furthermore, food quality may be more important than food quantity, and this is not incorporated into these models.

Turchin and Batzli (2001) related the general theory that we developed to one specific system about which we have enough data to arrive at reasonable estimates for most of the parameters: the brown lemmings of Point Barrow, Alaska. They adopted model 6 of table 13.1, and added seasonality to it to mimic winter and summer dynamics. They used this model to see whether the data they had available fit the food hypothesis and would generate population cycles. Their Barrow model exhibited oscillations approximately of the correct period and amplitude, thus giving some support to the food hypothesis. Nevertheless, they suggested that this result should be treated cautiously because key events explaining the population cycle in the model occur during winter, but the winter biology of lemmings is still poorly understood.

Other mechanisms for generating population fluctuations from food quality have been suggested, but none has been explored in detailed mathematical models. For example, Massey et al. (2008) suggested that silica plant defenses might cause vole population cycles, but this idea has not been explored theoretically in a formal model.

Models of the Predation Hypothesis

Since the early models of Lotka (1925) and Volterra (1926), many authors have suggested that the regular multiannual population oscillations of voles and lemmings are caused by predation, as we explored in chapter 9. Hanski et al. (2001) reviewed the research that has been done on the predation hypothesis primarily in Fennoscandia, and they have constructed models of predator-prey interaction. They concentrated on the interaction between least weasels (*Mustela nivalis*) and field voles (*Microtus agrestis*) because those species has been postulated to be the key specialist predator and key prey in the boreal forest region of Fennoscandia. The list of predators of field voles in Fennoscandia is large (table 13.3), and both generalist and specialist predators are common. Including all these predators in a model has not been attempted for a lack of specific data; predator-prey models generally have included at most only a few predators and one or two prey items.

Predator-prey dynamics have been modeled in many different ways, as reviewed by Hanski et al. (2001), who chose to model the prey and predators with two differential equations. For the rodent prey population,

$$\frac{dN}{dt} = rN\left(1 - \frac{N}{K}\right) - \frac{\alpha NP}{N + D} \qquad (13.1)$$

TABLE 13.3 **Approximate densities of the common small rodent predators at three latitudes in Fennoscandia. Predator density is much higher in southern Fennoscandia. From Hanski et al. 2001.**

Species	Density (individuals/km^2)		
	56°N	59°N	68°N
a) Individual species density			
Resident specialists			
Least weasel, *Mustela nivalis*	n.a.	n.a.	1–20
Stoat, *Mustela eriminea*	0.8–1.4	n.a.	0.5–2.0
Resident generalists			
Red fox, *Vulpes vulpes*	0.9–2.0	0.4	< 0.1
Badger, *Meles meles*	0.8–0.9	0.4	abs.
Domestic cat, *Felis catus*	1.8–2.5	0.2–0.4	abs.
Polecat, *Mustela putorius*	0.6–1.4	abs.	abs.
Pine marten, *Martes martes*	abs.	0.4	0.1
Common buzzard, *Buteo buteo*	1.5–2.5	0.2	abs.
Tawny owl, *Strix aluco*	0.7–1.4	< 0.1	abs.
Nomadic specialists			
Short-eared owl, *Asio flammeus*	abs.	abs.	< 0.1
Long-eared owl, *Asio otus*	0.2–1.3	0.0–0.3	abs.
Hawk owl, *Surnia ulula*	abs.	< 0.1	0.1–0.2
Tengmalm's owl, *Aegolius funereus*	abs.	0.0–0.6	< 0.1
Kestrel, *Falco tinnunculus*	0.1–1.3	< 0.1	< 0.1
Rough-legged hawk, *Buteo lagopus*	abs.	abs.	0.1–0.4
Long-tailed skua, *Stercorarius longicaudus*	abs.	abs.	0.1–0.6
b) Pooled density of resident generalists and nomadic specialists	7–13	2–3	0.3–1.5

n.a. = data not available; abs = absent

where N = density of the prey rodent population,
r = the prey's intrinsic rate of increase,
K = the prey's carrying capacity,
α = the maximum predation rate,
P = the predators' population density, and
D = the predator's attack rate.

For the predator population they assumed a type 2 functional response, so that

$$\frac{dP}{dt} = sP\left(1 - \frac{QP}{N}\right) \tag{13.2}$$

where P = the density of the predator population,
s = the predator's intrinsic rate of increase, and
Q = a constant relating to the carrying capacity for the predator population.

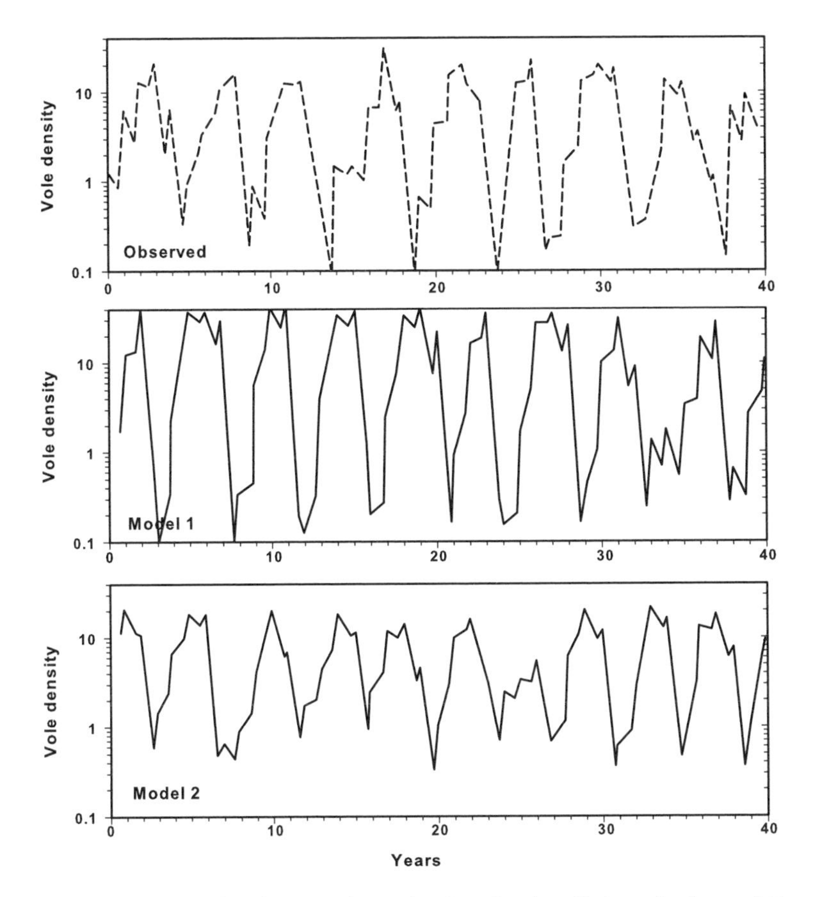

FIGURE 13.2 Comparison between observed and predicted oscillations of vole populations. The "vole density index" is a trapping index, a measure of relative vole density. *Top*: observed population oscillations at Kilpisjarvi, Finnish Lapland, from 1952 to 1992. Two data points (early summer and fall) are plotted for each year. *Middle and bottom*: sample outputs of two versions of the vole-weasel model for the best estimates of the parameter values, described in (*middle*) Turchin and Hanski (1997) and (*bottom*) Hanski et al. (1993). From Turchin and Hanski 1997.

The predator equation assumes logistic growth for the predator population so that the predators' increasing density reduces their growth rate. This might happen if, for example, they show territorial behavior.

The models for prey and predators have been parameterized with independent data as explained in detail in Hanski et al. (1993), Hanski and Korpimäki (1995), and Turchin and Hanski (1997). The parameters that are

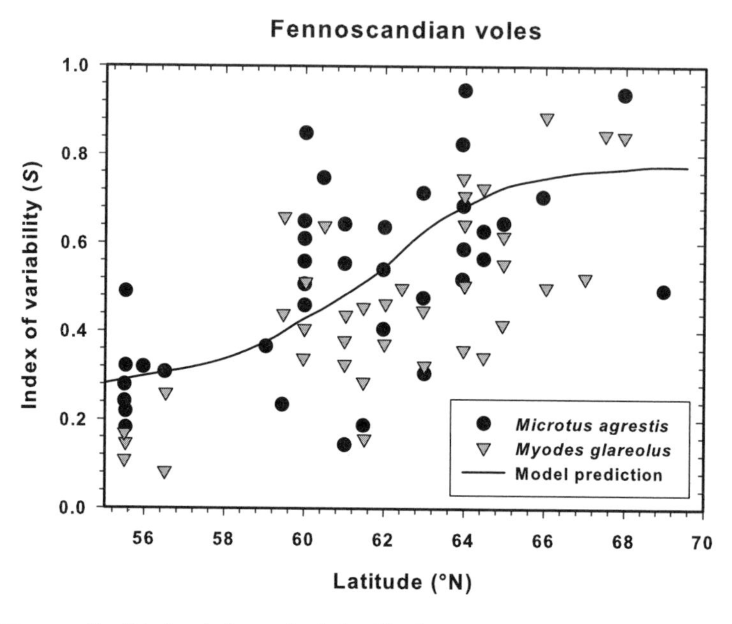

FIGURE 13.3 Predicted and observed relationships between latitude and the amplitude of population oscillations in Fennoscandia. The variability or amplitude of population fluctuations, S, was measured by the standard deviation of log-transformed vole density. The empirical data are for the field vole *Microtus agrestis* and the bank vole *Myodes glareolus* (from Hansson and Henttonen 1985). The line indicates the amplitude predicted by the model of Turchin and Hanski (1997), and includes the stabilizing effects produced by generalist predators, the density of which decreases as ones moves north (see table 13.3). From Hanski et al. 2001.

most difficult to estimate are Q, D, and the carrying capacity, K. Hanski and Korpimäki (1995) and Turchin and Hanski (1997) have investigated the sensitivity of the model-predicted dynamics to changes in the values of these parameters. Note that all the predator-prey models include a term for a density-dependent reduction in the rate of population growth, and this term is required because unless rodents' population growth is slowed as their density rises, their predation mortality can never catch up with the rapid rate of reproduction of which they are capable.

These predator-prey models produce population traces for the prey that are superficially similar to actual changes in vole indices reported by Heikki Henttonen from northern Finland (figure 13.2). The models are an incomplete description of Fennoscandian small-rodent dynamics, because they include only the resident specialist predators and not the generalists listed in table 13.3. Generalist predators produce direct density-dependent losses on their rodent prey; this tends to stabilize the rodents' population density, and thereby reduce the amplitude of their oscillations. By adding generalist predators to these models, Hanski et al. (2001) were able to mimic the reduced fluctuations shown by small rodents as one moves south in Fennoscandia (figure 13.3).

The conclusion of Hanski et al. (2001) was that their weasel-*Microtus* model, parameterized with field data, predicted well the broad patterns in vole population oscillations, including the amplitude of the fluctuations and the dominant period. An extension of the model incorporating generalist predators predicts well the observed latitudinal trends in vole dynamics: most notably the decreasing amplitude and the shortening of the cycle with decreasing latitude in Fennoscandia. Finally, a version of the model with two prey species corresponding to *Microtus* and *Myodes* (*Clethrionomys*) voles provides a plausible explanation for three patterns observed in the multispecies small-rodent communities in Fennoscandia. Hanski et al. (2001) concluded that although these results do not suffice to conclusively demonstrate the validity of the predation hypothesis, the breadth of observations that can be resolved under it is impressive.

Models of Other Mechanisms

Other models that will generate population fluctuations in small rodents have been published. One of the simplest is that of Smith et al. (2006), which requires as a mechanism only the timing of the onset of reproduction in voles and lemmings. Several field studies of cyclic vole and lemming populations have suggested a relationship between a given year's population density and the timing of the start of reproduction in the subsequent spring (e.g., Krebs 1964; Hansen et al. 1999; Norrdahl and Korpimäki 2002). Figure 3.14 illustrates data from T. Ergon showing that this delay is predictable in *Microtus agrestis* in England. Smith et al. (2006) mathematically examined the implications of variation in reproductive season length caused by delayed density-dependent changes in its start date. They demonstrated that when reproductive season length is a function of past population densities, it is possible to get realistic population cycles without invoking any changes in birth rates or survival. When parameterized for field voles (*Microtus agrestis*) in Kielder Forest in northern England, their most realistic model predicts population cycles of similar periodicity to the Kielder populations (figure 13.4). Reproductive timing could be

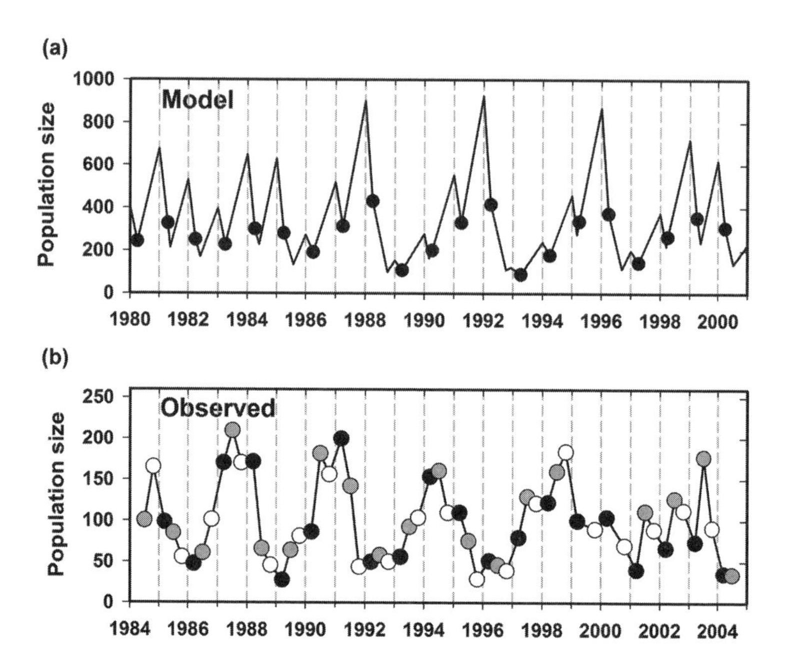

FIGURE 13.4 (a) Population changes predicted by Smith et al. (2006) from their model, which includes a delayed density-dependent onset of breeding. The dots represent the population density measured at the same point (the start of April) in each year. Clearly, sampling at only this fixed time in each year would fail to reveal the variable onset of reproductive season length. These dynamics appear qualitatively similar to those in the Kielder Forest *Microtus agrestis* data shown in (b). The data in (b) are average field vole densities between 1984 and 2004 for spring (black dots), summer (gray dots), and autumn (white dots), for the 14 to 18 sites in Kielder Forest. From Smith et al. 2006.

important in providing a delayed density-dependent effect on microtine populations. The exact mechanisms underlying this reproductive shift could be social (Drickamer 2007), or could possibly involve recovering food supplies. Investigations into the mechanisms underlying this reproductive delay are needed.

A variety of intrinsic explanations for population fluctuations, from the original stress hypothesis of Christian (1950) to the suggestions of Chitty (1960, 1967), have suggested that variations in behavior and individual quality could be a necessary component of cyclic population changes. None of these early intrinsic models were formulated mathematically, and many ecologists expressed considerable doubt that they could in fact produce population cycles in rodents. In an unpublished PhD thesis Judith

Anderson (1975) constructed a model of the Chitty hypothesis based on the heritability of aggressive behavior and showed that under high heritability of such behavior, one could generate a population fluctuation with an intrinsic mechanism based on genetic inheritance or maternal effects.

Inchausti and Ginzburg (1998) discuss the intrinsic hypothesis that has been implicit in discussions of social interactions by Dennis Chitty and John Christian: the maternal effects model. This model has the interesting history of having been proposed formally in 1996 at the meeting described briefly by Batzli (1996), but then sidelined by the modelers' conclusion that it was impossible to formulate a model of maternal effects that could generate population cycles. This conclusion turned out to be incorrect, as we now know, and as Anderson had pointed out in 1975.

Inchausti and Ginzburg (1998) formulated a discrete time model based on maternal effects to explain the density fluctuation patterns of microtine rodents. The idea of a maternal hypothesis is simple: there is a phenotypic transmission of quality from mothers to offspring, and this nongenetic transmission generates delayed density dependence, which in turn produces cyclic population fluctuations in the model.

The simplest model of maternal effects involves two linked dynamic variables:

$$N_{t+1} = N_t r_m F(X_t) \qquad (13.3)$$
$$X_{t+1} = X_t G(N_{t+1})$$

where N_t = rodents' population abundance,
r_m = their intrinsic rate of increase, and
X_t = average individual quality at time t,

and F and G are nonlinear, monotonically increasing functions describing the effects of individual quality on population growth rate, and of concurrent population density on the temporal changes in individual quality (figure 13.5). Individual quality X_t is left deliberately vague in biological terms, but quality is both the result of the interactions between individuals and their environment and a constraint on their reaction to changes in food and predators (Inchausti and Ginzburg 2009). By transmitting individual life history responses to the environment across generations, maternal effects effectively modify the survival, growth, movements, and reproduction of the progeny, depending on the environmental conditions that faced the parental generation, thus introducing a delayed density-dependent response in population growth rate. The Smith et al. (2006)

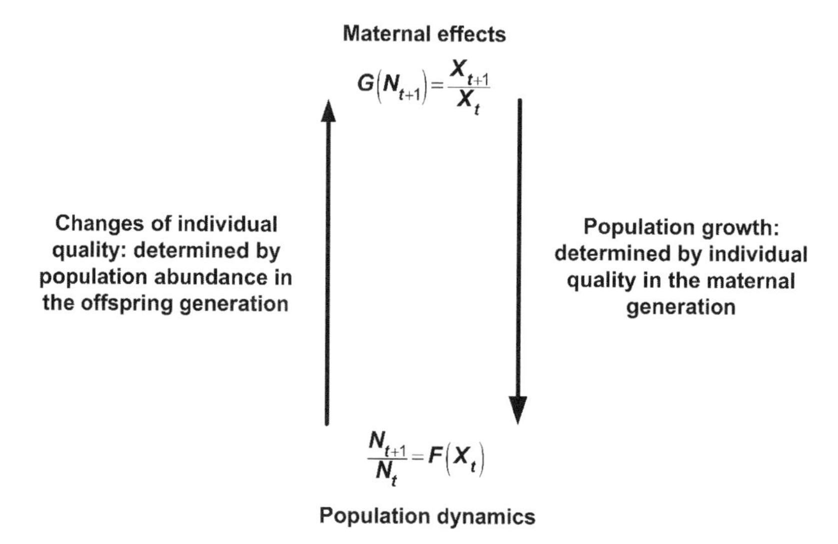

FIGURE 13.5 Feedbacks between population growth and changes in individual quality implied in the maternal effects hypothesis. In contrast to traditional kinetic approaches in population ecology, population abundance affects the growth rate through changes in individual quality, which in turn determines the abundance of offspring generation. These dependencies induce an inertial effect, whereby the current population growth rate depends on the growth rate at the previous generation. After Inchausti and Ginzburg 2009.

model described above could be viewed as a special case of this more general maternal effects model.

One interesting consequence of the maternal effects hypothesis (equation 13.3 above) is a focus on generations as the biological scale for viewing population dynamics. The maternal effects view effectively proposes a time scale that is defined by each species' demography, rather than the annual or semiannual scale often used in sampling schemes that are chosen for our convenience. The maternal effects hypothesis entails an inertial view of population dynamics according to which the cross-generational transmission of individual quality gives to every natural population an intrinsic tendency to oscillate with a period greater than six generations (Ginzburg 1998). This prediction of the maternal effects hypothesis provides an explanation for why similar species cycle in similar periods, why small mammals characteristically have three- to four-year cycles (Ginzburg and Taneyhill 1995; Krebs and Myers 1974; Stenseth 1999), and why wolves and moose have longer cycles (Peterson et al. 1984).

The maternal effects model is a general model; it does not specify the biological origin of the maternal effects. A wide variety of agents can produce maternal effects, and the next step in evaluating this idea is to determine a mechanism. Stress is clearly one such mechanism (Boonstra et al. 1998), which could arise from predation risk (Sheriff et al. 2009, 2011) or from aggressive interactions between territorial rodents (Wolff 1997; Krebs et al. 2007). The key point is that the generality of the maternal effects idea, while useful, must lead to detailed experimental work in different rodent species to see whether it can be confirmed.

An interesting footnote to the many attempts to construct numerical models of fluctuating populations has come from an analysis by Ginzburg and Jensen (2004), who argued that in ecology consumer-resource models were overparameterized—that is, the models contained too many parameters. For example, they pointed out that the Turchin and Batzli (2001) model has seven parameters, of which four have no empirically determined values, and that the Hanski et al. (2001) model has nine parameters, of which three have no empirically measured values. They argued that although these models may fit ecological field data, that may be misleading; and they provide four rules of thumb for judging ecological theories (table 13.4). This paper by Ginzburg and Jensen (2004) is an assessment of models by theoreticians, and is aimed at empirical ecologists trying to assess the utility of numerical models.

TABLE 13.4 **Rules of thumb for judging ecological theories. From Ginzburg and Jensen 2004, box 4.**

1. Compare the number of parameters with the number of data points. When a model uses 10 parameters to fit into a time series of 25 data points, chances are that it can fit almost any 25 data points.
2. Compare the complexity of the proposed model with the complexity of the phenomenon it seeks to explain. Often, proposed models turn out to be dramatically more complex than the ecological problems they seek to solve. If one can state the ecological phenomenon in fewer words than it takes to formulate the model, the theory is probably not useful.
3. Beware of meaningless caveats confessing oversimplification. Eager for their work to be embraced by ecologists, theoreticians like to conclude that their models are oversimplified. An already complex model that "admits" that there are more mechanisms (read: more parameters) to be taken into account betrays a tendency towards further unjustified complexity.
4. Beware of being given what you expect. As ecologists, we have come to expect that our data will be "messy," and many theoreticians will go out of their way to meet this expectation. One way to make the curves look "less perfect" is to simply add environmental noise and observational error (each variance adding one more parameter). Suspect that rhetoric is at work when models that are fully capable of producing a perfect fit are tweaked to show a more palatable near-perfect fit.

Conclusion

Few topics in population ecology are more discussed and argued about than the role of mathematical models. A sobering assessment of the role of models in science was given by Oreskes et al. (1994), who argued that *verification* and *validation* of numerical models of natural systems is not possible, because natural systems are not closed. More than one model construction can produce the same result. Models, in their view, can only be *confirmed*, by showing an agreement between observation and prediction. But a problem arises, because typically some data agree with model outputs and some data do not, so that confirmation of a particular model is always a matter of degree. Oreskes et al. (1994) concluded that any predictions coming from numerical models must always be open to question, and that the primary value of numerical models is heuristic. They focus our thoughts on which variables might be important and what possible range of values in them might be effective. They key feedback must be between models, observations, and experiments (Pielou 1981). Data from natural populations are always needed for testing model assumptions, and this is often the limiting process in determining whether a model is confirmed.

Mathematical models that describe ecological phenomena like rodent population fluctuations must be susceptible to empirical tests, and this interface of models with reality will often generate controversy (e.g., Oli 2003). The important point is to encourage a continual feedback between models and existing data, so that advice and understanding can flow in both directions.

CHAPTER FOURTEEN

Key Studies Yet to Be Done

Key Points:

- There is no resolution of the issue of whether we are searching for a general explanation for rodent population fluctuations or more specific explanations applying to only some species in some regions.
- All viable explanations for rodent population fluctuations involve multiple factors.
- The food hypothesis now involves food supplies, predation, and intrinsic processes, and three-factor experiments are now needed to evaluate this model.
- All the predation hypotheses involve both predators and at least one additional density-dependent process that could be intrinsic intraspecific interactions, generalized predation, or food resource interactions.
- How climatic changes will affect rodent population fluctuations is far from clear. Top-down models make no clear predictions of what might happen, and the constraints to accurate prediction are great.
- The general issue of what determines the average density of any particular rodent population is largely unresolved, with many simple, untested generalizations available for precise testing.

Given that in the last 13 chapters I have tried to summarize the state of knowledge of the demography of population changes in fluctuating small rodents, the next important issue is to construct a list of desirable studies that need to be done to resolve outstanding issues. Because I am not a mathematical modeler, I will not attempt to suggest further modeling but will instead leave it to others who are more highly qualified. My emphasis is on empirical studies that could add critical data on uncertain points of knowledge.

Demographic Studies

Detailed demographic studies of rodent populations are badly needed, particularly long-term studies that extend for 5 to 10 years. Unfortunately, these are beyond the time scope of most graduate theses, and clever arrangements will be required to carry them out. Lowell Getz's studies at Urbana, Illinois, are a model for this, and they need to be repeated elsewhere. There is a basic limitation to these studies in that they require a landscape that is protected from disturbance. Small patches of farmland or woodland are useful for studies that try to determine the effect of habitat fragmentation on small rodent dynamics. But as Sinclair (1998) has pointed out, we need extensive, undisturbed habitats for studies of ecological processes in order to have the "control" values for comparison with our human-fragmented landscapes. The advances in analysis techniques, typified by Program MARK (White 2008; Efford et al. 2009; Cooch and White 2010) were scarcely available when most of the field data I have discussed in this book were gathered.

Descriptive studies need to be combined with concurrent experimental studies. For temperate and polar species we need much more data from the winter period. For lemmings in particular, we know all too little about what happens to populations under the snow (Turchin and Batzli 2001; Krebs 2011). The use of radiotelemetry has greatly increased our ability to follow individuals and gain more understanding of space use (e.g., Lambin 1994b; Steen et al. 1997). When this information can be combined with DNA methods to identify relatives and possibly dispersers, we will have more powerful tools for understanding spacing behavior and dispersal (Rousset 2001).

The breeding season of voles and lemmings is typically extended during phases of population growth. We have a very poor understanding of why breeding seasons start and stop in small rodents. The great variability shows that it is clearly not photoperiod that is driving breeding seasons. Winter reproduction in voles and lemmings, when it occurs, must operate under the standard physiological constraints on small mammal reproduction with respect to temperature limitations and food restrictions (Speakman and Król 2005). Winter litter sizes are low in lemmings (Millar 2001), but there are no detailed data that permit us to determine the temperature threshold at which winter reproduction becomes impossible, assuming that food is not limiting in winter (yet another problem that needs data). There is unequivocal evidence that voles and lemmings can breed under the snow in winter and increase in numbers, but exactly how they achieve this remains to be determined (Millar 2001). Winter breeding in voles is more readily documented in temperate climates (e.g., Krebs et al. 1969) but the demographic consequences are rarely quantitatively analyzed. Millar (2007) pointed out that in many small mammal studies the observed recruitment rate exceeded the estimated reproductive rate—an impossible demographic situation. Some of these cases must be due to autumn or winter reproduction that was missed in the census program.

Except for laboratory studies, we have very little information on the severity of infanticide mortality in voles and virtually no information in lemmings. Mallory and Brooks (1978) showed that female collared lemmings committed infanticide in the laboratory while male lemmings did not. The remaining question is how often infanticide operates in the field, a topic on which we have very little data. Millar (2007) reviewed nest mortality in small mammals, which he found ranged from 30% to 96% in the first three weeks of life, but he concluded that these losses could rarely be partitioned as to the exact time or the exact cause of death. Infanticide in the field is very difficult to detect. Lambin and Krebs (1993) found very little infanticide in a *Microtus townsendii* population, although they could not rule out that juveniles were killed by strange adults immediately after leaving the nest. Ylönen et al. (1997) estimated infanticide mortality of 37% in field populations of bank voles. The sole estimate of infanticide losses in field populations of collared lemmings from Reid et al. (1995) ranged from zero to 5% for a low-density population in which most losses of nestlings (43%–70%) were due to predation. Determining whether infanticide in the nest is a serious mortality will require many more detailed studies with radiotelemetry.

The common assumption in demographic studies of voles and lemmings is that death is caused by predation. But disease and food shortage, as well as direct aggression by conspecifics or other rodent species, could also cause death. There is a schism of views on this matter. Chitty (1960) argued that the immediate cause of death was immaterial, and that the important thing to determine was whether there were predisposing conditions that made death more likely. The contrary opinion is that the immediate cause of death is worth determining, particularly if it is predation or disease, and also that one should be aware of conditions that predispose individuals to dying. For example, food shortage may reduce body condition, which permits more severe parasitic infections, which could make individuals move more slowly, which may allow predators to catch them more easily. These kinds of causal chains are nearly impossible to decipher, and the best evidence for that is the difficulty of specifying the cause of death of humans in developed countries. But I think that it would be useful to try to pin down the exact timing of death in small rodents with intensive radiotelemetry studies to see, for example, how often individuals are killed by predators or by other rodents.

Immigration and emigration are typically ignored in small rodent studies, yet methods now exist in Program MARK for estimating these components of demography. More analysis is needed to determine the behavioral ecology behind the emigration and immigration movements of individuals in species where the fence effect discussed in chapter 5 occurs. Such studies will also be important for small habitat patches that are increasingly being produced by the habitat fragmentation that is associated with human activities, particularly agriculture. It is commonly found, for example, that grassy fence rows a few meters wide between cropped fields contain *Microtus* populations at densities that are very high in comparison to those of open field populations (Krebs, unpublished). Why this should be is a mystery, but it must have something to do with vole territoriality and dispersal in linear habitats.

The bottom line is that much remains to be done on the dynamics of small rodent populations. The design of each study must be carefully thought out, the hypotheses clearly stated, and the decision structure declared before fieldwork is begun. Enough is already known about small rodent populations to do this, and new studies should make maximum use of the accumulated knowledge about how best to proceed.

Food Studies

It is a truism to say that we do not know the diet breadth of many species of voles and lemmings, even those we have studied intensively. Diet breadth may be strongly affected by habitat, so that a single study of diet, or a study confined to one season of the year or one location, is never enough. Voles and lemmings typically feed on a variety of plant species and are rarely tied to a single food plant. But they may be generalists in summer and specialists in winter, or vice versa. Techniques for determining diet vary from cafeteria studies and isotope studies to analyses of stomach contents or fecal pellet plant fragments. Pitfalls abound, and again much attention must be paid to details of methodology. For example, Hwang et al. (2007) showed that in several species of voles the isotopic composition varied from the stomach to the feces, because isotopes were differentially absorbed as food passed through the gut. Isotopic composition studies will not replace the older methods of cafeteria trials and stomach analysis of plant fragments.

Given that the food habits are well documented, the next important step, as outlined by Turchin and Batzli (2001) in chapter 13 of this book, is to determine the growth trajectory of the critical vegetation quantitatively (cf. figure 13.1). Primary production measurements are difficult to make, and they must be specific to a particular study area, rather than estimated from satellite data. Given that primary production can be measured, simple ecosystem models like ECOPATH can be used to explore the relative offtake of various herbivores feeding on a suite of plants. For example, Legagneux et al. (2012) showed that on Bylot Island in the Canadian Arctic, herbivores, including the brown and collared lemmings, took less than 10% of annual primary production, This suggests that unless lemmings are tied to a few scarce food plants, they cannot be limited by absolute food shortage. Another key item of interest is whether grazing by voles or lemmings in year *t* affects vegetative production in year *t+1*. If there is no delayed effect of grazing, it is unlikely that food supplies by themselves can drive population fluctuations.

Food supplementation experiments on small rodents have typically shown that adding extra food of high quality increases the average density or carrying capacity of the habitat but does not seem to affect the overall pattern of population collapse in the decline phase (e.g., Taitt and Krebs 1981; Taitt et al. 1981). In *Microtus townsendii* food addition reduced the rate of, but did not eliminate, the spring decline at the start of the summer breeding period (Taitt et al. 1981). The suggestion from these and other studies is that food addition experiments for small rodents need to add high-quality food for sustained periods of time in open populations that are intensively studied so that immigration and emigration can be measured. Doing these kinds of experiments in enclosures is attractive, but it risks artifacts caused by the scale of the enclosures.

Another approach to food supply studies is to use a landscape approach, as suggested by Lidicker (1995). This type of study would have to be extensive across a landscape with clearly distinct habitats, and should at the same time be intensive with detailed demographic data available for the vole or lemming population in each habitat. To date, the study closest to this ideal has been that of Delattre et al. (1999) in an agricultural area of eastern France, discussed in chapter 12. This study could not obtain the detailed demographic data from each habitat type, which understandably would be a large task. But if we are to be serious about landscape effects on population dynamics, we need experimental studies with extensive resources and person-power. The general problem with landscape studies is that they are entirely descriptive and rarely articulate a clear hypothesis that is being tested.

Predator Studies

The bane of many predator studies has been the acquisition of detailed data on the predators without comparable data on prey dynamics, or vice versa. The most desirable studies on the role of predators in affecting small rodent population changes need to acquire detailed demography of both predators and prey. This shortage of data is most clearly illustrated by attempts to assess the role of least weasel (*Mustela nivalis*) and ermine (*Mustela erminea*) predation on vole and lemming cycles. In most cases we do not know the precise density of mustelids; so when we prepare models of the predation hypothesis, as discussed in chapter 12, we do not know whether to make weasel density in the model 2 to 20/km^2 or 0.1 to 2/km^2. Arguments about weasel predation often boil down to the question of how many there are, rather than to their feeding rate. A very high priority in areas where weasels are considered a key predator is to determine their density accurately, a task that is not easy because of their large movements. Studies of weasels in patchy landscapes, like Kielder Forest in northern England, will be particularly useful in this type of study (Brandt and Lambin 2007). The key question in predators, as we have seen, is whether they have a numerical response to changes in prey density. If there is no numerical response, there is no way in which predators like weasels can regulate prey numbers, as shown in figure 9.2.

Another issue that clouds interpretation of predators' role in causing population changes in prey is intraguild predation: the tendency of predators to kill other predators (Thompson and Gese 2007; Sergio and Hiraldo 2008). Weasels and arctic foxes in the arctic, for example, are taken by snowy owls (Pitelka et al. 1955). Birds of prey are well known for killing other birds of prey (Sergio and Hiraldo 2008). The key question is how intraguild predation might restrict predators of voles or lemmings from being able to strongly impact on the rodent population. To answer this question, detailed data are required on the diets and densities of all the predators in a study region.

Predator models of cyclic rodents have always indicated that by themselves, predators cannot stop population growth in voles or lemmings because of the time lag in their numerical response (Hanski et al. 2001). A second factor must operate in a density-dependent manner to reduce the population growth rate to near zero at the peak of the fluctuation; this factor could be food shortage or intrinsic factors in social behavior. A consequence of this finding is that any future studies of the role of predation would be most useful if they include attempts to find the missing density-dependent factor in the existing models.

Removal experiments on predators are the best way of testing their role in rodent population changes (e.g., Korpimäki et al. 2002). By and large, such experiments are impossible because of the large scale at which most predators operate, and because of ethical constraints on harming wildlife. Some experiments can be carried out in predator-proof exclosures if the problems of spatial scale can be alleviated. Klemola et al. (2000) reported on *Microtus* populations in 0.5-hectare enclosures where predators were excluded and food was added; they concluded that predation, rather than food shortage, was the key factor limiting populations. These experiments produced much greater effects on *Microtus* population density than similar experiments conducted in open populations because predators were excluded more effectively and could not immigrate into the enclosures.

Intrinsic Factor Studies

Two lines of evidence have suggested further studies that can shed light on the role of social behavior in affecting population rates of change. The Smith et al. (2006) model suggests that delayed onset of the reproductive cycle can by itself generate a population cycle. Since it is already well known that reproduction often starts slowly in the decline phase of vole and lemming fluctuations, it is imperative to find out what is causing this. Can the delay be removed if high-quality food is added? Is the delay due to chronic stress passed down as a maternal effect (e.g., Sheriff et al. 2011)? If it is a maternal effect, is it produced by social interactions involving aggression or by exposure to predation risk? Is it associated with diseases that show delayed density dependence?

The Chitty hypothesis and the maternal effects hypothesis of Inchausti and Ginzburg (2009) both suggest that individual quality is compromised as population density increases. The problem is to define what this quality might be. Klemola et al. (2002) attempted to test the individual quality hypothesis by introducing *Microtus* voles into 0.5-hectare enclosures from the low phase of the cycle in one year and from the increase phase of the cycle in the following year. Both populations increased at the same rate in the enclosures, and the researchers concluded that individual quality did not change in a way that affected population rates of increase. There is one statistical problem: the rates of population growth were measured in two different years, so that year is a confounding variable in the analysis. Two other issues are unresolved. The first issue is whether enclosures are a sufficient analog of open fields, in which dispersal is not constrained by fences and somewhat arbitrary rules of deciding what a dispersing individual is. Wolff et al. (1996), for example, rejected the use of barrier strips in enclosures to identify individual voles that were dispersing. The second issue is the origin and small number of voles introduced to each enclosure. Given the problems with this study, it can be interpreted either as rejecting the individual quality hypothesis or as being inadequate for study of these intrinsic effects.

The unsolved problem with all the intrinsic hypotheses involves the question of exactly which biological characteristics are affecting quality. Originally it was thought that selection for aggressive behavior might be the key. Aggressive behavior could be genetic in origin (a possibility for which there is at present no evidence); physiologically driven from social interactions; or situational, as in kin-selection models. If individuals that are either realted or familiar with each other are aggressive only against unrelated or unfamiliar animals, introducing a group of unrelated individuals into an enclosure could affect social organization in ways quite different from what happens in established field populations. In their enclosure experiments, Klemola et al. (2002) found a density-dependent reduction in population growth, which they attributed to delayed maturation and reduced reproductive success—both variables that are socially conditioned when predation has been eliminated. Individual quality can be rejected as a component of population change only on the assumption that quality is a constant in the life span of an individual.

Chitty and others who have suggested that intrinsic mechanisms are a necessary part of population fluctuations have always admitted that they do not know exactly what aspects of social behavior might be involved. There is considerable room for additional studies of the social behavior of voles and lemmings, and especially of whether the well-known aspects of that behavior, such as delayed maturation, operate only in a directly density-dependent manner or might have a delayed component, as has been suggested for snowshoe hares by Sheriff et al. (2009). Techniques are now available for measuring the stress levels of wild mammals (Boonstra 2004; Sheriff et al. 2010a), and, given that stress may arise from a variety of sources, this is one effective entry into measuring individual quality. Measuring immunocompetence is another approach that could be useful (Lochmiller 1996).

One way to test for undesirable effects of enclosures is to construct larger and larger enclosures to see whether results similar, for example, to those found by Klemola et al. (2000) occur in all sizes of enclosures. Unfortunately, this research would require a large and expensive series of experiments. But I can see no other way to resolve the problem of artifacts produced by small enclosure studies, which mislead us in our quest to understand population changes in rodents.

Miscellaneous Problems

An underlying issue in small rodent studies is the scale issue: the question whether the results we get from observational studies or manipulative experiments depend on the landscape structure. Are *Microtus* studies informative if carried out in agricultural landscapes in which most of the original vole habitat is now used for agricultural crops and little is left as native grassland? Of course such studies are informative for these particular kinds of landscapes, but are they comparable to those done in open grasslands that occupy hundreds to thousands of hectares? In a sense this is the issue that underlies the varied thoughts about the utility of small enclosures for population studies of rodents. What is clear is that rodent enclosure studies have in some cases misled population ecologists trying to understand population processes. The best examples come from house mice (*Mus musculus*) studies. Small enclosure studies have shown the power of territoriality to restrict population growth (e.g., Crowcroft 1996), but these and other aspects of social organization of house mice have not been found in field populations and do not appear to restrict house mouse population outbreaks in Australia (Singleton et al. 2005). Whether the utility of enclosures varies with the species is itself an interesting question. Wolff (2003b) described a set of behavioral conclusions developed from laboratory studies of rodents that were found to be artifacts, and he cautioned against accepting results without adequate field testing.

If small enclosures produce artifacts, what size of enclosure will not produce them? This question is still unanswered. It could be investigated by using islands of different size to investigate dynamics. Tamarin (1978b) described the unusual dynamics of a population of *Microtus* on an island off Cape Cod on the east coast of the United States. Muskeget Island is 2.6 km^2 in size, has no mammalian predators, and supports a population of *Microtus breweri* that does not fluctuate even though its close relative *M. pennsylvanicus*, on the adjacent mainland, shows normal three- to four-year population fluctuations. Tamarin argues that the absence of population fluctuations on Muskeget Island is caused by the absence of an effective dispersal sink. If voles cannot disperse on an island, he argues, they will develop *K*-selected traits and persist at high density. More studies like this are now needed, along with long-term studies on islands of different sizes with and without mammalian predators. This is a challenging order, but it would be an important test case of patch size and population fluctuations in small rodents.

Climate Change Effects

Climate change and its implications for biodiversity are a key focus of much current research. The first panic button regarding population cycles was pushed when rodent fluctuations began to disappear in the 1980 and 1990s in Fennoscandia. The first warnings were sounded in 1991 by Birger Hörnfeldt in his PhD thesis (Hörnfeldt 1991). As illustrated in figure 14.1, Hörnfeldt (2004) presented eight hypotheses that might explain the collapse. The simplest hypothesis, that continued sampling of the same plots over many years had decimated the population, had been rejected by

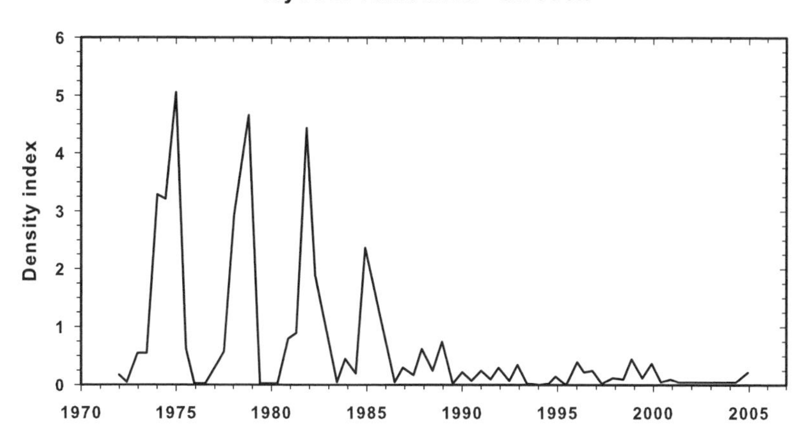

FIGURE 14.1 Collapse of population cycles in the grey-sided vole *Myodes (Clethrionomys) rufocanus* in northern Sweden from 1970 to 2005. Data from Hörnfeldt et al. 2005.

Christensen and Hörnfeldt (2003). The problem of the disappearance of rodent population cycles was partially solved by Ecke et al. (2006), who showed that continued forest harvesting had reduced the landscape occupied by voles to a series of patches that were too small for occupancy—too small, at least, for gray-sided voles. Nevertheless, the question about collapsing cycles continued to be raised. Ims et al. (2008) concluded that fading cycles in rodents were part of an overall northern ecosystem collapse related to climate change. Extensive analyses by Xavier Lambin (pers. comm.) suggest that the amplitude of cycles in Europe has been decaying since the 1980s, and that this might be caused by changes in climate affecting seasonality. There is much to be done on these climatic effects, and long-term data are needed for critical testing of various hypotheses.

Some models have been presented to explore the impact of climatic variables on fluctuating rodent populations. Gilg et al. (2009) produced a model that demonstrated mathematically that climate changes could produce lemming cyclic collapse. No one except Hörnfeldt (2004, p. 386) seems to have explored the possibility that strong cyclic fluctuations might return as they have in 2010. More investigations need to be carried out on climatic influences on rodent populations (e.g., Reid et al. 2011), but models such as the one with 29 parameters produced by Gilg et al. (2009) seem

TABLE 14.1 **Constraints encountered by ecologists attempting to project the effects of climate change on mammals. The first column indicates the type of model used and the type of constraint encountered. References give an example of each constraint. From Berteaux et al. 2006.**

Constraint	Example	Reference
Forecasts		
1. Values of unobserved variables potentially changing	Our model assumes that vegetation is the only driving variable in the current distribution of each species and therefore ignores potentially important anthropogenic affects that are independent of climate effects.	Johnston and Schmidt (1997)
2. Need to extrapolate beyond the range of observations	Forecasting models can project future population dynamics of three African ungulates, provided that observed climatic conditions recur but cannot make safe predictions beyond the range of observed climatic conditions.	Ogutu and Owen-Smith (2003)
Predictions		
3. Inherent noise is large	Predictions as to which arctic species will evolve fast enough to adapt to new climatic and ecological conditions are currently difficult, because there is a lack of data on the quantitative genetics of many species.	Berteaux et al. (2004)
4. Need for better definition of boundary conditions	The model assumes the length and severity of seasonal energetic bottlenecks, and not the distribution of prey, predators, or caves, determines the winter range limits of little brown bats.	Humphries et al. (2002)
5. Need for additional fine tuning regarding aggregation and representation	Despite it being one of the best-understood mammalian species distributional limits, mechanisms underlying the northern limit of Virginia opossums remain enigmatic. There are three major areas of inadequate understanding: the microclimates actually experienced by opossums, the exact relationship of opossum foraging behavior with ambient temperature, and the role of human-related resources operating to mitigate restrictive climatic effects.	Kanda (2005)
Both		
6. Uncertainties of climatic scenarios	A different climatic scenario with a faster rate of temperature change would result in an increased mismatch between the rate of habitat change and the rate at which species can relocate to occupy new habitats in appropriate climate envelopes.	Callaghan et al. (2004)

to represent modeling carried to some logical extreme (cf. Ginzburg and Jensen 2004). Fading population cycles need continuous study, but much of the data used to investigate long-term trends in rodent abundance are of poor quality by current standards, and need validation and improvement.

Climate change impacts can be studied only in part by manipulative experiments, but this is no excuse for failing to specify hypotheses about the ways in which climate might act. If the conclusion is that food resources are rarely limiting for many rodent populations, then climate change will have little bottom-up impact, and the key effects will come from top-down changes in predator abundance and feeding opportunities. Insofar as social behavior processes cause changes in reproduction or mortality, there seems to be no clear way in which climate change will affect social dynamics directly. Berteaux et al. (2006) have explored the problems of relating climate change to mammalian geographical distributions and population dynamics, and they make recommendations for the constraints under which we must operate (table 14.1). The problems are large, but the challenge is even greater in view of anticipated scenarios.

Conclusion

There is an interesting contrast between the excellent historical overview of brown lemming cycles at Barrow, Alaska, by Pitelka and Batzli (2007) and the summary by Chitty (1996) of his unsuccessful attempts to find a general solution to population cycles. Pitelka and Batzli (2007) review the effects of three key factors at Barrow: overgrazing of food resources, predation by raptors and owls as well as weasels, and winter conditions of snow cover and meltoff in spring. Chitty (1996) argues, by contrast, that overgrazing and severe vegetation damage do not occur in all population fluctuations, that predators are sometimes at low abundance, and that many *Microtus* populations in temperate latitudes fluctuate in cycles in the absence of snow. The difference between these two visions of the particular and the general has not yet been resolved in the study of small rodent population changes. For the present, we can proceed only by developing more and more case- and location-specific studies, by doing experimental manipulations where possible, and by striving to achieve a general hypothesis as far as possible. It remains to be resolved whether this will mean that population cycles of a particular species in Britain are explained differently than population cycles of the same species in Fennoscandia or of similar species in Illinois. As scientists we strive for generality, but our efforts must be tempered by reality. At present, ecology is in a phase of attention to particular details rather than of generality, and many ecologists seem convinced that the "general" rules, once uncovered, will be less general than many have hoped. We shall see, as we do more research.

CHAPTER FIFTEEN

Synthesis of Rodent Population Dynamics

Key Points:

- Two problems must be distinguished: what sets the average density of a population, and what determines the population growth rate. Both these questions have simple modeling explanations, yet complex empirical explanations.
- Average density is most likely explained by food resources, or by primary productivity in general. For the solution of a conservation or management problem, the hypothesis concerning food resources should be tested first.
- Population rates of growth are affected by many factors, and for each population fluctuation a complex of multiple factors must be involved. At present, predation and social behavior appear to be the dominant mechanisms of regulation.
- If we wish to determine which of the multiple factors are necessary and which are contingent, we must use experiments that compare experimental populations to controls in the normal scientific manner.
- A good approach for such experimentation would be to seek simplicity and distrust it. A dose of humility would also be advisable since no one knows exactly where we will finish, and so much still remains to be done.

In this chapter I will suggest an outline of a synthesis for rodent populations that is sketchy because many of the details are yet to be filled in. As such, you can use this chapter as a set of guideposts for how we might increase our understanding of rodent population systems. Such an understanding may be useful if we have a rodent pest that must be controlled or an endangered species that should be protected.

Two Distinct Problems

Empirical population analyses are plagued by a misunderstanding of the two fundamental problems of population dynamics. This confusion has arisen because the problems are solved so simply with a graphical or mathematical model that many ecologists assume that the empirical solution should also be simple (cf. Krebs 2009a, figure 13.2). The two distinct problems involve what determines the population's average density, and what determines the population's growth rate. Models confound these two issues by dividing the effect of mechanisms into density-dependent effects and density-independent effects. Density-dependent effects can be further subdivided into direct density dependence or delayed density dependence.

The problem of what determines average density can be attacked by conventional statistical experimental designs in which a particular factor is present in one population and absent in another, or is at a high level in one population and at a low level in another. Comparative or manipulative studies are available, and there is little controversy over their results. The only difficulty at present is that there are too few of these kinds of studies for small rodents.

Determining what controls the population growth rate is much more difficult, because in a fluctuating population the growth rate changes from year to year and, within a year, from season to season. Again it is essential to use the comparative method to answer questions about controls. If a population does not breed during an extended winter period, it will not increase in size unless there is immigration, and the answer to what controls its rate of decline over that winter can be determined by comparing its rates of predation, disease, or other factors with those of one or more other populations. As one extends this approach over longer time periods, such as multiple years, difficulties occur because of compensatory mechanisms. For example, if predation is higher in one winter in area A, reproduction may start earlier the following spring in that same area, so that when one compares the predation in one area over a winter with with that in another area that has had less predation over that same winter, one may find that by midsummer both areas have identical population densities, so that the overall annual rate of population growth is the same at both sites in spite of their different predation levels. Whether this kind of compensatory effect exists in the real world needs, of course, to be determined.

What Determines Average Density in Small Rodents?

Changing the quantity and quality of the food supply increases average density in virtually all studies of small rodents. Only a few examples need to be cited. Figure 15.1 shows the average September density of two *Microtus* species in central Illinois, averaged over 25 years. *Microtus ochrogaster* is at higher density in alfalfa fields, while *M. pennsylvanicus* is at higher density in tallgrass prairie. Taitt and Krebs (1981) found that adding extra food to *Microtus townsendii* populations doubled population density. Desy and Batzli (1989) found that prairie vole populations with added food achieved about three times the density of unfed populations over a 24-week period. Bujalska (1975) reported that an island population of *Myodes glareolus* in Poland increased about 50% when supplemental food was added for one year. Boutin (1990) reviewed 138 food supplementation experiments in terrestrial vertebrates and reported that the typical population response was a two- to threefold increase in density. The evidence is very strong that food supplies limit average density.

Cover manipulation experiments can be a useful way of changing food supplies, cover from predators, and habitat complexity for social interactions—if, for example, the plants are fertilized, or if the food factor can be eliminated through the use of some form of inert cover. Two good examples of cover manipulation are the studies by Birney et al. (1976), discussed in chapter 12, and the fertilization experiment at Barrow, Alaska, mentioned in Pitelka and Batzli (2007). Unfortunately the details of the Barrow experiment were never reported, but the summary was that plant growth and plant quality increased under the addition of fertilizer, and that winter use of the fertilizer patch by lemmings also increased greatly (as measured by winter nest count), but that the lemming population then collapsed in the following spring because of concentrated predation (Pitelka and Batzli 2007, p. 327). Taitt et al. (1981), on the other hand, used straw as inert cover to manipulate populations of Townsend's vole.

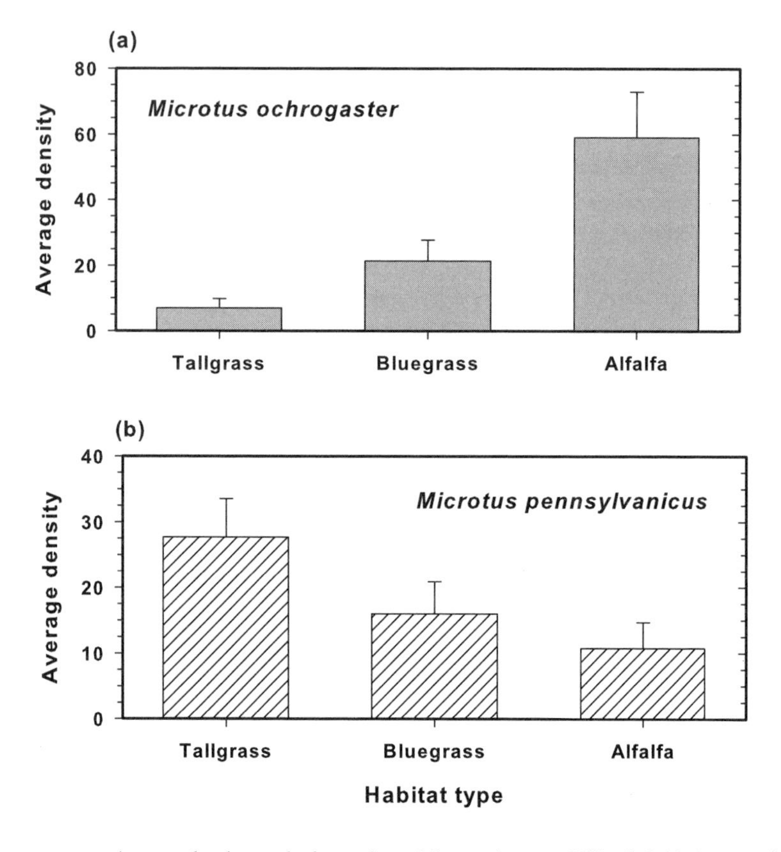

FIGURE 15.1 Average density per hedtare of two *Microtus* in central Illinois fields from 1972 to 1996. The preferred habitats have higher densities on average, and these differences are directly caused by the preferred food plants of each vole species. Data courtesy of Getz et al. 2005.

Dissecting the mechanisms behind these cover manipulations would give us more information about the nexus of factors affecting vole and lemming numbers.

There is little evidence showing a similar response when predators are removed from a small rodent population. Enclosure studies indicate a strong response by rodents to predator exclusion (e.g., Klemola et al. 2000), but the problem of unknown enclosure effects makes comparison with field populations difficult. In chapter 9 of this book I have already reviewed the lack of response of field voles to the removal of least weasels in experiments carried out by Graham and Lambin (2002), as well as the modest effects observed by Korpimäki and Norrdahl (1998).

If there is a general deficiency in most experiments that explore the variables affecting average rodent density, it is in their time frame. Press experiments become more valuable the longer they operate. In some environments it may be impossible to continue press experiments over more than one growing season, but longer-term press experiments would assist in answering the question of population limitation most directly.

What Determines Population Growth Rates in Small Rodents?

This question is typically addressed to populations whose size fluctuates in cycles or at least oscillates irregularly. Stable populations are rarely described in voles and lemmings, so for the present I will not discuss them. Rates of population change in fluctuating rodents are typically broken down into phases—increase, peak, decline, and low—and each phase is at least a season long and often a year in length. The first problem is always to obtain a good description of the quantitative demography of each phase, and an equally good description of the chosen mechanisms that operate during that phase. Our ability to obtain this information has been limited, particularly in lemmings, by the difficulty of gathering data in winter. The consequent lack of information can result in some impossible demographic statements about, for example, how much population growth can occur over winter, or how much population decline over winter may be caused by resident predators.

The increase phase of a population fluctuation typically begins at or near the maximum possible rate of increase, but may already be slowed in its later stages by losses of nestlings or juveniles immediately after they have emerged from the natal nest. The competing hypotheses here concern predation and social mortality. The use of remote cameras could answer this question, but there are no data available at present for deciding among these two hypotheses. Dispersal of subadults as they enter breeding condition may predispose individuals to mortality from predators or social mortality, and again we do not know the importance of either factor.

The peak phase of a fluctuation can be strongly affected by both specialist and generalist predators—and in general, mobile specialist predators like birds of prey can have strong effects, as the experiments of Erkki Korpimäki and his colleagues have demonstrated. But the dominant features of the peak phase are demographic: shortening of the breeding season, delayed maturation of subadult young-of-the-year, and increased movements. These demographic changes are social in origin, and factors such as kin-group structuring (Boyce and Boyce 1988b) can be of critical importance. There could be shortages of high-quality food in this phase, but they seem unlikely, at least in the summer growing season. The key processes in the peak phase of a population fluctuation are predation and social behavior.

As the peak phase of a fluctuation gives way to the decline phase, the data become fewer and the controversies more numerous. If breeding ceases or is greatly reduced at the end of the peak phase, populations must decline. This time period is the transition between the twin predictions of the breeding suppression hypothesis of Ylönen et al. (1997; Ylönen and Brown 2007). This hypothesis predicts that female voles should favor survivorship over reproduction in the decline phase of a population fluctuation, and fecundity over survivorship during the increase phase. Once breeding has stopped at the end of the peak phase, a variety of factors could begin to reduce numbers: bad weather, parasitic disease, predation, or food shortage. It seems unlikely that social interactions can play any part in this early part of the decline in which there is no breeding. Social aggression in rodents is a breeding-season phenomenon.

The start of the breeding season in the decline phase is a critical time. Breeding in the decline phase often starts late in spring and ends early in autumn, results that are consistent with the predictions of the breeding suppression hypothesis. Infanticide may or may not be a serious cause of juvenile mortality in this phase. For example, Ylönen et al. (1997) found that about 30% of male and female bank voles from a laboratory stock committed infanticide when encountering unguarded pups, and this result was similar to what was observed both in the field and in enclosures. Specialist predators like weasels and ermine often concentrate on patches of small rodents and cause additional mortality.

Social mortality at the start of the breeding season has been highlighted as a window on the events that drive numbers down. The spring decline at the start of the breeding period occurs in many species of voles and lemmings, and would seem an ideal candidate for short-term experiments. Krebs and Boonstra (1978) described a series of spring declines in Townsend's voles and pointed out the great variation in their intensity and duration, which on circumstantial evidence they suggested was partly due to aggressive spacing behavior over territories. Taitt and Krebs (1983) followed up on this analysis by manipulating predation, cover, and food during a spring decline in Townsend's vole, and concluded that predation was the dominant cause of the early part of the spring decline, but that spacing behavior causing dispersal was the dominant mechanism in the latter part of the decline as reproduction was starting. Dispersing voles could die from a number of causes, and the key is that an adult vole without a breeding territory has no chance of survival.

Both specialist and generalist predators are probably responsible for much of the losses during the decline phase, and the only outstanding question is whether the kill rate is operating on specific age or sex groups, and whether the individuals killed can be considered "surplus" animals in the Errington (1956) context—that is, animals excluded from the territorial social system. The common assumption is that predators are taking not only the "surplus" but also the "capital," and that the collapse of the population in the decline phase is sufficient evidence for rejecting Errington's idea that predators kill only "surplus" rodents. But caution would suggest that more data on this question would be useful, along with detailed studies of social organization in declining populations. Boyce and Boyce (1988b) reported a severe decline in *Microtus arvalis* in Germany that had no social mortality component—no infanticide in this species, no aggressive attacks on unrelated juveniles—and they concluded that the decline might be due to food shortage or predation, but was certainly not associated with social interactions. We are led back to the question of the mechanisms that are a necessary component of the decline phase, and predation seems the best candidate to explain the largest part of the decreased population growth rate. Social factors could be involved only as part of the mechanism setting the start and perhaps the stop of breeding.

The low phase of population fluctuations is controversial because of two difficulties. First, because populations are at low density, data are sparse and often imprecise, so that a description of the low phase may not be accurate (figure 15.2). For some rodent fluctuations this is clearly not the case (figure 15.3). Boonstra et al. (1998) suggested two major types of hypotheses to account for the low phase. The first proposes that something is "wrong" with the extrinsic environment. The most promising of these extrinsic explanations is that predation, acting either directly or indirectly, has delayed density-dependent effects on prey populations during the low phase. A second hypothesis, which involves delayed effects of grazing on plant quality, has been tested—for example, by Klemola et al. (2000)—but appears unlikely to be correct. The second class of hypotheses proposes that something is "wrong" with the animals themselves. The most likely intrinsic factors are maternal effects or age effects on fitness during the

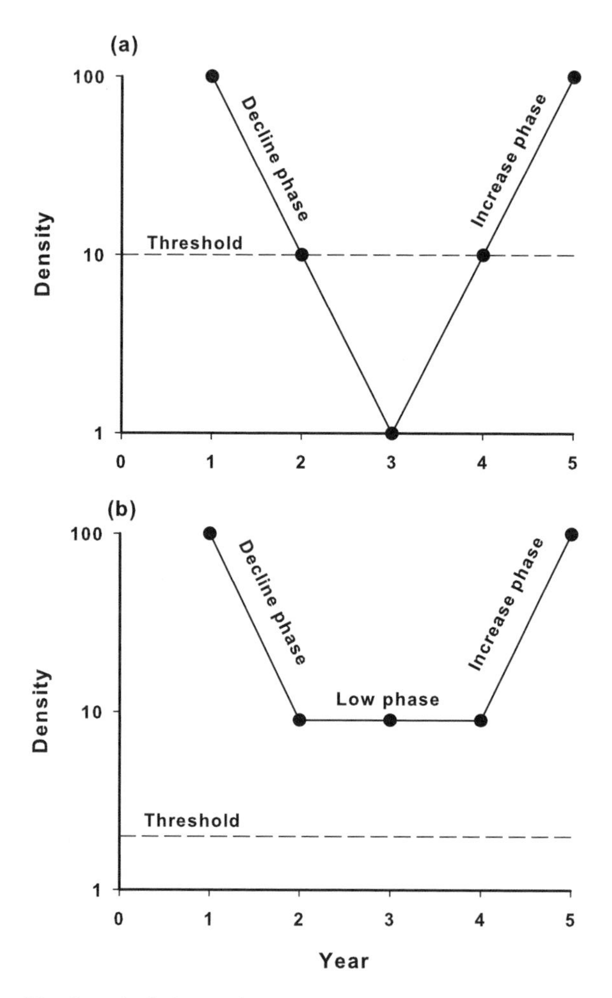

FIGURE 15.2 Two hypothetical scenarios of the low phase in fluctuating small rodents. In the simpler model (a), no low phase exists; we perceive a low in year 3 only because there is a detection threshold below which normal trapping methods cannot measure density. In the more complex model (b), the detection threshold is lower, and we can be sure that a low phase of more or less stable numbers persists. Field data support scenario (b) in some well-studied cases. From Boonstra et al. 1998.

low phase. There are many possible explanations for the low phase, ranging from generalist predation to genetic mechanisms of spacing behavior traits. Boonstra et al. (1998) reviewed these mechanisms and pointed out that speculation greatly exceeds data for this part of the population cycle, and that experimental tests for each of these hypotheses are needed.

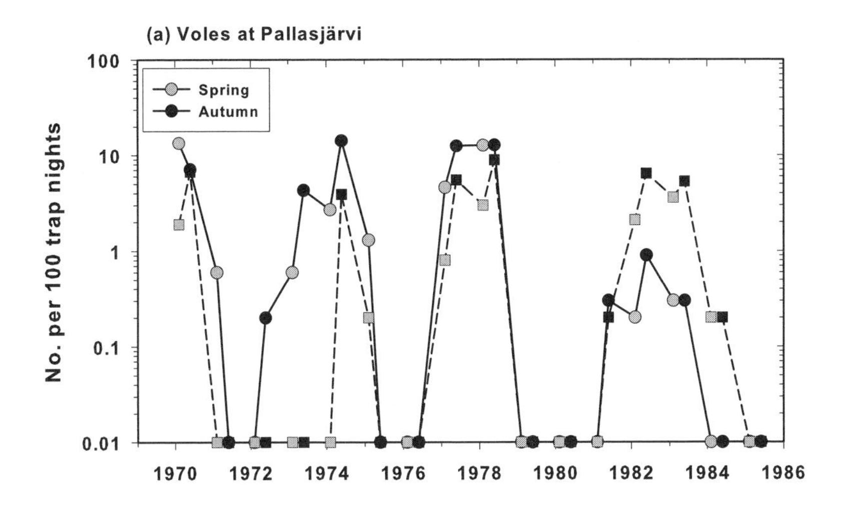

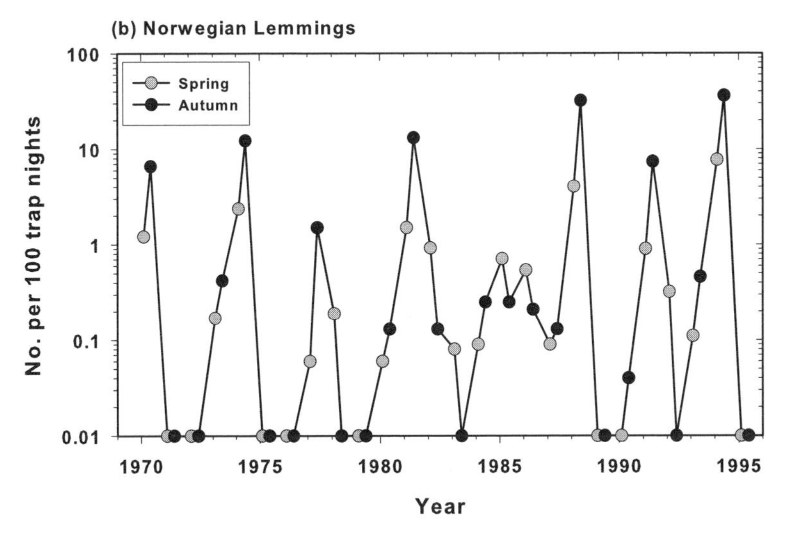

FIGURE 15.3 Two examples of low phases in populations of small rodents. (a) Population cycles in field voles (*Microtus agrestis*, circles) and root voles (*Microtus oeconomus*, squares) from Pallasjärvi, northern Finland. Spring and autumn snap trapping were done on permanent small quadrats. Values of 0.01 are equivalent to zero. Note the low phases in 1971–73 and 1984–85, and the extended low phase in 1979–81 (Henttonen et al. 1987). (b) Population cycles in Norwegian lemmings (*Lemmus lemmus*) from Finse, Norway. Data were snap-trap samples obtained from a one-hectare grid from the most productive site (grid M). Data from Framstad et al. 1993, 1997.

Points of Debate

I draw together here six points of disagreement that have arisen in the study of small rodent populations with an indication of the importance I attach to each of these issues. There is no need to agree with my assessment of these points, which are always in need of discussion and reassessment.

1. *We require a general hypothesis that will apply to all rodent population fluctuations.*

   This viewpoint can be considered a long-term goal; while useful for debates, it should not affect the observations and experiments that need to be carried out now. Detailed evidence is essential, no matter how general the hypotheses.

2. *We need to do many more single-factor studies to make sure there is not a simple hypothesis that will explain all the features of rodent fluctuations.*

   This viewpoint is obsolete in view of all the research that has been done, and it is essential that all future studies consider and test multiple factors.

3. *We need to do long-term, large-scale studies to further our understanding of population fluctuations.*

   This view is correct, but it needs to recognize that not all studies can be done over the long term or on a large scale. Small, carefully designed studies can help us to understand any of the mechanisms that could be involved in population change.

4. *Factors not implicated as being necessary for population regulation are not important items for study.*

   This view is pernicious and must be rejected completely. If we should find out that disease or spacing behavior is not an essential part of an explanation for population cycles, it does not mean this factor is not important for many other aspects of life-history theory. We are dealing here with one aspect of the ecology of rodents—population limitation and regulation—but many other aspects are also vital to investigate.

5. *A mathematical model is an essential component of every study of rodent population dynamics.*

   This view is certainly not correct, as is evidenced by many elegant and useful studies carried out on the basis of verbal models of population dynamics. It is certainly worth debating the role of specific types of mathematical models in population studies, but most comprehensive models to date have had so many unmeasurable parameters as to be not very useful empirically.

6. *Rodents are a specialized group of mammals; their cyclic dynamics are taxa-specific and do not inform general advances in ecological understanding.*

   This view is incorrect, as we shall see in the next chapter—but rodents comprise a large group in the class Mammalia have led the way in advancing our understanding of population dynamics in general. Ecologists can ignore them, but only at their peril.

Conclusion

We seek generality in our explanations of why small-rodent populations fluctuate in size, and why they are on average more abundant in some habitats than in others. The level of generality we will ultimately achieve is unknown, and discussions about generality should not impede progress in sorting out these questions because studies must be time- and place-specific, and thus must involve only one or a few species. What is clear to me after reviewing our current state of knowledge is that multiple-factor explanations for population fluctuations are necessary and that all observational and manipulative studies should operate with a detailed set of hypotheses about the factors thought to be part of the explanation. In my judgment the most likely critical factors are predation and social behavior, and I would concentrate on them. Others may wish to investigate disease and food resources in more detail, along with social behavior. In addition to these large-scale, long-term studies that directly address the population regulation issue, valuable studies can be done on very specific mechanisms and processes that can be studied in the short term. Examples might include predator selectivity with respect to age and sex, variability in food habits with season, individual differences in parasite loads and immunocompetence, and the heritability of aggressive behavior. All these kinds of studies, which are more suitable to graduate thesis topics, can give insight into the mechanisms of population change.

On an even broader scale, we need much more information about the landscape context of population fluctuations. There are almost no rigorous data, for example, on the extent of synchrony in lemming cycles across the arctic regions of Canada and Alaska. Detailed studies of landscape effects, coupled with experimental manipulations or observations on predators or food supplies, would form an ideal nexus to advance our appreciation of landscape effects. Because small-rodent habitat in agricultural areas is being fragmented, it is particularly important to determine whether fragmentation at several scales can affect dynamics, as Lidicker (1995) has proposed.

Any synthesis of information is only as solid as the data on which it is based. Ecological data gathering has many pitfalls that have been addressed through the development of increasingly robust methods over the last 50 years. But many classical studies of small-rodent populations have lacked what are now defined as essential methods of estimation, simply because methods have improved over time and good computer programs are available. Methodology is even more important than replication in developing and testing evidence-based hypotheses.

CHAPTER SIXTEEN

Comparative Dynamics of Rodents and Other Mammals

Key Points:

- If the factors affecting mammal populations can be separated into extrinsic and intrinsic factors, one conjecture originating from Graeme Caughley was that large mammals are extrinsically regulated, and small mammals intrinsically regulated.
- Wolff (1997) suggested from evolutionary first principles that intrinsic population regulation was possible only for mammals that had altricial young, so that territoriality evolved as a strategy to prevent infanticide.
- Large mammals typically have precocial young, and their populations are primarily regulated by predation or by food supplies if they have no effective predators, with human harvesting an important concern often overriding everything else.
- For small rodents with altricial young, social behavior and predation are the suggested control factors, with the exception of desert rodents and agricultural pests, which have boom-bust dynamics largely driven by food resources.
- Comparative population dynamics will be improved if everyone will clearly state which factors they have considered and which they were unable to study. By constructing precise multiple-factor hypotheses we can progress rapidly in our understanding of what causes population changes in all mammals.

How does the research on small rodents interface with that of other mammals? Are small rodents, particularly those that show cyclic population dynamics, an unusual case with little general applicability, or do they inform more general theory? The problems of population dynamics have been addressed with vigor for more than 50 years, and some ecologists have argued that they are now solved and we can move on. Unfortunately, the solutions offered by various ecologists differ and yet it is important for the management of our natural resources that useful generalizations are carefully analyzed and presented. In this chapter, I attempt to generalize across all mammals to address the question of which factors produce population change.

Caughley's Conjecture

Graeme Caughley recognized in the early 1980s that small-mammal and large-mammal ecologists were working with two separate paradigms of population regulation, and he attempted to explain why this might be. Caughley and Krebs (1983) presented a set of arguments about why we should expect large and small herbivorous mammals to have different mechanisms of population dynamics based on first principles. They argued that large herbivorous mammals (intrinsic rate of increase <0.45 per year, body mass >30 kg) ought to have extrinsic controls while small herbivorous mammals would be expected to show intrinsic controls. Given that extrinsic control is typically identified with predators, diseases, and food shortage, these ought to be the critical factors that managers dealing with large mammals need to consider. For small mammals intrinsic, control is typically identified with territoriality, infanticide, physiological stress, and other social interactions, and small mammal ecologists should thus deal with a completely different set of mechanistic forces in trying to understand population changes. I will refer to this set of large- and small-mammal generalizations as Graeme Caughley's conjecture because it satisfied his lifelong quest for mechanistic generality in population dynamics. Have these generalizations turned out to be correct in the light of further research?

Several different challenges have been made to the Caughley and Krebs (1983) generalizations. Within the large herbivorous mammal group of ecologists, individuals have fallen into two clusters, identified as those arguing for top-down control via predation versus those arguing for bottom-up control via food supplies. Sinclair and Krebs (2002) proposed that primary control in all herbivorous vertebrate populations was bottom-up from food supplies, and that this control could be secondarily modified by top-down natural enemy effects, social interactions, or stochastic weather events. As we saw in chapter 9, Korpimäki et al. (2002) argued in a series of papers that top-down control was also the norm for small-mammal populations, and implicitly argued against the dichotomy proposed by Caughley and Krebs (1983). The evidence presented in the previous 15 chapters of this book argues against a simple dichotomy and favors multiple-factor explanations for population control in small rodents.

Many factors interact to affect the change in numbers of any population, and the job of the population ecologist could be (1) to catalog all the factors affecting changes in population size, or (2) to determine the major factors driving changes in numbers. I have adopted the second approach, with the assumption that it will provide the most useful information for natural resource management and conservation. I have described this approach elsewhere as the mechanistic paradigm (Krebs 2002), and I wish to identify the mechanistic factors that affect population growth rates. My concern here is not with the density-dependent paradigm, which, I have argued, is less useful for understanding population dynamics (Krebs 2002). Much analysis has been made of population time series to determine whether first-order (density-dependent) dynamics occur, and if, in addition, second-order dynamics (delayed density-dependence) are present (Stenseth 1999). This approach is a useful first step in analyzing population changes, but it does not arrive at a mechanistic explanation of what causes population change. It is the understanding of mechanistic factors that will illuminate natural resource management and conservation problems (Caughley 1994).

Evolution of Intrinsic Population Regulation

Population control by extrinsic factors like predation and food shortage is easy to understand in evolutionary terms, since it makes the best of a bad situation. But under what conditions should we expect a population to be regulated by intrinsic processes involving spacing behavior? We might expect that mammals, with their relatively complex behavior, would be the species that might most obviously show intrinsic regulation. Wolff (1997)

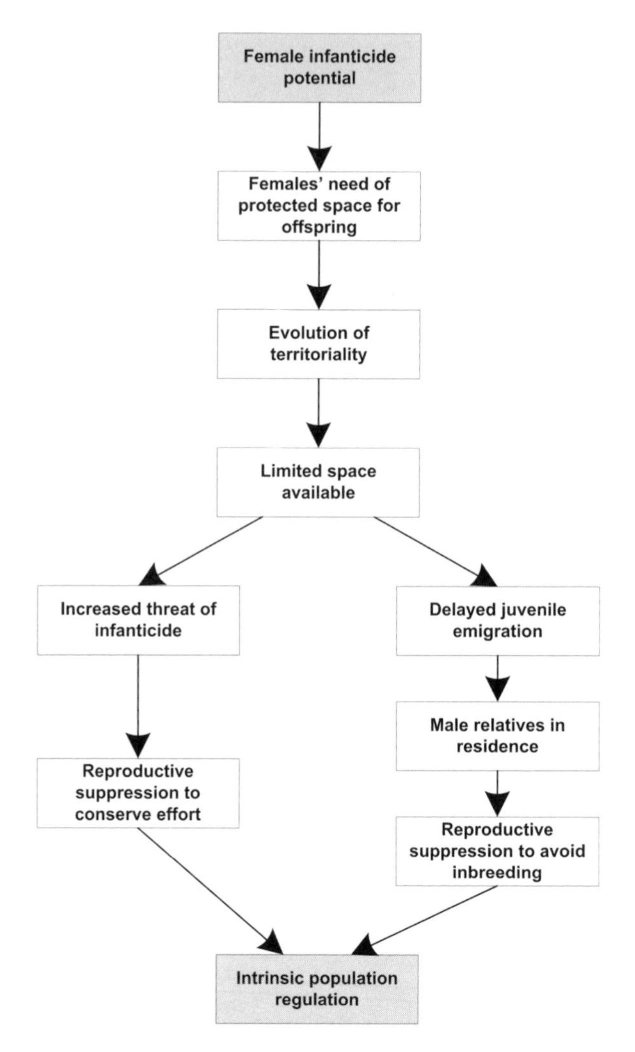

FIGURE 16.1 Wolff's hypothesis for the evolution of intrinsic population regulation in mammals. Spacing behavior is the key mechanism to evolve in species competing for space away from infanticidal individuals. Species in which infanticide does not occur should never show intrinsic regulation. From Wolff 1997.

has suggested a conceptual model that predicts which mammals have the potential for intrinsic regulation. Wolff's model (figure 16.1) suggests an evolutionary path to intrinsic regulation. The key assumption in the model is that territoriality in female mammals has evolved as a counterstrategy to infanticide committed by strange females. Infanticide is a mechanism of competition by which intruders usurp the breeding space of residents and increase their fitness by killing the offspring of resident females.

Female mammals should evolve territorial behavior to defend their young from infanticide only if the young are not mobile at birth. Females with precocial young, which have their eyes open and can move very soon after birth, will not be susceptible to infanticide and will not defend territories. These predictions from Wolff's model are consistent with most of what is known about mammalian social systems (Wolff and Sherman 2007). For example, hares have precocial young while rabbits have altricial young. Infanticide is unknown in hares but is known to occur in rabbits. Many carnivores (for example, lions) have altricial young, are subject to infanticide, and are territorial. By contrast, kangaroos have altricial young but carry them around in a pouch so that they are not vulnerable to infanticide. None of the kangaroo species are territorial.

Another feature of self-regulation in mammals is reproductive suppression of juveniles (Wolff 1997). If juveniles do not disperse from their natal area, they risk the possibility of breeding with close relatives. Selection against inbreeding has molded the dispersal pattern of mammals such that male juveniles will emigrate while female offspring remain near their natal sites (Clobert et al. 2001; Lambin et al. 2001). But high density may make dispersal costly due to aggressive encounters such that all juveniles stay near the birthplace. At high density, adults may suppress sexual maturation of their offspring through pheromones in order to prevent inbreeding, especially if space for breeding is limited. The result can be that a large fraction of the population does not breed, as has been observed in many rodents, primates, and wolves. This reproductive suppression of juveniles at high density acts as a density-dependent factor to potentially regulate the population.

Most large mammal species are not subject to potential infanticide, and these species would be expected to be subject to extrinsic regulation by predators, food shortage, disease, or weather. Wolff (1997) pointed out that intrinsic regulation is not in itself an evolved strategy. What evolves are behavioral strategies such as territoriality, dispersal, and reproductive inhibition, and these individual strategies can result in population regulation at the level of the population. Wolff has pointed out that these behavioral strategies may influence population changes in small rodents, but may also be ineffective in many populations. Wolff's model has effectively replaced the Caughley and Krebs (1983) model, and it forms the basis for our current understanding of population regulation in mammals.

Large Mammal Population Dynamics

Caughley (1981) argued that for large herbivores we should first thoroughly study the large mammal-food supply nexus before we try to understand how predators and competitors affect at balance. Sinclair and Krebs (2002) agreed with this approach. They suggested that food supply is the primary factor determining population growth rate in all animal populations, and thus postulated bottom-up control as the universal primary standard for mammals. They pointed out, however, that bottom-up control by food shortage can be overridden or severely modified by three secondary processes: top-down processes from predators, social interactions within the species, and stochastic disturbances. Interactions between these four controls produce the variety of complex, nonlinear effects on population growth that we see in nature.

Sinclair and Krebs (2002) suggested some generalizations or working hypotheses for classifying animal populations with regard to regulatory processes as follows. Bird populations are driven by primary food limitations coupled with social interactions over territories. Food supply drives changes in large-mammal populations, and top-down processes rarely intervene. Small mammals may be affected more by top-down controls in combination with social interactions, and their population growth rarely seems limited by food. The population growth of fish and many invertebrates, by contrast, seems affected more by stochastic disturbances affecting recruitment processes through primary food limitation.

To evaluate the Caughley (1981) and Sinclair and Krebs (2002) conjectures in the light of recent studies, I begin with the dynamics of large herbivorous mammals, many of which have been strongly affected by human activities (figure 16.2). I divide all large mammal populations into two classes: (1) large herbivorous mammals with no predators (e.g., elephants [*Loxodonta Africana*], and white-tailed deer [*Odocoileus virginianus*] in eastern North America), and (2) large mammals with effective natural predators (e.g., red kangaroos [*Macropus rufus*] in central Australia, woodland caribou [*Rangifer tarandus caribou*] in Canada, and zebra [*Equus burchellii*] in Africa). Inevitably there are populations that fall between these two groups, such as eastern gray kangaroos (*Macropus giganteus*) in eastern New South Wales, where most but not all of the predators have been suppressed by poisoning (Banks et al. 2000; Fletcher 2006), so that they operate like the first group, large mammals with no predators.

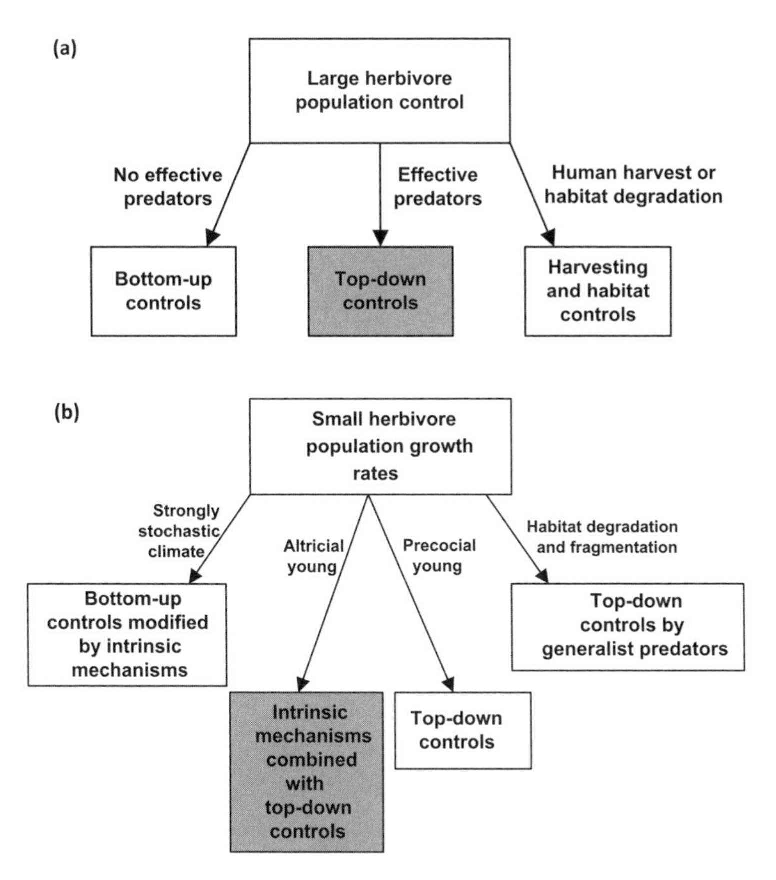

FIGURE 16.2 A global model of population control in large and small herbivorous mammals. Typical pathways are shown in grey shading. For large mammals (a), the key questions are whether numbers are driven by human harvest and whether effective predation is impossible. For small mammals (b), the three key variables are whether the species has altricial or precocial young, the strength of climatic variability, and the severity of habitat disturbances associated with agriculture and land management. Harvesting controls and habitat degradation can be viewed as top-down controls, with humans as the apex predator. From Krebs 2009b.

Clearly the first group of large herbivorous mammals with no predators is the simplest to analyze. Caughley et al.'s (1987) study of kangaroos at Kinchega in New South Wales serves as a paradigm for this approach, since those kangaroo populations existed in the virtual absence of dingo predation. If predators are absent or ineffective, population changes must be caused by change in food resources (in the absence of disease or climatic limitation). The general message for these kinds of large mammal populations is to analyze carefully plant-herbivore dynamics as the key to understanding population changes, a point made effectively by Caughley (1976) and shown well by the Soay sheep (*Ovis aries*) study on St. Kilda (Crawley et al. 2004). Johnson (2010) pointed out the massive changes in plant communities that followed the extinction of the megafauna, large herbivores that went extinct around the world during the last 50,000 years. These large herbivores were nearly immune to predation (but see Rule et al. 2012) and had a large effect on the vegetation communities in which they lived (Zimov et al. 1995). One caveat to these conclusions about large herbivores and food limitation is that we do not have extensive data on juvenile survival and what factors cause juvenile losses.

A special case exists for the existing large mammals in which human hunting or poaching is the main cause of population change. Conservation for these species becomes a matter of setting proper harvesting quotas along with effectively enforcing the law, often easier said than done. The black rhinoceros (*Diceros bicornis*) in east Africa is a good example, showing the complications introduced when poaching drives the dynamics (Metzger et al. 2007). The hunting of elephants for ivory is another example (Caughley et al. 1990;, Milner-Gulland and Beddington 1993).

When natural predator communities are more or less intact, it is tempting to reverse the onus of proof and begin with the assumption of top-down control by predators. A good example is the woodland caribou in Ontario. Woodland caribou have been declining in southern Canada for more than 30 years, presumably as a result of predation by wolves, bears, and lynx. Bergerud et al. (2007) obtained the data shown in figure 16.3 on two populations of woodland caribou: one isolated from predators on offshore islands in Lake Superior, and one with predators on mainland sites that continues to slowly decline. No studies were carried out on the food supplies of these populations, but the presumption is that the island population is food-limited. From a management perspective, however, predation is the key process threatening mainland populations, and the conservation question concerns what can be done to reduce predator kills of caribou.

White-tailed deer populations in eastern North America are an example of what was formerly top-down control of ungulate numbers. A trophic cascade has resulted from the killing of wolves and bears in the highly settled areas of eastern United States. Trophic cascades are a form

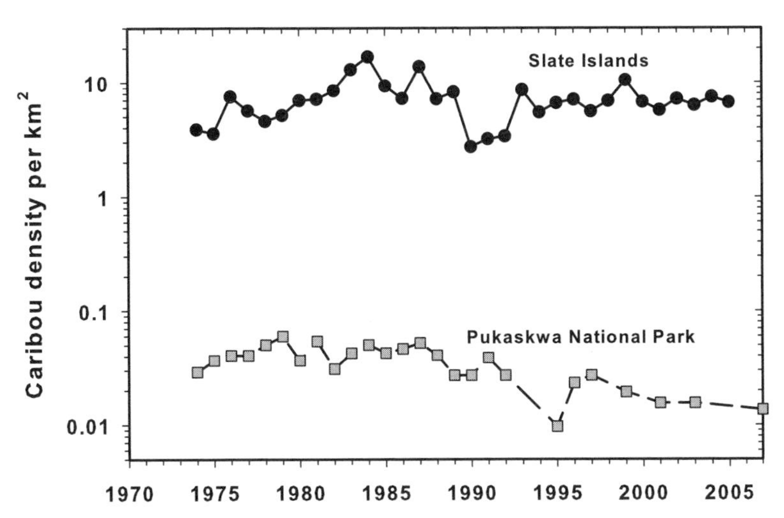

FIGURE 16.3 Woodland caribou densities on the Slate Islands of Lake Superior and on the adjacent mainland in Pukaskwa National Park, Ontario, from 1974 to 2007. Island populations free of predators exist at 100 times the density of mainland populations that coexist with predators. Predators are driving the mainland populations of woodland caribou toward extinction. From 1987 to 2007 the mainland herd showed a significant decline of 6% per year. Data from Bergerud et al. 2007.

of strong top-down control in which impacts flow down the food chain as a series of positive and negative effects on successive trophic levels in the community. White-tailed deer are now overabundant, and the resulting highway collisions and damage to regenerating forests (Mladenoff and Stearns 1993;, Stromayer and Warren 1997) are an ongoing conservation dilemma. A trophic cascade is not a necessary consequence in a system with top-down control; it will occur only when the herbivores involved are dominant species in a community. Rare species may well be controlled from the top down yet not be abundant enough to affect plant resources.

Fryxell et al. (1988) pointed out that large-mammal migrations were one strategy that reduced predator effectiveness and permitted some migratory species to escape predator limitation and thus be controlled from the bottom up by food supplies. The wildebeest of the Serengeti are the classic example of this strategy, which is a version of a spatial refuge for the prey. Group formation among large mammals also helps to reduce the effectiveness of predators (Fryxell et al. 2007). Migration typically evolves to track seasonal food supplies, and one secondary consequence is that predators may become less effective. Migration and group formation, however, will not be effective as predator-escape strategies if the predators themselves are migratory, as is the case with wolves and barren ground caribou in North America.

Large herbivore mammals are not greatly affected directly by weather, with a few exceptions. Peary caribou in the Canadian Arctic islands are periodically reduced by autumn episodes of freezing rain that render their food plants unavailable and cause starvation (Caughley and Gunn 1993). Weather may also be a driving variable of population changes in reindeer on Svalbard (Solberg et al. 2001). Soay sheep winter mortality is strongly affected by wet, windy weather interacting with per-capita food supplies (Hone and Clutton-Brock 2007). More typically, weather affects vegetation growth and affects herbivores via food shortage, a classic response demonstrated by Caughley et al. (1987) for kangaroos. Direct weather effects on large mammals operate as an interaction with bottom-up controls via food supplies, and climate change effects will operate via changes in plant productivity.

Some tension in the literature about both large and small herbivore population dynamics could be avoided if the objectives of each study were carefully outlined. The dichotomy in all mechanistic population studies is between finding out which variables are the main causes of population change (so that management activities can be assigned accordingly) and, on the other hand, understanding how the nexus of food supplies, predation, disease, weather, social interactions, and habitat structure operates to control change in numbers. These different objectives demand different experimental designs. The key is to specify exactly the objectives of each population study so that an end point for understanding can be specified. There is no doubt that factor interactions need to be analyzed—for example, if poor food supplies permit increased parasite loads that contribute to higher predation rates. Management agencies typically have only a few major variables they can manipulate within their budgets to change population trends, and knowing the details of how factors interact may not be necessary for effective management action.

For large mammals, I propose to revise the conjecture of Sinclair and Krebs (2002) that for mammalian herbivores the primary control is bottom-up, with a more complex tripartite separation for those populations with and without natural predator assemblages, and for those completely subject to human control via harvesting and habitat changes (figure 16.2).

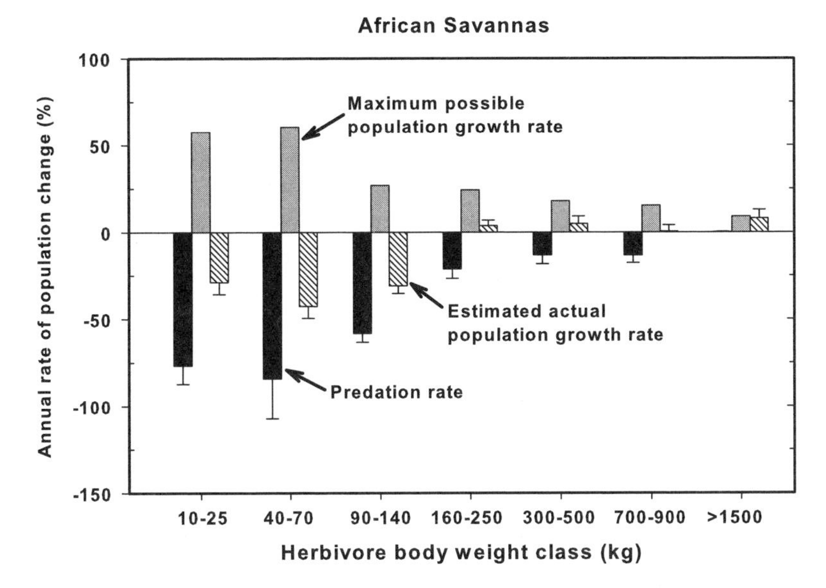

FIGURE 16.4 Estimated average rates of population growth in mammals of four African savanna ecosystems, and their observed predation rates. Herbivores are grouped by body size. The maximum possible rate of increase was estimated from the allometric equation in Sinclair (1996). Predation rates are averages based on field studies. The actual population growth rate is the difference between the maximum rate and the predation rate. Species with body mass below about 150 kg are more likely to be limited by predation in these ecosystems. Means ± S.D. After Fritz et al. 2011, figure 2.

I suggest that primary control for large mammalian herbivores in intact ecosystems is top-down, with secondary modifications produced by food supplies, disease, and weather. Top-down control in large mammals can be size-related; Fritz et al. (2011) argued that in African savannahs herbivores smaller than 150 kg were controlled by predation, while those above this size limit were food-limited (figure 16.4). The Serengeti wildebeest (*Connochaetes taurinus*, body mass 170–270 kg) shows another way to escape predation limitation, because its migration strategy sheds potential predator controls on abundance (Fryxell et al. 2007). By contrast, small resident populations of wildebeest in Kruger National Park appear to be regulated by predation (Hopcraft et al. 2010).

The interactions of climate, soils, and plant production in African savannas has provided a complex picture of how top-down versus bottom-up regulation can be affected by physical factors (Hopcraft et al. 2010). Figure 16.5 qualitatively illustrates the synthesis of how these factors operate

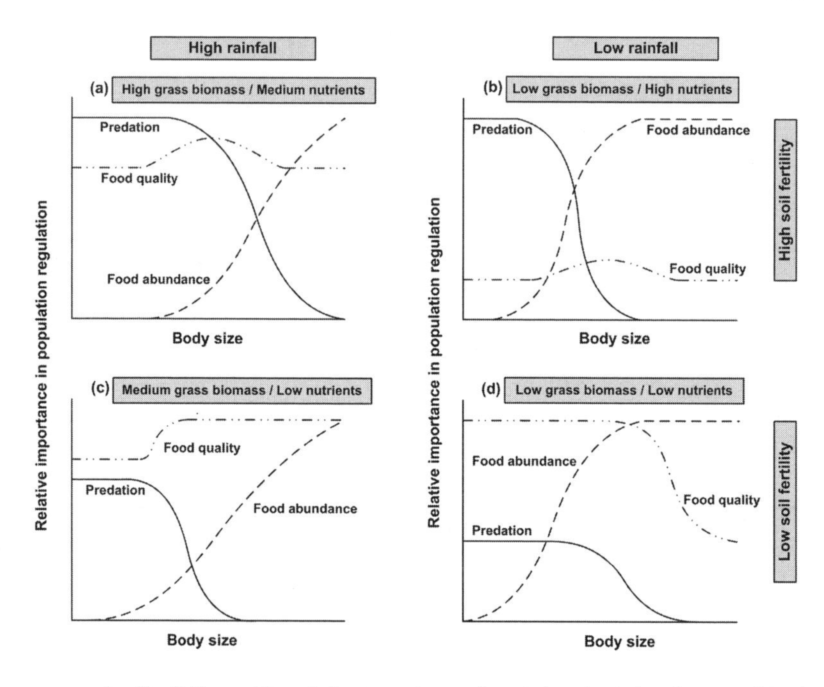

FIGURE 16.5 Predictions of the relative importance of predation, food abundance, and food quality in limiting herbivore populations of increasing body size across rainfall and soil fertility gradients in African savannas. (a) High rainfall and high soil nutrients produce high grass biomass with plants of medium nutrient levels. In these cases, large herbivores like elephants are limited by food abundance, while small herbivores are limited by predation. (b) If rainfall is low in areas of good soil, the predation limitation zone shrinks and food abundance is most limiting. (c) In areas of low soil fertility, predation becomes less important and food quality and food abundance are limiting; this effect is even stronger in areas of low rainfall and poor soils (d). After Hopcraft et al. 2010.

in the savannas of Africa. Areas of high rainfall and good soils support more top-down regulation (figure 16.5a), while similar areas with low rainfall (figure 16.5b) support much less primary production and bottom-up limitation tends to dominate. Climate change may shift ecosystems along this gradient, with changing consequences for population regulation.

The results of the syntheses given in Fritz et al. (2011) and Hopcraft et al. (2010) also suggest that the massive decline of very large herbivores that occurred at the end of the Pleistocene (Brook and Bowman 2004), as well as the persecution of large predators by humans (Brasheres et al. 2004), may have drastic effects on the dynamics and functioning of terrestrial ecosystems, similar to those of the simpler predator-prey system in North America (Ripple and Beschta 2006). Thus, Fritz et al. (2011) suggest that temperate terrestrial ecosystems that maintained a large diversity of predators before their extirpation function very differently today than they did in the past (Johnson et al. 2007; Letnic et al. 2011). Further study of these ecosystems as climate change advances would help to test the predictions illustrated in figure 16.5.

Almost no large herbivore studies consider the potential role of intrinsic processes in population dynamics. The assumption is that these social processes are not important for populations of large herbivores, as predicted by Wolff (1997).

Small Mammal Population Dynamics

As we have seen in the previous 15 chapters, small-mammal population changes illustrate all too well the potential confusion of correlation and causation in population dynamics. Many factors change as numbers rise and fall. For example, a population regulated from the top down by predators that show a time lag in their numerical response will of necessity show a correlation of population changes with food supply per capita. Observing such a correlation is not proof of bottom-up control. Factor interactions make these kinds of observations even more difficult when, for example, parasite infections change the individual susceptibility of prey to predation or extreme weather events. The solution, which has been known for some time, is to adopt an experimental approach.

If social factors such as territoriality and infanticide are necessary components causing small mammal population fluctuations, and predation is also an essential component, as I have argued in chapter 14, this generalization will not automatically translate into methods for the control of pest rodents in agriculture, where they are typically injected into an environment in which food is periodically produced in great excess and then taken away, and in which predators are typically greatly reduced by humans. House mice in grain crops of southeastern Australia are a good example of this (Brown et al. 2010), as are ricefield rats in Southeast Asia (Singleton et al. 2005). The role of social behavior in population control of agricultural pests cannot be ignored, but the message is clear that social factors alone cannot prevent destructive outbreaks (Eccard et al. 2011).

Social interactions among house mice in grain fields have important effects on population growth (Sutherland and Singleton 2006), and knowledge of these interactions can assist a program of fertility control. Similarly, rats in Asian rice fields illustrate too well the failure of predation and social behavior to restrict densities, but damage can be reduced by cultural controls and crop management (Singleton et al. 2005). Knowing the limiting factors can help to avoid the setting up of pest management programs that cannot succeed. The classic examples of unsuccessful pest rodent control programs in tropical countries are those that encourage predators like cats and barn owls in situations where their offtake is minimal in relation to the rodents' rates of increase in crops (Singleton et al. 2003). Pest problems in agriculture are a graphic illustration of how the removal of predators and the provisioning of high-quality food can generate destructive population outbreaks that are not prevented by intrinsic social controls of territoriality, infanticide, and emigration.

For small herbivorous mammals I have suggested a multiple-factor explanation of population changes, based on direct and delayed density-dependent control from social behavior and predation. Following Wolff (1997), I suggest that population change in species with precocial young will be subject usually to top-down controls, depending on the ecosystem, and will in effect operate like population change in large mammals (figure 16.2). There are two important exceptions for small mammals with altricial young. In severe environments such as deserts (Dickman et al. 1999; Letnic and Dickman 2010), bottom-up control is enforced by pulsed climatic events driving food production so that social interactions, although present, are not the main cause of population changes. Desert rodents increase in number after rain because of improved food supplies, and their numbers collapse in subsequent drought, partly because of predation and partly due to food shortage (Letnic and Dickman 2010). If their habitats are sufficiently broken up by agriculture or forestry operations, generalized predators, supported by human activities, can become major drivers of their population dynamics as suitable habitat patches become minor parts of the landscape. Many results of Scandinavian studies of small rodents must be viewed in the light of these landscape effects.

A common problem in discussions of the factors responsible for population change in small rodents is that some authors completely ignore the possible impact of social processes on events. The implicit assumption is often made that top-down or bottom-up factors like food resources, predation, or disease yield a complete explanation of population change. This error is further propagated by adopting the exclusion method of analysis: to say, for example, that only predation mortality or food shortage can control population changes, and if my results show that there is no food shortage, then this population change must have been caused by predation, even if predation was not explicitly studied. It is important to recognize the limitations we are working under for all populations that we study. If you wish to assume that social factors are not a relevant factor for a particular population, that wish ought to be stated and justified, if only to make it clear that you had insufficient time or money to study additional factors. Similarly, if you wish to assume that predation is not a relevant factor for your population, that should be stated and justified. This kind of transparency would prevent some arguments in the literature that seem to be at cross-purposes. It is too rare that authors list the deficiencies of their own studies in their published papers, even while they gladly admit them in conversation.

Conclusion

A fundamental problem in analyzing the literature on population regulation is that only rarely are hypotheses stated clearly with predictions, and with the limitations of the studies cleanly pointed out. The fact that one particular factor affects population size or population growth rate is rarely of much utility unless it is part of a general attempt to be comprehensive. The typical approach of locating the density-dependent factors in population dynamics has been rather less useful than the simple models in ecology textbooks would suggest, and so we have moved on to look directly at the mechanisms behind change: predators, parasites, weather, food and social interactions.

The wealth of research into the population dynamics of both large and small mammals has complicated the search for simple generalizations. Evolutionary history and changing climate are too rarely brought into discussions of population limitation and regulation. As we gain more understanding of how intact ecosystems operate to determine the size of mammal populations, we recognize more and more that human disruptions of food webs have consequences we cannot anticipate without further study. The Serengeti ecosystem of east Africa is possibly the most intensively studied terrestrial ecosystem to date, yet we lack much understanding of the details of large-mammal, small-mammal, and bird population dynamics for the purposes of conservation (Estes et al. 2011).

It is a major triumph of population ecologists to have reached a general understanding of both large- and small-mammal population changes, as I have tried to summarize briefly in this final chapter. There is more work to be done in population dynamics to consolidate and extend this understanding. But the biggest challenge now is to integrate the information we have on population dynamics into community- and ecosystem-level studies so that we can understand how communities operate, and which connections within them are key to their structures and functions. This information is critical for the conservation of biodiversity, a major goal of ecological research in this century.

Bibliography

Abe, H. 1976. "Age determination of *Clethrionomys rufocanus bedfordiae*." *Japanese Journal of Ecology* 26:221–227.

Aber, J. D. 1997. "Why don't we believe the models?" *Bulletin of the Ecological Society of America* 78 (3): 232–233.

———. 1998. "Mostly a misunderstanding, I believe." *Bulletin of the Ecological Society of America* 79 (4): 256–257.

Abrams, P. A. 1993. "Effect of increased productivity on the abundances of trophic levels." *American Naturalist* 141:351–371.

Abramsky, Z., and C. R. Tracy. 1979. "Population biology of a "noncycling" population of prairie voles and a hypothesis on the role of migration in regulating microtine cycles." *Ecology* 60 (2): 349–361.

Adams, C. E. 1960. "Studies on prenatal mortality in the rabbit, *Oryctolagus cuniculus*: The amount and distribution of loss before and after implantation." *Journal of Endocrinology* 19:325–344.

Agrell, J., J. O. Wolff, and H. Ylonen. 1998. "Counter-strategies to infanticide in mammals: costs and consequences." *Oikos* 83 (3): 507–517.

Allen, P., F. W. R. Brambell, and I. H. Mills. 1947. "Studies on sterility and prenatal mortality in wild rabbits. I. The reliability of estimates of prenatal mortality based on counts of corpora lutea, implantation sites and embryos." *Journal of Experimental Biology* 23:312–331.

Anderson, D. R. 2003. "Index values rarely constitute reliable information." *Wildlife Society Bulletin* 31 (1): 288–291.

Anderson, D. R., K. P. Burnham, and W. L. Thompson. 2000. "Null hypothesis testing: Problems, prevalence, and an alternative." *Journal of Wildlife Management* 64 (4): 912–923.

Anderson, J. L. 1975. "Phenotypic correlates among relatives, and variability in reproductive performance in populations of the vole, *Microtus townsendii*." PhD thesis, University of British Columbia.

Anderson, J. L., and R. Boonstra. 1979. "Some aspects of reproduction in the vole *Microtus townsendii*." *Canadian Journal of Zoology* 57 (1): 18–24.

Anderson, R. M., and R. M. May. 1979. "Population biology of infectious diseases: Part I." *Nature* 280:361–367.

Andersson, M., and S. Jonasson. 1986. "Rodent cycles in relation to food resources on an alpine heath." *Oikos* 46 (1): 93–106.

Andrewartha, H. G., and L. C. Birch. 1954. *The Distribution and Abundance of Animals.* Chicago: University of Chicago Press.

Angerbjörn, A., M. Tannerfeldt, and H. Lundberg. 2001. "Geographical and temporal patterns of lemming population dynamics in Fennoscandia." *Ecography* 24 (3): 298–308.

Aunapuu, M., J. Dahlgren, T. Oksanen, D. Grellmann, L. Oksanen, J. Olofsson, Ü. Rammul, M. Schneider, B. Johansen, and H. O. Hygen. 2008. "Spatial patterns and dynamic responses of arctic food webs corroborate the Exploitation Ecosystems Hypothesis (EEH)." *American Naturalist* 171 (2): 249–262. doi:10.1086/524951.

Banks, P. B., A. E. Newsome, and C. Dickman. 2000. "Predation by red foxes limits recruitment in populations of eastern grey kangaroos." *Austral Ecology* 25 (3): 283–291.

Batzli, G. O. 1983. "Responses of arctic rodent populations to nutritional factors." *Oikos* 40:396–406.

Batzli, G. O. 1985. "Nutrition." In *Biology of New World Microtus*, edited by R. H. Tamarin, 779–811. Shippensburg, PA: American Society of Mammalogists.

———. 1992. "Dynamics of small mammal populations: A review." In *Wildlife 2001: Populations*, edited by D. R. McCullough and R. H. Barrett, 831–850. New York: Elsevier Applied Science.

———. 1996. "Population cycles revisited." *Trends in Ecology and Evolution* 11 (12): 488–489.

Batzli, G. O., R. G. White, S. F. Maclean, F. A. Pitelka, and B. D. Collier. 1980. "The herbivore-based trophic system." In *An Arctic Ecosystem*, edited by J. Brown, P. C. Miller, L. L. Tiezen and F. L. Bunnell, 335–410. Stroudsburg, PA: Dowden, Hutchinson, and Ross.

Beacham, T. D. 1980a. "Dispersal during population fluctuations of the vole, *Microtus townsendii*." *Journal of Animal Ecology* 49:867–877.

———. 1980b. "Survival of cohorts in a fluctuating population of the vole *Microtus townsendii*." *Journal of Zoology* 191 (1): 49–60. doi: 10.1111/j.1469-7998.1980.tb01448.x.

Beer, J. R., C. F. MacLeod, and L. D. Frenzel. 1957. "Prenatal survival and loss in some cricetid rodents." *Journal of Mammalogy* 38 (3): 392–402.

Beldomenico, Pablo M., Sandra Telfer, Stephanie Gebert, Lukasz Lukomski, Malcolm Bennett, and Michael Begon. 2008. "Poor condition and infection: a vicious circle in natural populations." *Proceedings of the Royal Society of London, B: Biological Sciences* 275 (1644): 1753–1759. doi: 10.1098/rspb.2008.0147.

Bergeron, J. M., and L. Jodoin. 1987. "Defining "high quality" food resources of herbivores: the case for meadow voles (*Microtus pennsylvanicus*)." *Oecologia* 71 (4): 510–517.

———. 1993. "Intense grazing by voles (*Microtus pennsylvanicus*) and its effect on habitat quality." *Canadian Journal of Zoology* 71:1823–1830.

Bergerud, A. T., W. J. Dalton, H. Butler, L. Camps, and R. Ferguson. 2007. "Woodland caribou persistence and extirpation in relic populations on Lake Superior." *Rangifer* 17 (special issue): 57–78.

Bernard, N., D. Michelat, F. Raoul, J.-P. Quere, P. Delattre, and P. Giraudoux. 2010. "Dietary response of barn owls (*Tyto alba*) to large variations in populations of common voles (*Microtus arvalis*) and European water voles (*Arvicola terrestris*)." *Canadian Journal of Zoology* 88 (4): 416–426.

Berryman, A. 2002. *Population Cycles: The Case for Trophic Interactions*. New York: Oxford University Press.

Berteaux, D., M. M. Humphries, C. J. Krebs, M. Lima, A. G. McAdam, N. Pettorelli, D. Reale, T. Saitoh, E. Tkadlec, R. B. Weladji, and N. C. Stenseth. 2006. "Constraints to projecting the effects of climate change on mammals." *Climate Research* 32:151–158.

Berteaux, D., D. Réale, A. G. McAdam, and S. Boutin. 2004. "Keeping pace with fast climate change: Can arctic life count on evolution?" *Integrative and Comparative Biology* 44:140–151.

Bierman, S. M., J. P. Fairbairn, S. J. Petty, D. A. Elston, D. Tidhar, and X. Lambin. 2006. "Changes over time in the spatiotemporal dynamics of cyclic populations of field voles (*Microtus agrestis* L.)." *American Naturalist* 167 (4): 583–590.

Birney, E. C., W. E. Grant, and D. D. Baird. 1976. "Importance of vegetative cover to cycles of Microtus populations." *Ecology* 57:1043–1051.

Bjørnstad, O. N., S. Champely, N. C. Stenseth, and T. Saitoh. 1996. "Cyclicity and stability of grey-sided voles, *Clethrionomys rufocanus*, of Hokkaido: Spectral and principal components analysis." *Philosophical Transactions of the Royal Society of London, Series B* 351:867–875.

Bjørnstad, O. N., W. Falck, and N. C. Stenseth. 1995. "A geographic gradient in small rodent density fluctuations: A statistical modelling approach." *Proceedings of the Royal Society of London, Series B* 262:127–133.

Bjørnstad, O. N., and B. T. Grenfell. 2001. "Noisy clockwork: Time series analysis of population fluctuations in animals." *Science* 293 (27 July 2001): 638–643.

Bjørnstad, O. N., N. C. Stenseth, and T. Saitoh. 1999. "Synchrony and scaling in dynamics of voles and mice in northern Japan." *Ecology* 80 (2): 622–637.

Bjørnstad, O. N., N. C. Stenseth, T. Saitoh, and O. C. Lingjaerde. 1998. "Mapping the regional transition to cyclicity in *Clethrionomys rufocanus*: Spectral densities and functional data analysis." *Researches on Population Ecology* 40 (1): 77–84.

Blachford, A. M. 2011. *Five Studies in Life History Evolution*. Department of Zoology, University of British Columbia, Vancouver, BC.

Blank, B. F., J. Jacob, A. Petri, and A. Esther. 2011. "Topography and soil properties contribute to regional outbreak risk variability of common voles (*Microtus arvalis*)." *Wildlife Research*. doi: 10.1071/WR10192.

Bondrup-Nielsen, S., and R. A. Ims. 1986. "Comparison of maturation of female *Clethrionomys glareolus* from cyclic and noncyclic populations." *Canadian Journal of Zoology* 64:2099–2102.

Boonstra, R. 1985. "Demography of *Microtus pennsylvanicus* in Southern Ontario: enumeration versus Jolly-Seber estimation compared." *Can J Zool* 63 (5): 1174–1180.

———. 2004. "Coping with changing northern environments: The role of the stress axis in birds and mammals." *Integrative and Comparative Biology* 44 (2): 95–108).

Boonstra, R., and P. T. Boag. 1987. "A test of the Chitty Hypothesis: Inheritance of life-history traits in meadow voles *Microtus pennsylvanicus*." *Evolution* 41 (5): 929–947.

Boonstra, R., D. Hik, G. R. Singleton, and A. Tinnikov. 1998. "The impact of predator-induced stress on the snowshoe hare cycle." *Ecological Monographs* 68 (3): 371–394.

Boonstra, R., and I. Hogg. 1988. "Friends and strangers: A test of the Charnov-Finerty hypothesis." *Oecologia* 77:95–100.

Boonstra, R., and C. J. Krebs. 1977. "A fencing experiment on a high-density population of *Microtus townsendii*." *Canadian Journal of Zoology* 55:1166–1175.

———. 2012. "Population dynamics of red-backed voles (*Myodes*) in North America." *Oecologia* 168 (3): 601–620. doi:10.1007/s00442–011–2120-z.

Boonstra, R., C. J. Krebs, and T. D. Beacham. 1980. "Impact of botfly parasitism on *Microtus townsendii* populations." *Canadian Journal of Zoology* 58 (9): 1683–1692.

Boonstra, R., C. J. Krebs, and N. C. Stenseth. 1998. "Population cycles in mammals: the problem of explaining the low phase." *Ecology* 79:1479–1488.

Boonstra, R., and F. H. Rodd. 1983. "Regulation of breeding density in *Microtus pennsylvanicus*." *Journal of Animal Ecology* 52 (3): 757–780.

Boratynski, Z., E. Koskela, T. Mappes, and T. A. Oksanen. 2010. "Sex-specific selection on energy metabolism: Selection coefficients for winter survival." *Journal of Evolutionary Biology* 23 (9):1969–1978. doi: 10.1111/j.1420–9101.2010.02059.x.

Borchers, D. L., and M. G. Efford. 2008. "Spatially explicit maximum likelihood methods for capture-recapture studies." *Biometrics* 64 (2): 377–385.

Boutin, S. 1990. "Food supplementation experiments with terrestrial vertebrates: Patterns, problems, and the future." *Canadian Journal of Zoology* 68:203–220.

Boyce, C. C. K., and J. L. Boyce III. 1988a. "Population biology of *Microtus arvalis*. I. Lifetime reproductive success of solitary and grouped breeding females." *Journal of Animal Ecology* 57:711–722.

Boyce, C. C. K., and J. L. Boyce III. 1988b. "Population biology of *Microtus arvalis*. III. Regulation of numbers and breeding dispersion of females." *Journal of Animal Ecology* 57:737–754.

Brambell, F. W. R., and I. H. Mills. 1948. "Studies on sterility and prenatal mortality in wild rabbits. IV. The loss of embryos after implantation." *Journal of Experimental Biology* 25:241–269.

Brandt, M. J., and X. Lambin. 2007. "Movement patterns of a specialist predator, the weasel *Mustela nivalis* exploiting asynchronous cyclic field vole *Microtus agrestis* populations." *Acta Theriologica* 52 (1): 13–25.

Brashares, J. S., P. Arcese, M. K. Sam, P. B. Coppolillo, A. R. E. Sinclair, and A. Balmford. 2004. "Bushmeat hunting, wildlife declines, and fish supply in West Africa." *Science* 306 (5699): 1180–1183.

Brommer, J. E., H. Pietiainen, K. Ahola, P. Karell, T. Karstinen, and H. Kolunen. 2010. "The return of the vole cycle in southern Finland refutes the generality of the loss of cycles through 'climatic forcing.'" *Global Change Biology* 16 (2): 577–586. doi: 10.1111/j.1365-2486.2009.02012.x.

Brook, B. W., and D. M. J. S. Bowman. 2004. "The uncertain blitzkrieg of Pleistocene megafauna." *Journal of Biogeography* 31 (4): 517–523.

Brown, P. R., G. R. Singleton, R. Pech, L. H. Hinds, and C. J. Krebs. 2010. "Rodent outbreaks in Australia: Mouse plagues in cereal crops" In *Rodent Outbreaks: Ecology and Impacts*, edited by G. R. Singleton, S. R. Belmain, P. R. Brown and B. Hardy. Manila: International Rice Research Institute.

Bujalska, G. 1970. "Reproductive stabilizing elements in an island population of *Clethrionomys glareolus* (Schreber 1780)." *Acta Theriologica* 15 (25): 381–412.

———. 1973. "The role of spacing behaviour among females in the regulation of reproduction in the bank vole." *Journal of Reproduction and Fertility* Supplement 19:465–474.

———. 1975. "The effect of supplementary food on some parameters in an island population of *Clethrionomys glareolus*." *Bulletin de l'Academie Polonaise des Sciences, Serie des Sciences Biologiques* 13 (1): 23–28.

Callaghan, T. V., L. O. Björn, Y. Chernov, T. Chapin, T. R. Christensen, B. Huntley, R. A. Ims, M. Johansson, D. Jolly, S. Jonasson, N. Matveyeva, N. Panikov, W. Oechel, and G. R. Shaver. 2004. "Uncertainties and recommendations." *Ambio* 33 (7): 474–479.

Caswell, H. 2001. *Matrix Population Models: Construction, Analysis, and Interpretation* 2nd ed. Sunderland, MA: Sinauer Associates.

Caughley, G. 1976. "Plant-herbivore systems." In *Theoretical Ecology*, edited by R. M. May, 94–113. Philadelphia: Saunders.

———. 1981. "What we do not know about the dynamics of large mammals." In *Dynamics of Large Mammal Populations*, edited by C. W. Fowler and T. D. Smith, 361–372. New York: John Wiley and Sons.

———. 1994. "Directions in conservation biology." *Journal of Animal Ecology* 63:215–244.

Caughley, G., H. Dublin, and I. Parker. 1990. "Projected decline of the African elephant." *Biological Conservation* 54:157–164.

Caughley, G., and A. Gunn. 1993. "Dynamics of large herbivores in deserts: Kangaroos and caribou." *Oikos* 67:47–55.

Caughley, G., and C. J. Krebs. 1983. "Are big mammals simply small mammals writ large?" *Oecologia* 59:7–17.

Caughley, G., N. Shepherd, and J. Short. 1987. *Kangaroos: Their Ecology and Management in the Sheep Rangelands of Australia*. Cambridge: Cambridge University Press.

Cerqueira, D., P. Delattre, B. De Sousa, C. Gabrion, S. Morand, and J. P. Quere. 2007. "Numerical response of a helminth community in the course of a multi-annual abundance cycle of the water vole (*Arvicola terrestris*)." *Parasitology* 134:705–711.

Charbonnel, N., Y. Chaval, K. Berthier, J. Deter, S. Morand, R. Palme, and J.-F. Cosson. 2008. "Stress and demographic decline: a potential effect mediated by impairment of reproduction and immune function in cyclic vole populations." *Physiological and Biochemical Zoology* 81 (1): 63–73. doi: 10.1086/523306.

Charnov, E. L., and J. P. Finerty. 1980. "Vole population cycles: A case for kin-selection?" *Oecologia* 45:1–2.

Charnov, E. L., and W. M. Schaffer. 1973. "Life-history consequences of natural selection: Cole's result revisited." *American Naturalist* 107:791–793.

Chitty, D. 1938. "A laboratory study of pellet formation in the short-eared owl (*Asio flammeus*)." *Proceedings of the Zoological Society of London, Series A*, 108:267–287.

———. 1952. "Mortality among voles (*Microtus agrestis*) at Lake Vyrnwy, Montgomeryshire in 1936–9." *Philosophical Transactions of the Royal Society of London* 236:505–552.

———. 1954. "Tuberculosis among wild voles: with a discussion of other pathological conditions among certain mammals and birds." *Ecology* 35 (2): 227–237.

———. 1957. "Self-regulation of numbers through changes in viability." *Cold Spring Harbor Symposia on Quantitative Biology* 22:277–280.

———. 1960. "Population processes in the vole and their relevance to general theory." *Canadian Journal of Zoology* 38 (1): 99–113.

———. 1967. "The natural selection of self-regulatory behaviour in animal populations." *Proceedings of the Ecological Society of Australia* 2:51–78.

———. 1996. *Do Lemmings Commit Suicide? Beautiful Hypotheses and Ugly Facts*. New York: Oxford University Press.

Chitty, D., and M. Nicholson. 1942. "Canadian Arctic Wild Life Enquiry, 1940–41." *Journal of Animal Ecology* 11 (2): 270–287.

Chitty, H. 1961. "Variations in the weight of the adrenal glands in the field vole, *Microtus agrestis*." *Journal of Endocrinology* 22:387–393.

Christensen, P. , and B. Hörnfeldt. 2003. "Long-term decline of vole populations in northern Sweden: a test of the destructive sampling hypothesis." *Journal of Mammalogy* 84 (4): 1292–1299.

Christian, J. J. 1950. "The adreno-pituitary system and population cycles in mammals." *Journal of Mammalogy* 31:247–259.

———. 1956. "Adrenal and reproductive responses to population size in mice from freely growing populations." *Ecology* 37 (2): 258–273.

———. 1970. "Social subordination, population density, and mammalian evolution." *Science* 168:84–90.

Christian, J. J., and D. E. Davis. 1964. "Endocrines, behavior, and population." *Science* 146:1550–1560.

Clarke, J. R. 1955. "Influence of numbers on reproduction and survival in two experimental vole populations." *Proceedings of the Royal Society of London, Series B* 144:68–85.

———. 1977. "Long and short term changes in gonadal activity of field voles and bank voles." *Oikos* 29:457–467.

Clobert, J., E. Danchin, A. A. Dhondt, and J. D. Nichols. 2001. *Dispersal*. Oxford: Oxford University Press.

Cockburn, A., and W. Z. Lidicker, Jr. 1983. "Microhabitat heterogeneity and population ecology of an herbivorous rodent, *Microtus californicus*." *Oecologia* 59:167–177.

Cole, F. R., and G. O. Batzli. 1979. "The influence of supplemental feeding on a vole population." *Journal of Mammalogy* 59 (4): 809–819.

Cole, L. C. 1954. "The population consequences of life history phenomena." *Quarterly Review of Biology* 29 (2): 103–137.

Coley, P. D., J. P. Bryant, and F. S. Chapin III. 1985. "Resource availability and plant antiherbivore defense." *Science* 230:895–899.

Collett, R. 1895. "*Myodes lemmus*, its habits and migrations in Norway." *Christiana, Videnskabs-Selskabs Forhandlinger 1895* 3:1–62.

Cooch, E., and G.C. White. 2010. *Program MARK: A Gentle Introduction*. 9th ed: http://www.phidot.org/software/mark/docs/book/.

Crawley, M. C., S. D. Albon, D. R. Bazely, and J. M. Milner. 2004. "Vegetation and sheep population dynamics." In *Soay Sheep: Dynamics and Selection in an Island Population*, edited by T. H. Clutton-Brock and J. M. Pemberton, 89–112. Cambridge: Cambridge University Press.

Crowcroft, P. 1966. *Mice All Over*. London: G.T. Foulis.

Dahlgren, J., L. Oksanen, T. Oksanen, J. Olofsson, P. A. Hambäck, and Å. Lindgren. 2009. "Plant defences to no avail? Responses of plants of varying edibility to food web manipulations in a low arctic scrubland." *Evolutionary Ecology Research* 11:1189–1203.

Dalton, C. L. 2000. "Effects of female kin groups on reproduction and demography in the gray-sided vole, *Microtus canicaudus*." *Oikos* 90 (1): 153–159.

Dambacher, J. M., H. W. Li, J. O. Wolff, and P. A. Rossignol. 1999. "Parsimonious interpretation of the impact of vegetation, food, and predation on snowshoe hare." *Oikos* 84 (3): 530–532.

Danielson, B. J., M. L. Johnson, and M. S. Gaines. 1986. "An analysis of a method for comparing residents and colonists in a natural population of *Microtus ochrogaster*." *Journal of Mammalogy* 67:733–736.

Davis, S., M. Begon, L. De Bruyn, V.S. Ageyev, N. L. Klassovskly, S. B. Pole, H. Viljugrein, N.C. Stenseth, and H. Leirs. 2004. "Predictive thresholds for plague in Kazakhstan." *Science* 304 (30 April 2004): 736–738.

Delattre, P., B. De Sousa, E. Fichet-Calvet, J. P. Quere, and P. Giraudoux. 1999. "Vole outbreaks in a landscape context: Evidence from a six year study of *Microtus arvalis*." *Landscape Ecology* 14:401–412.

Delattre, P., N. Morellet, P. Codreanu, S. Miot, J.-P. Quere, F. Sennedot, and J. Baudry. 2009. "Influence of edge effects on common vole population abundance in an agricultural landscape of eastern France." *Acta Theriologica* 54 (1): 51–60.

Desy, E. A., and G. O. Batzli. 1989. "Effects of food availability and predation on prairie vole demography a field experiment." *Ecology* 70 (2): 411–422.

Desy, E. A., G. O. Batzli, and J. Liu. 1990. "Effects of food and predation on behaviour of prairie voles: A field experiment." *Oikos* 58 (2): 159–168.

Deter, J., Y. Chaval, M. Galan, K. Berthier, A. Ribas Salvador, J. C. C. Garcia, S. Morand, J.-F. Cosson, and N. Charbonnel. 2007. "Linking demography and host dispersal to *Trichuris arvicolae* distribution in a cyclic vole species." *International Journal for Parasitology* 37 (7): 813–824. doi: 10.1016/j.ijpara.2007.01.012.

Dhondt, A. A. 1988. "Carrying capacity: A confusing concept." *Acta Oecologica* 9:337–346.

Diamond, J. 1986. "Overview: Laboratory experiments, field experiments, and natural experiments." In *Community Ecology*, edited by J. Diamond and T. J. Case, 3–22. New York: Harper and Row.

Dickman, C., P. S. Mahon, P. Masters, and D. F. Gibson. 1999. "Long-term dynamics of rodent populations in arid Australia: The influence of rainfall." *Wildlife Research* 26:389–403.

Diffendorfer, J. E., M. S. Gaines, and R. D. Holt. 1995. "Habitat fragmentation and movements of three small mammals (*Sigmodon*, *Microtus*, and *Peromyscus*)" *Ecology* 76 (3): 827–839.

Diffendorfer, J. E., N. A. Slade, M. S. Gaines, and R. D. Holt. 1995. "Population dynamics of small mammals in fragmented and continuous old-field habitats." In *Landscape Approaches in Mammalian Ecology and Conservation*, edited by W. Z. Lidicker Jr., 175–199. Minneapolis: University of Minnesota Press.

Dobson, F. S., and M. K. Oli. 2008. "The life histories of orders of mammals: Fast and slow breeding." *Current Science* 95 (7): 862–865.

Drickamer, L. C. 1984. "Seasonal variation in acceleration and delay of sexual maturation in female mice by urinary chemosignals." *Journal of Reproduction and Fertility* 72:55–58.

———. 2007. "Acceleration and delay of reproduction in rodents." In *Rodent Societies: An Ecological and Evolutionary Perspective*, edited by J. O. Wolff and P. W. Sherman, 106–114. Chicago: University of Chicago Press.

Dueser, R. D., M. L. Wilson, and R. K. Rose. 1981. "Attributes of dispersing meadow voles in open grid populations." *Acta Theriologica* 26:139–162.

Dufty, A. M., Jr., and J. R. Belthoff. 2001. "Proximate mechanisms of natal dispersal: the role of body condition and hormones." In *Dispersal*, edited by J. Clobert, E. Danchin, A. A. Dhondt, and J. D. Nichols, 217–229. Oxford: Oxford University Press.

Duncan, R., D. Forsyth, and J. Hone. 2007. "Testing the metabolic theory of ecology: Allometric scaling exponents in mammals." *Ecology* 88:324–333.

Dupuy, G., P. Giraudoux, and P. Delattre. 2009. "Numerical and dietary responses of a predator community in a temperate zone of Europe." *Ecography* 32 (2): 277–290.

Eberhardt, L. L. 2003. "What should we do about hypothesis testing?" *Journal of Wildlife Management* 67 (2): 241–247.

Eccard, J. A., I. Jokinen, and H. Ylönen. 2011. "Loss of density-dependence and incomplete control by dominant breeders in a territorial species with density outbreaks." *BMC Ecology* 11:16. doi: 10.1186/1472-6785-11-16.

Ecke, F., P. Christensen, P. Sandström, and B. Hörnfeldt. 2006. "Identification of landscape elements related to local declines of a boreal grey-sided vole population." *Landscape Ecology* 21 (4): 485–497.

Efford, M. 2004. "Density estimation in live-trapping studies." *Oikos* 106 (3): 598–610.

Efford, M. G., D. L. Borchers, and A. E. Byrom. 2009. "Density estimation by spatially explicit capture-recapture: Likelihood-based methods." In *Modeling Demographic Processes in Marked Populations*, edited by E. G. Cooch, D. L. Thomson, and M. J. Conroy, 255–269. New York: Springer.

Ehrich, D., P. E. Jorde, C. J. Krebs, A. J. Kenney, J. E. Stacy, and N. C. Stenseth. 2001. "Spatial structure of lemming populations (*Dicrostonyx groenlandicus*) fluctuating in density." *Molecular Ecology* 10:481–495.

Elton, C. 1942. *Voles, Mice and Lemmings: Problems in Population Dynamics*. Oxford: Clarendon Press.

Elton, C. S. 1924. "Periodic fluctuations in the numbers of animals: Their causes and effects." *British Journal of Experimental Biology* 2:119–163.

Engeman, R. M. 2003. "More on the need to get the basics right: population indices." *Wildlife Society Bulletin* 31:286–287.

Erb, J., M. S. Boyce, and N. C. Stenseth. 2001. "Population dynamics of large and small mammals." *Oikos* 92 (1): 3–12.

Ergon, T. 2003. "Fluctuating life-history traits in overwintering field voles (*Microtus agrestis*)." PhD thesis, Department of Zoology, University of Oslo, Norway.

———. 2007. "Optimal onset of seasonal reproduction in stochastic environments: when should overwintering small rodents start breeding?" *Ecoscience* 14 (3): 330–346.

Ergon, T., X. Lambin, and N. C. Stenseth. 2001. "Life-history traits of voles in a fluctuating population respond to the immediate environment." *Nature* 411:1043–1045.

Ergon, T., N. G. Yoccoz, and J. D. Nichols. 2009. "Estimating latent time of maturation and survival costs of reproduction in continuous time from capture-recapture data." In *Modelling Demographic Processes in Marked Populations*, edited by D. L. Thomson, E. G. Cooch and M. J. Conroy, 173–197. New York: Springer.

Erlinge, S. 1987. "Predation and noncyclicity in a microtine population in southern Sweden." *Oikos* 50:347–352.

Erlinge, S., G. Göransson, G. Högstedt, G. Jansson, O. Liberg, J. Loman, I. N. Nilsson, T. von Schantz, and M. Sylvén. 1984. "Can vertebrate predators regulate their prey?" *American Naturalist* 123 (1): 125–133.

Erlinge, S., D. Hasselquist, M. Svensson, P. Frodin, and P. Nilsson. 2000. "Reproductive behaviour of female Siberian lemmings during the increase and peak phase of the lemming cycle." *Oecologia* 123 (2): 200–207.

Ernest, E. S. M. 2003. "Life history characteristics of placental nonvolant mammals: Ecological Archives E084–093." *Ecology* 84 (12): 3402.

Errington, P. L. 1956. "Factors limiting higher vertebrate populations." *Science* 124:304–307.

Estes, J. A., J. Terborgh, J. S. Brashares, M. E. Power, J. Berger, W. J. Bond, S. R. Carpenter, T. E. Essington, R. D. Holt, J. B. C. Jackson, R. J. Marquis, L.

Oksanen, Tarja Oksanen, Robert T. Paine, Ellen K. Pikitch, William J. Ripple, Stuart A. Sandin, M. Scheffer, T. W. Schoener, J. B. Shurin, A. R. E. Sinclair, M. E. Soulé, R. Virtanen, and D. A. Wardle. 2011. "Trophic downgrading of Planet Earth." *Science* 333 (6040): 301–306. doi: 10.1126/science.1205106.

Ferraz, G., J. D. Nichols, J. E. Hines, P. C. Stouffer, R. O. Bierregaard, Jr., and T. E. Lovejoy. 2007. "A large-scale deforestation experiment: Effects of patch area and isolation on Amazon birds." *Science* 315 (5809): 238–241.

Fisher, D. O. , X. Lambin, and S. Yletyinen. 2009. "Experimental translocation of juvenile water voles in a Scottish lowland metapopulation." *Population Ecology* 51 (2): 289–295.

Fisher, R. A. 1930. *The Genetical Theory of Natural Selection*. Oxford: Oxford University Press.

Fletcher, D. 2006. *Population dynamics of Eastern grey kangaroos in temperate grasslands* PhD, Applied Science, University of Canberra, Canberra.

Framstad, E., N. C. Stenseth, O. N. Bjørnstad, and W. Falck. 1997. "Limit cycles in Norwegian lemmings: Tensions between phase-dependence and density-dependence." *Proceedings of the Royal Society of London, Series B* 264:31–38.

Framstad, E., N. C. Stenseth, and E. Østbye. 1993. "Demography of *Lemmus lemmus* through five population cycles." In *The Biology of Lemmings*, edited by N. C. Stenseth and R. A. Ims, 117–133. London: Academic Press.

Frank, F. 1954. "Beitrage zur Biologie der Feldmaus, *Microtus arvalis* (Pallas). I. Gehegeversuche." *Zoologische Jahrbucher* 82 (3/4): 353–404.

Freeland, W. J. 1974. "Vole cycles: Another hypothesis." *American Naturalist* 108 (960): 238–245.

Fritz, H., M. Loreau, S. Chamaillé-Jammes, M. Valeix, and J. Clobert. 2011. "A food web perspective on large herbivore community limitation." *Ecography* 34 (2): 196–202. doi: 10.1111/j.1600–0587.2010.06537.x.

Fryxell, J. M., J. Greever, and A. R. E. Sinclair. 1988. " Why are migratory ungulates so abundant?" *American Naturalist* 131:781–798.

Fryxell, J. M., A. Mosser, A. R. E. Sinclair, and C. Packer. 2007. "Group formation stabilizes predator-prey dynamics." *Nature* 449 (25 Oct 2007): 1041–1043.

Fujimaki, Y. 1981. "Reproductive activity in *Clethrionomys rufocanus bedfordiae*. 4. Number of embryos and prenatal mortality." *Japanese Journal of Ecology* 31 (3): 247–256.

Gaines, M. S., and L. R. McClenaghan, Jr. 1980. "Dispersal in small mammals." *Annual Review of Ecology and Systematics* 11:163–196.

Gaines, M. S., and R. K. Rose. 1976. "Population dynamics of *Microtus ochrogaster* in eastern Kansas." *Ecology* 57 (6): 1145–1161.

Gaines, M. S., N. C. Stenseth, M. L. Johnson, R. A. Ims, and S. Bondrup-Nielsen. 1991. "A response to solving the enigma of population cycles with a multifactorial perspective." *Journal of Mammalogy* 72 (3): 627–631.

Gaines, M. S., A. M. Vivas, and C. L. Baker. 1979. "An experimental analysis of dispersal in fluctuating vole populations: Demographic parameters." *Ecology* 60:814–828.

Galindo, C., and C. J. Krebs. 1985. "Evidence for competition in small rodents." *Oikos* 46:116–119.

Gauthier, G., D. Berteaux, C. J. Krebs, and D. Reid. 2009. "Arctic lemmings are not simply food limited: A comment." *Evolutionary Ecology Research* 11 (3): 483–484.

Gauthier, G., J. Bety, J. F. Giroux, and L. Rochefort. 2004. "Trophic Interactions in a high arctic snow goose colony." *Integrative and Comparative Biology* 44:119–129.

Getz, L. L., B. McGuire, T. Pizzuto, J. E. Hofmann, and B. Frase. 1993. "Social organization of the prairie vole (*Microtus ochrogaster*)." *Journal of Mammalogy* 74 (1): 44–58.

Getz, L. L., J. E. Hofmann, B. McGuire, and T. W.III Dolan. 2001. "Twenty-five years of population fluctuations of *Microtus ochrogaster* and *M. pennsylvanicus* in three habitats in east-central Illinois." *Journal of Mammalogy* 82 (1): 22–34.

Getz, L. L., L. E. Simms, and B. McGuire. 2000. "Nestling survival and population cycles in the prairie vole, *Microtus ochrogaster*." *Canadian Journal of Zoology* 78 (10): 1723–1731.

Getz, L. L., M. K. Oli, J. E. Hofmann, and B. McGuire. 2005. "Habitat-specific demography of sympatric vole populations over 25 years." *Journal of Mammalogy* 86:561–568.

Getz, L. L., N. G. Solomon, and T. M. Pizzuto. 1990. "The effects of predation of snakes on social organization of the prairie vole, *Microtus ochrogaster*." *American Midland Naturalist* 123 (20): 365–371.

Gilbert, B. S., C. J. Krebs, D. Talarico, and D. B. Cichowski. 1986. "Do *Clethrionomys rutilus* females suppress maturation of juvenile females?" *Journal of Animal Ecology* 55:543–552.

Gilbert, N. E. 1989. "What use are population models?" *Bulletin of the British Ecological Society* 20 (2): 89–92.

Gilg, O., B. Sittler, and I. Hanski. 2009. "Climate change and cyclic predator-prey population dynamics in the high arctic." *Global Change Biology* 15:2634–2652.

Ginzburg, L. R. 1998. "Inertial growth, population dynamics based on maternal effects." In *Maternal Effects as Adaptations*, edited by T. A. Mousseau and C. W. Fox, 42–53. New York: Oxford University Press.

Ginzburg, L. R., and C. X. J. Jensen. 2004. "Rules of thumb for judging ecological theories." *Trends in Ecology and Evolution* 19 (3): 121–126. doi: 10.1016/j.tree.2003.11.004.

Ginzburg, L. R., and D. E. Taneyhill. 1995. "Higher growth rate implies shorter cycle, whatever the cause: A reply to Berryman." *Journal of Animal Ecology* 64:294–195.

Ginzburg, L. R., C. X. J. Jensen, and J. V. Yule. 2007. "Aiming the 'unreasonable effectiveness of mathematics' at ecological theory." *Ecological Modelling* 207 (2–4): 356–362.

Goddard, T. R. 1935. "A census of short-eared owls (*Asio f. flammeus*) at Newcastleton, Roxburghshire, 1935." *Journal of Animal Ecology* 4 (2): 289–290.

Goswami, V. R., L. L. Getz, J. A. Hostetler, A. O., and M. K. Oli. 2011. "Synergistic influences of phase, density, and climatic variation on the dynamics of fluctuating populations." *Ecology* 92 (8): 1680–1690.

Goundie, T. R., and S. H. Vessey. 1986. "Survival and dispersal of young white-footed mice born in nest boxes." *Journal of Mammalogy* 67 (1): 53–60.

Graham, I. M. 2001. "Weasels and vole cycles: An experimental test of the specialist predator hypothesis. PhD thesis, Department of Zoology, University of Aberdeen, Aberdeen.

Graham, I. M., and X. Lambin. 2002. "The impact of weasel predation on cyclic field-vole survival: The specialist predator hypothesis contradicted." *Journal of Animal Ecology* 71:946–956.

Greenwald, G.S. 1956. "The reproductive cycle of the field mouse, *Microtus californicus*." *Journal of Mammalogy* 37 (2): 213–222.

Grenfell, B. T., and B. F. Finkenstadt. 1998. "Seasonality, stochasticity and population cycles." *Researches on Population Ecology* 40:141–143.

Gruyer, N., G. Gauthier, and D. Berteaux. 2010. "Demography of two lemming species on Bylot Island, Nunavut, Canada." *Polar Biology* 33 (6): 725–736.

Hagen, A., N. C. Stenseth, E. Ostbye, and H. J. Skar. 1980. "The eye lens as an age indicator in the root vole." *Acta Theriologica* 25:39–50.

Hairston, N. G., F. E. Smith, and L. B. Slobodkin. 1960. "Community structure, population control, and competition." *American Naturalist* 94:421–425.

Hamilton, W. J., Jr. 1937. "The biology of microtine cycles." *Journal of Agricultural Research* 54:779–790.

Hanley, T. A., and J. C. Barnard. 1999. "Spatial variation in population dynamics of Sitka mice in floodplain forests." *Journal of Mammalogy* 80 (3): 866–879.

Hansen, T. F., N. C. Stenseth, and H. Henttonen. 1999. "Multiannual vole cycles and population regulation during long winters: An analysis of seasonal density dependence." *American Naturalist* 154 (2): 129–139.

Hanski, I., H. Henttonen, E. Korpimäki, L. Oksanen, and P. Turchin. 2001. "Small-rodent dynamics and predation." *Ecology* 82 (6): 1505–1520.

Hanski, I., and E. Korpimäki. 1995. "Microtine rodent dynamics in northern Europe: Parameterized models for the predator-prey interation." *Ecology* 76:840–850.

Hanski, I., P. Turchin, E. Korpimaki, and H. Henttonen. 1993. "Population oscillations of boreal rodents: Regulation by mustelid predators leads to chaos." *Nature* 364:232–235.

Hansson, L. 1984. "Winter reproduction of small mammals in relation to food conditions and population dynamics." In *Winter Ecology of Small Mammals*, edited by J. F. Merritt, 225–234. Pittsburgh, Pennsylvania: Carnegie Museum of Natural History, Special Publication 10.

Hansson, L., and H. Henttonen. 1985. "Gradients in density variations of small rodents: The importance of latitude and snow cover." *Oecologia* 67:394–402.

Hasler, J. F. 1975. "A review of reproduction and sexual maturation in the microtine rodents." *The Biologist* 57 (2): 52–86.

Henden, J.-A., R. A. Ims, and N. G. Yoccoz. 2009. "Nonstationary spatiotemporal small rodent dynamics: Evidence from long-term Norwegian fox bounty data." *Journal of Animal Ecology* 78 (3): 636–645. doi: 10.1111/j.1365–2656.2008.01510.x.

Henttonen, H. 1986. "Causes and geographic patterns of microtine cycles." PhD thesis, University of Helsinki, Finland.

Henttonen, H., A. D. McGuire, and L. Hansson. 1985. "Comparisons of amplitudes and frequencies (spectral analyses) of density variation in long-term data sets of *Clethrionomys* species." *Annales Zoologici Fennici* 22:221–227.

Henttonen, H., T. Oksanen, A. Jortikka, and V. Haukisalmi. 1987. "How much do weasels shape microtine cycles in the northern Fennoscandian taiga?" *Oikos* 50:353–365.

Hestbeck, J. B. 1982. "Population regulation of cyclic mammals: The social fence hypothesis." *Oikos* 39:157–163.

Hilborn, R., and S. C. Stearns. 1982. "On inference in ecology and evolutionary biology: The problem of multiple causes." *Acta Biotheoretical* 31:145–164.

Hoffmann, R. S. 1958. "The role of reproduction and mortality in population fluctuations of voles (*Microtus*)." *Ecological Monographs* 28:79–109.

Hoffmeister, D. F., and L. L. Getz. 1968. "Growth and age classes in the prairie vole, *Microtus ochrogaster*." *Growth* 32:57–69.

Holling, C. S. 1959. "The components of predation as revealed by a study of small mammal predation of the European pine sawfly." *Canadian Entomologist* 91:293–320.

Holmes, J. C. 1995. "Population regulation: A dynamic complex of interactions." *Wildlife Research* 22:11–19.

Holmes, W. G., and J. M. Mateo. 2007. "Kin recognition in rodents: issues and evidence." In *Rodent Societies: An Ecological and Evolutionary Perspective*, edited by J. O. Wolff and P. W. Sherman, 216–228. Chicago: University of Chicago Press.

Hone, J., and T. H. Clutton-Brock. 2007. "Climate, food, density and wildlife population growth rate." *Journal of Animal Ecology* 76:361–367.

Hopcraft, J. G. C., H. Olff, and A. R. E. Sinclair. 2010. "Herbivores, resources and risk: Alternating regulation along primary environmental gradients in savannas." *Trends in Ecology & Evolution* 25:119–128.

Hörnfeldt, B. 1991. "Cycles of voles, predators, and alternative prey in boreal Sweden." PhD thesis, Department of Animal Ecology, University of Umeå, Umeå, Sweden.

———. 2004. "Long-term decline in numbers of cyclic voles in boreal Sweden: analysis and presentation of hypotheses." *Oikos* 107 (2): 376–392.

Hörnfeldt, B., T. Hipkiss, and U. Eklund. 2005. "Fading out of vole and predator cycles?" *Proceedings of the Royal Society of London, Series B* 272:2045–2049.

Hudson, P. J., A. P. Dobson, and D. Newborn. 1998. "Prevention of population cycles by parasite removal." *Science* 282:2256–2258.

Huffaker, C. B. 1958. "Experimental studies on predation: dispersion factors and predator-prey oscillations." *Hilgardia* 27 (14): 343–383.

Huitu, O., I. Jokinen, E. Korpimäki, E. Koskela, and T. Mappes. 2007. "Phase dependence in winter physiological condition of cyclic voles." *Oikos* 116 (4): 565–577. doi: 10.1111/j.0030-1299.2007.15488.x.

Huitu, O., N. Kiljunen, E. Korpimäki, E. Koskela, T. Mappes, H. Pietiäinen, H. Pöysä, and H. Henttonen. 2009. "Density-dependent vole damage in silviculture and associated economic losses at a nationwide scale." *Forest Ecology and Management* 258 (7): 1219–1224. doi: 10.1016/j.foreco.2009.06.013.

Huitu, O., M. Koivula, E. Korpimäki, T. Klemola, and K. Norrdahl. 2003. "Winter food supply limits growth of northern vole populations in the absence of predation." *Ecology* 84 (8): 2108–2118. doi: 10.1890/02–0040.

Humphries, M. M., D. W. Thomas, and J. R. Speakman. 2002. "Climate-mediated energetic constraints on the distribution of hibernating mammals." *Nature* 418:313–316.

Hwang, Y. T., J. S. Millar, and F. J. Longstaffe. 2007. "Do δ^{15}N and δ^{13} values of feces reflect the isotopic composition of diets in small mammals?" *Canadian Journal of Zoology* 85:388–396. doi: doi:10.1139/Z07–019.

Imholt, C., A. Esther, J. Perner, and J. Jacob. 2011. "Identification of weather parameters related to regional population outbreak risk of common voles (*Microtus arvalis*) in eastern Germany." *Wildlife Research* 38(7): 551–559.

Ims, R. A., J.-A. Henden, and S. T. Killengreen. 2008. "Collapsing population cycles." *Trends in Ecology & Evolution* 23 (2): 79–86.

Ims, R.A., and D. Ø. Hjermann. 2001. "Condition-dependent dispersal." In *Dispersal*, edited by J. Clobert, E. Danchin, A. A. Dhondt and J. D. Nichols, 203–216. Oxford: Oxford University Press.

Ims, R. A., and H. P. Andreassen. 2005. "Density-dependent dispersal and spatial population dynamics." *Proceedings of the Royal Society of London, Series B: Biological Sciences* 272 (1566): 913–918. doi: 10. 1098/rspb.2004.3025.

Inchausti, P., and L. R. Ginzburg. 1998. "Small mammals cycles in northern Europe: Patterns and evidence for a maternal effect hypothesis." *Journal of Animal Ecology* 67 (2): 180–194. doi: 10.1046/j.1365–2656.1998.00189.x.

———. 2009. "Maternal effects mechanism of population cycling: A formidable competitor to the traditional predator-prey view." *Philosophical Transactions of the Royal Society B: Biological Sciences* 364:1117–1124.

Innes, D. G., and J. S. Millar. 1990. "Numbers, of litters, litter size and survival in two species of microtines at two elevations." *Holarctic Ecology* 13 (3): 207–216.

Jett, D. A., and J. D. Nichols. 1987. "A field comparison of nested grid and trapping web density estimators." *Journal of Mammalogy* 68:888–892.

Johnson, C. N. 2010. "Ecological consequences of Late Quaternary extinctions of megafauna." *Proceeding of the Royal Society of London, Series B* 276 (1667): 2509–2519. doi: 10.1098/rspb.2008.1921.

Johnson, C. N., J. L. Isaac, and D. O. Fisher. 2007. "Rarity of a top predator triggers continent-wide collapse of mammal prey: Dingoes and marsupials in Australia." *Proceedings of the Royal Society of London, Series B* 274:341–346.

Johnston, K. M., and O. J. Schmidt. 1997. "Wildlife and climate change: assessing the sensitivity of selected species to simulated doubling of atmospheric CO_2." *Global Change Biology* 3:531–544.

Jongbloet, P. H., H. M. M. Groenewoud, S. Huber, M. Fieder, and N. Roeleveld. 2007. "Month of birth related to fecundity and childlessness among contemporary women." *Human Biology* 79 (5): 479–490. doi: 10.1353/hub .2008.0006.

Kalela, O. 1957. "Regulation of reproductive rate in subarctic populations of the vole, *Clethrionomys rufocanus* (Sund.)." *Annales Academiae Scientiarum Fennicae, Series A, IV Biologica* 34:1–60.

———. 1962. "On the fluctuations in the numbers of arctic and boreal small rodents as a problem of production biology." *Annales Academiae Scientiarum Fennicae, Series A, IV* 66:1–38.

Kallio, E. R., M. Begon, H. Henttonen, E. Koskela, T. Mappes, A. Vaheri, and O. Vapalahti. 2009. "Cyclic hantavirus epidemics in humans: Predicted by rodent host dynamics." *Epidemics* 1 (2): 101–107. doi: 10.1016/j.epidem.2009.03.002.

Kallio, E. R., L. Voutilainen, O. Vapalahti, A. Vaheri, H. Henttonen, E. Koskela, and T. Mappes. 2007. "Endemic hantavirus infection impairs the winter survival of its rodent host." *Ecology* 88 (8): 1911–1916.

Kanda, L. L. 2005. "Winter energetics of Virginia opossums *Didelphis virginiana* and implications for the species' northern distributional limit." *Ecography* 28:731–744.

Kareiva, P. 1990. "Population dynamics in spatially complex environments: Theory and data." *Philosophical Transactions of the Royal Society of London* 330 (Series B): 175–190.

Kasparian, K., and J. S. Millar. 2004. "Effects of extra food on nestling growth and survival in red-backed voles (*Clethrionomys gapperi*)." *Canadian Journal of Zoology* 82 (8): 1219–1224.

Keller, B. L., and C. J. Krebs. 1970. "Microtus population biology. III. Reproductive changes in fluctuating populations of *M. ochrogaster* and *M. pennsylvanicus* in southern Indiana, 1965–67." *Ecological Monographs* 40 (3): 263–294.

Keller, B. L. 1985. "Reproductive patterns." In *Biology of New World Microtus*, edited by R. H. Tamarin, 725–778. Lawrence, Kansas: American Society of Mammalogists.

Klemola, T., M. Koivula, E. Korpimäki, and K. Norrdahl. 2000. "Experimental tests of predation and food hypotheses for population cycles of voles." *Proceedings of the Royal Society of London, Series B* 267:1–6.

Klemola, T., E. Korpimaki, and M. Koivula. 2002. "Rate of population change in voles from different phases of the population cycle." *Oikos* 96 (2): 291–298.

Klemola, T., K. Norrdahl, and E. Korpimäki. 2000. "Do delayed effects of overgrazing explain population cycles in voles?" *Oikos* 90 (3): 509–516.

Kokorev, Y. I., and V. A. Kuksov. 2002. "Population dynamics of lemmings, *Lemmus sibirica* and *Dicrostonyx torquatus*, and arctic fox *Alopex lagopus* on the Taimyr peninsula, Siberia, 1960–2001." *Ornis Svecica* 12:139–143.

Korpimaki, E. 1993. "Regulation of multiannual vole cycles by density-dependent avian and mammalian predation?" *Oikos* 66 (2): 359–363.

———. 1994. "Rapid or delayed tracking of multi-annual vole cycles by avian predators?" *Journal of Animal Ecology* 63:619–628.

Korpimäki, E., T. Klemola, K. Norrdahl, L. Oksanen, T. Oksanen, P. B. Banks, G. O. Batzli, and H. Henttonen. 2003. "Vole cycles and predation." *Trends in Ecology and Evolution* 18 (10): 494–495.

Korpimäki, E., and K. Norrdahl. 1998. "Experimental reduction of predators reverses the crash phase of small-rodent cycles." *Ecology* 79 (7): 2448–2455.

Korpimäki, E., K. Norrdahl, T. Klemola, T. Pettersen, and N. C. Stenseth. 2002. "Dynamic effects of predators on cyclic voles: Field experimentation and model extrapolation." *Proceedings of the Royal Society of London, Series B* 269:991–997. doi: 10.1098/rspb.2002.1972.

Koshkina, T. V. 1965. "Population density and its importance in regulating the abundance of the red vole (*Clethrionomys rutilus*)." *Byulleten' Moskovskogo Obshchestva Ispytatelei Prirody Otdel Biologicheskii* 70 (1): 5–19.

Koskela, E., T. Mappes, and H. Ylonen. 1999. "Experimental manipulation of breeding density and litter size: Effects on reproductive success in the bank vole." *Journal of Animal Ecology* 68 (3): 513–521. doi: 10.1046/j.1365-2656.1999.00308.x.

Krebs, C. J. 1964. "The lemming cycle at Baker Lake, Northwest Territories, during 1959–62." *Arctic Institute of North America Technical Paper 15*:104 pp.

———. 1966. "Demographic changes in fluctuating populations of *Microtus californicus*." *Ecological Monographs* 36:239–273.

———. 1978. "A review of Chitty's hypothesis of population regulation." *Canadian Journal of Zoology* 56:2463–2480.

———. 1988. "The experimental approach to rodent population dynamics." *Oikos* 52:143–149.

———. 1996. "Population cycles revisited." *Journal of Mammalogy* 77 (1): 8–24.

———. 1999. *Ecological Methodology*. 2nd ed. Menlo Park, CA: Addison Wesley Longman.

———. 2002. "Two complementary paradigms for analyzing population dynamics." *Philosophical Transactions of the Royal Society of London, Series B* 357: 1211–1219.

———. 2006. "Ecology after 100 years: Progress and pseudo-progress." *New Zealand Journal of Ecology* 30 (1): 3–11.

———. 2009a. *Ecology: The Experimental Analysis of Distribution and Abundance*. 6th ed. San Francisco: Benjamin Cummings.

———. 2009b. "Population dynamics of large and small mammals: Graeme Caughley's grand vision." *Wildlife Research* 36 (1): 1–7.

———. 2011. "Of lemmings and snowshoe hares: The ecology of northern Canada." *Proceedings of the Royal Society of London, Series B* 278 (1705): 481–489. doi: 10.1098/rspb.2010.1992.

Krebs, C. J., and R. Boonstra. 1978. "Demographic attributes of the spring decline in populations of the vole *Microtus townsendii*." *Journal of Animal Ecology* 47:1007–1015.

Krebs, C. J., R. Boonstra, K. Cowcill, and A. J. Kenney. 2009. "Climatic determinants of berry crops in the boreal forest of the southwestern Yukon." *Botany* 87:401–408.

Krebs, C. J., R. Boonstra, B. S. Gilbert, D. G. Reid, A. J. Kenney, and E. J. Hofer. 2011. "Density estimation for small mammals from live trapping grids: Rodents in northern Canada." *Journal of Mammalogy* 92 (5): 974–981. doi: 10.1644/10-MAMM-A-313.1

Krebs, C. J., K. Cowcill, R. Boonstra, and A. J. Kenney. 2010. "Do changes in berry crops drive population fluctuations in small rodents in the southwestern Yukon?" *Journal of Mammalogy* 91 (2): 500–509.

Krebs, C. J., Z. T. Halpin, and J. N. M. Smith. 1977. "Aggression, testosterone, and the spring decline in populations on the vole *Microtus townsendii*." *Canadian Journal of Zoology* 55 (2): 430–437.

Krebs, C. J., B. L. Keller, and R. H. Tamarin. 1969. "*Microtus* population biology: demographic changes in fluctuating populations of *M. ochrogaster* and *M. pennsylvanicus* in southern Indiana." *Ecology* 50 (4): 587–607.

Krebs, C. J., A. J. Kenney, S. Gilbert, K Danell, A. Angerbjörn, S. Erlinge, R. G. Bromley, C. Shank, and S. Carriere. 2002. "Synchrony in lemming and vole populations in the Canadian arctic." *Canadian Journal of Zoology* 80 (8): 1323–1333. doi:10.1139/z02–120.

Krebs, C. J. , X. Lambin, and J. O. Wolff. 2007. "Social behavior and self-regulation in murid rodents." In *Rodent Societies: An Ecological and Evolutionary Perspective*, edited by J. O. Wolff and P. W. Sherman, 173–181. Chicago: University of Chicago Press.

Krebs, C. J., and J. H. Myers. 1974. "Population cycles in small mammals." *Advances in Ecological Research* 8:267–399.

Labov, J. B., U. W. Huck, R. W. Elwood, and R. J. Brooks. 1985. "Current problems in the study of infanticidal behavior of rodents." *Quarterly Review of Biology* 60 (1): 1–20.

Lack, D. 1954. *The Natural Regulation of Animal Numbers*. Oxford: Clarendon Press.

Lakatos, I., and A. Musgrave. 1970. *Criticism and the Growth of Knowledge*. Cambridge: Cambridge University Press.

Lambin, X. 1994a. "Natal philopatry, competition for resources, and inbreeding avoidance in Townsend's voles (*Microtus townsendii*)." *Ecology* 75:224–235.

———. 1994b. "Territory acquisition and social facilitation by litter-mate Townsend's voles (*Microtus townsendii*)." *Ethology, Ecology & Evolution* 6:213–220.

Lambin, X., J. Aars, and S. B. Piertney. 2001. "Dispersal, intraspecific competition, kin competition and kin facilitation: A review of the empirical evidence." In *Dispersal*, edited by J. Clobert, E. Danchin, A. A. Dhondt and J. D. Nichols, 110–122. Oxford: Oxford University Press.

Lambin, X., D. A. Elston, S. J. Petty, and J. L. MacKinnon. 1998. "Spatial asynchrony and periodic travelling waves in cyclic populations of field voles." *Proceedings of the Royal Society of London, Series B* 265:1491–1496.

Lambin, X., and I. M. Graham. 2003. "Testing the specialist predator hypothesis for vole cycles." *Trends in Ecology & Evolution* 18 (10): 493.

Lambin, X., and C. J Krebs. 1991. "Can changes in female relatedness influence microtine population dynamics?" *Oikos* 61 (1): 126–132.

———. 1993. "Influence of female relatedness on the demography of female Townsend's vole populations in the spring." *Journal of Animal Ecology* 62 (3): 536–550.

Lambin, X., and N. Yoccoz. 1998. "The impact of population kin-structure on nestling survival in Townsend's voles, *Microtus townsendii*." *Journal of Animal Ecology* 67 (1): 1–16.

Lantová, P., K. Zub, E. Koskela, K. Šíchová, and Z. Borowski. 2011. "Is there a linkage between metabolism and personality in small mammals? The root vole (*Microtus oeconomus*) example." *Physiology & Behavior* 104 (3): 378–383. doi: 10.1016/j.physbeh.2011.04.017.

Legagneux, P., G. Gauthier, D. Berteaux, J. Bêty, M-C. Cadieux, F. Bilodeau, E. Bolduc, L. McKinnon, A. Tarroux, J-F. Therrien, L. Morissette, and C. J. Krebs. 2012. "Disentangling trophic relationships in a high arctic tundra ecosystem through food web modeling." *Ecology* 93(7): 1707–1716.

Lemaître, J., D. Fortin, P.-O. Montiglio, and M. Darveau. 2009. "Bot fly parasitism of the red-backed vole: Host survival, infection risk, and population growth." *Oecologia* 159 (2): 283–294. doi: 10.1007/s00442–008–1219–3.

Leslie, P. H. 1945. "On the use of matrices in certain population mathematics." *Biometrika* 33:183–212.

Leslie, P. H., J. S. Perry, and J. S. Watson. 1945. "The determination of the median body-weight at which female rats reach maturity." *Proceedings of the Zoological Society of London* 115:473–488.

Letnic, M., and C. R. Dickman. 2010. "Resource pulses and mammalian dynamics: conceptual models for hummock grasslands and other Australian desert habitats." *Biological Reviews* 85 (3): 501–521. doi: 10.1111/j.1469–185X.2009.00113.x.

Letnic, M., A. Greenville, E. Denny, C. R. Dickman, M.Tischler, C. Gordon, and F. Koch. 2011. "Does a top predator suppress the abundance of an invasive mesopredator at a continental scale?" *Global Ecology and Biogeography* 20 (2): 343–353. doi: 10.1111/j.1466–8238.2010.00600.x.

Lewontin, R. C. 1965. "Selection of colonizing ability." In *The Genetics of Colonizing Species*, edited by H. G. Baker and G. L. Stebbins, 77–94. New York: Academic Press.

Lewontin, R. 1966. "On the measurement of relative variability." *Systematic Zoology* 15:141–142.

Lidicker, W. Z., Jr. 1962. "Emigration as a possible mechanism permitting the regulation of population density below carrying capacity." *American Naturalist* 96 (1): 29–33.

———. 1973. "Regulation of numbers in an island population of the California vole: A problem in community dynamics." *Ecological Monographs* 43:271–302.

———. 1975. "The role of dispersal in the demography of small mammals." In *Small Mammals: Their Productivity and Population Dynamics*, edited by F. B. Golley, K. Petrucewicz and L. Ryskowski, 103–134. Cambridge: Cambridge University Press.

———. 1978. "Regulation of numbers in small mammal populations: Historical reflections and a synthesis." In *Populations of Small Mammals under Natural Conditions*, edited by D. P. Snyder, 122–166. Pymatuning, PA: Pymatuning Laboratory of Ecology.

———. 1988. "Solving the enigma of microtine 'cycles.'" *Journal of Mammalogy* 69: 225–235.
———. 1991. "In defense of a multifactor perspective in population ecology." *Journal of Mammalogy* 72 (3): 631–635.
———. 1995. "The landscape concept: Something old, something new." In *Landscape Approaches in Mammalian Ecology and Conservation*, edited by W. Z. Lidicker Jr., 3–19. Minneapolis: University of Minnesota Press.
———. 2000. "A food web / landscape interaction model for microtine rodent density cycles." *Oikos* 91 (3): 435–445.
Lidicker, W. Z., Jr., and S. F. Maclean, Jr. 1969. "A method for estimating age in the California vole, *Microtus californicus*." *American Midland Naturalist* 82:450–470.
Lochmiller, R. L. 1996. "Immunocompetence and animal population regulation." *Oikos* 76:594–602.
Lochmiller, R. L., and C. B. Dabbert. 1993. "Immunocompetence, environmental stress, and the regulation of animal populations." *Trends in Comparative Biochemistry and Physiology* 1:823–855.
Loeb, S. C., and R. G. Schwab. 1987. "Estimation of litter size in small mammals bias due to chronology of embryo resorption." *Journal of Mammalogy* 68 (3): 671–675.
Loehle, C. 1987. "Errors of construction, evaluation, and inference: A classification of sources of error in ecological models." *Ecological Modelling* 36:297–314.
Loehle, C., and J. H. K. Pechmann. 1988. "Evolution: The missing ingredient in systems ecology." *American Naturalist* 132:884–899.
Lotka, A. J. 1925. *Elements of Physical Biology* (reprint). New York: Dover Publications.
Lummaa, V. 2003. "Early developmental conditions and reproductive success in humans: Downstream effects of prenatal famine, birthweight, and timing of birth." *American Journal of Human Biology* 15 (3): 370–379. doi: 10.1002/ajhb.10155.
Luque-Larena, J. J., F. Mougeot, J. Viñuela, D. Jareño, L. Arroyo, X. Lambin, and B. Arroyoc. 2012. "Recent large-scale range expansion and eruption of common vole (*Microtus arvalis*) outbreaks in NW Spain." *Pest Management Science*, submitted.
Mackinnon, J. L., S. J. Petty, D. A. Elston, C. J. Thomas, T. N. Sherratt, and X. Lambin. 2001. "Scale invariant spatio-temporal patterns in field vole density." *Journal of Animal Ecology* 70 (1): 101–111.
Madison, D. M., and W. J. McShea. 1987. "Seasonal changes in reproductive tolerance, spacing, and social organization in meadow voles: A microtine model." *American Zoologist* 27 (3): 899–908.
Madison, D. M. 1985. "Activity rhythms and spacing." In *Biology of New World Microtus*, edited by R. H. Tamarin, 373–419. Lawrence, KS: American Society of Mammalogists, Special Publication 8.
Maher, W. J. 1967. "Predation by weasels on a winter population of lemmings, Banks Island, Northwest Territories." *Canadian Field-Naturalist* 81 (4): 248–250.

———. 1970. "The pomarine jaeger as a brown lemming predator in northern Alaska." *Wilson Bulletin* 82 (2): 130–157.

Mallory, F. F., and R. J. Brooks. 1978. "Infanticide and other reproductive strategies in the collared lemming, *Dicrostonyx groenlandicus*." *Nature* 273:144–146.

———. 1980. "Infanticide and pregnancy failures: Reproductive strategies in the female collared lemming (*Dicrostonyx groenlandicus*)." *Biology of Reproduction* 22 (2): 192–196.

Mappes, T., M. Koivula, E. Koskela, T. A. Oksanen, T. Savolainen, and B. Sinervo. 2008. "Frequency and density-dependent selection on life-history strategies: A field experiment." *PLoS ONE* 3 (2):e1687. doi: 10.1371/journal.pone.0001687.

Maron, J. L., D. E. Pearson, and R. J. Fletcher. 2010. "Counterintuitive effects of large-scale predator removal on a midlatitude rodent community." *Ecology* 91 (12): 3719–3728. doi: 10.1890/10-0160.1.

Massey, F. P., M. J. Smith, X. Lambin, and S. E. Hartley. 2008. "Are silica defences in grasses driving vole population cycles?" *Biology Letters* 4:419–422.

Massey, F. P., and S. E. Hartley. 2006. "Experimental demonstration of the antiherbivore effects of silica in grasses: Impacts on foliage digestibility and vole growth rates." *Proceedings of the Royal Society of London, B: Biological Sciences* 273 (1599): 2299–2304. doi: 10.1098/rspb.2006.3586.

May, R. M. 1981. *Stability and Complexity in Model Ecosystems*. Princeton, NJ: Princeton University Press.

May, R. M., and R. M. Anderson. 1979. "Population biology of infectious diseases: Part II." *Nature* 280:455–461.

May, R. M., G. R. Conway, M. P. Hassell, and T. R. E. Southwood. 1974. "Time delays, density dependence and single-species oscillations." *Journal of Animal Ecology* 43:747–770.

McAdam, A. G., and J. S. Millar. 1999. "Dietary protein constraint on age at maturity: An experimental test with wild deer mice." *Journal of Animal Ecology* 68 (4): 733–740.

McArdle, B., and K. Gaston. 1993. "The temporal variability of populations." *Oikos* 67:187–191.

McArdle, B. H., and K. J. Gaston. 1992. "Comparing population variabilities." *Oikos* 64:610–612.

McGuire, B., L. L. Getz, and M. K. Oli. 2002. "Fitness consequences of sociality in prairie voles, *Microtus ochrogaster*: Influence of group size and composition." *Animal Behaviour* 64 (4): 645–654. doi: 10.1006/anbe.2002.3094.

McQueen, D. G., J. R. Post, and E. L. Mills. 1986. "Trophic relationships in freshwater pelagic ecosystems." *Canadian Journal of Fisheries and Aquatic Sciences* 43:1571–1581.

Merritt, J. F. 1984. *Winter Ecology of Small Mammals*. Special Publication 10, *Special Publications of the Carnegie Museum of Natural History*. Pittsburgh: Carnegie Museum of Natural History.

Messier, F. 1992. "On the functional and numerical responses of wolves to changing prey density." Paper read at Ecology and Conservation of Wolves in a Changing

World, Proceedings of the Second North American Symposium On Wolves, at Edmonton, Alberta.

———. 1994. "Ungulate population models with predation: A case study with the North American moose." *Ecology* 75:478–488.

Messier, F., and D. O. Joly. 2000. "Comment: Regulation of moose populations by wolf predation." *Canadian Journal of Zoology* 78 (3): 506–510.

Metzger, K. L., A. R. E. Sinclair, K. L. I. Campbell, R. Hilborn, J. Grant, C. Hopcraft, S. A. R. Mduma, and R. M. Reich. 2007. "Using historical data to establish baselines for conservation: The black rhinoceros (*Diceros bicornis*) of the Serengeti as a case study." *Biological Conservation* 139 (3–4): 358–374.

Millar, J. S. 2001. "On reproduction in lemmings." *Ecoscience* 8 (2): 145–150.

———. 2007. "Nest mortality in small mammals." *Ecoscience* 14 (3): 286–291. doi: 10.2980/1195-6860(2007)14[286:nmism]2.0.co;2.

Millar, J. S., and A. G. McAdam. 2001. "Life on the edge: The demography of short-season populations of deer mice." *Oikos* 93 (1): 69–76.

Millon, A., and V. Bretagnolle. 2008. "Predator population dynamics under a cyclic prey regime: Numerical responses, demographic parameters and growth rates." *Oikos* 117 (10): 1500–1510.

Milner-Gulland, E. J., and J. R. Beddington. 1993. "The exploitation of elephants for the ivory trade: An historical perspective." *Proceedings of the Royal Society of London, Series B* 252:29–37.

Mladenoff, D. J., and F. Stearns. 1993. "Eastern hemlock regeneration and deer browsing in the northern Great Lakes region: A re-examination and model simulation." *Conservation Biology* 7:889–900.

Moran, P. A. P. 1953. "The statistical analysis of the Canadian lynx cycle. II. Synchronization and meteorology." *Australian Journal of Zoology* 1:291–298.

Morris, D. W. 1989. "Density-dependent habitat selection: Testing the theory with fitness data." *Evolutionary Ecology* 3:80–94.

Moss, R., and A. Watson. 2001. "Population cycles in birds of the grouse family (Tetraonidae)." *Advances in Ecological Research* 32:53–111.

Murray, B. G., Jr. 2001. "Are ecological and evolutionary theories scientific?" *Biological Reviews* 76:255–289.

Nelson, J., J. Agrell, S. Erlinge, and M. Sandell. 1991. "Reproduction of different female age categories and dynamics in a non-cyclic field vole, *Microtus agrestis*, population." *Oikos* 61:73–78.

Norrdahl, K., and E. Korpimäki. 1995. "Effects of predator removal on vertebrate prey populations: Birds of prey and small mammals." *Oecologia* 103:241–248.

———. 2000. "Do predators limit the abundance of alternative prey? Experiments with vole-eating avian and mammalian predators." *Oikos* 91 (3): 528–540.

———. 2002. "Changes in population structure and reproduction during a 3-yr population cycle of voles." *Oikos* 96 (2): 331–345.

O'Connor, R. J. 2000. "Why ecology lags behind biology." *The Scientist* 14 (16 October 2000): 35.

Odum, E. P. 1971. *Fundamentals of Ecology*. Third ed. Philadelphia: W. B. Saunders.

Ogutu, J. O., and N. Owen-Smith. 2003. "ENSO, rainfall and temperature influences on extreme population declines among African savanna ungulates." *Ecology Letters* 6:412–419.

Oksanen, L., S. D. Fretwell, J. Arruda, and P. Niemelä. 1981. "Exploitation ecosystems in gradients of primary productivity." *American Naturalist* 118:240–261.

Oksanen, L., and T. L. Oksanen. 2000. "The logic and realism of the hypothesis of exploitation ecosystems." *American Naturalist* 155 (6): 703–723.

Oli, M. K. 2003. "Population cycles of small rodents are caused by specialist predators: Or are they?" *Trends in Ecology and Evolution* 18:105–107.

Oli, M. K., and F. S. Dobson. 1999. "Population cycles in small mammals: The role of age at sexual maturity." *Oikos* 86 (3): 557–565.

———. 2003. "The relative importance of life-history variables to population growth rate in mammals: Cole's prediction revisited." *American Naturalist* 161 (3): 422–440.

Oreskes, N. 1998. "Evaluation (not validation) of quantitative models." *Environmental Health Perspectives Supplements* 106 (S6): 1453–1460.

Oreskes, N., K. Shrader-Frechette, and K. Belitz. 1994. "Verification, validation, and confirmation of numerical models in the earth sciences." *Science* 263:641–646.

Ostfeld, R. S. 1994. "The fence effect reconsidered." *Oikos* 70:340–348.

———. 2008. "Parasites as weapons of mouse destruction." *Journal of Animal Ecology* 77 (2): 201–204. doi: 10.1111/j.1365-2656.2008.01364.x.

Ostfeld, R. S., C. D. Canham, and S. R. Pugh. 1993. "Intrinsic density-dependent regulation of vole populations." *Nature* 366:259–261.

Ostfeld, R. S., W. Z. Lidicker, Jr., and E. J. Heske. 1985. "The relationship between habitat heterogeneity, space use, and demography in a population of California voles." *Oikos* 45:433–442.

Paradis, E. 1995. "Survival, immigration and habitat quality in the Mediterranean pine vole." *Journal of Animal Ecology* 64 (5): 579–591.

Parmenter, R. R., T. L. Yates, D. R. Anderson, K. P. Burnham, J. L. Dunnum, A. B. Franklin, M. T. Friggens, B. C. Lubow, M. C. Miller, G. S. Olson, C. A. Parmenter, J. H. Pollard, E. Rexstad, T. M. Shenk, T. R. Stanley, and G. C. White. 2003. "Small-mammal density estimation: A field comparison of grid-based vs. web-based density estimators." *Ecological Monographs* 73 (1): 1–26.

Pedersen, A. B., and T. J. Greives. 2008. "The interaction of parasites and resources cause crashes in a wild mouse population." *Journal of Animal Ecology* 77 (2): 370–377. doi: 10.1111/j.1365-2656.2007.01321.x.

Pelikan, J. 1979. "Sufficient sample size for evaluating the litter size in rodents." *Folia Zoologica* 28:289–297.

Peters, R. H. 1991. *A Critique for Ecology*. Cambridge: Cambridge University Press.

Peterson, R. O., R. E. Page, and K. M. Dodge. 1984. "Wolves, moose, and the allometry of population cycles." *Science* 224:1350–1352.

Pielou, E. C. 1981. "The usefulness of ecological models: A stock-taking." *Quarterly Review of Biology* 56:17–31.

Pitelka , F.A. 1964. "The nutrient-recovery hypothesis for arctic microtine cycles.

I. Introduction." In *Grazing in Terrestrial and Marine Environments*, edited by D. J. Crisp, 55–56. Oxford: Blackwell Scientific Publications.

Pitelka, F. A., and G. O. Batzli. 2007. "Population cycles of lemmings near Barrow, Alaska: A historical review." *Acta Theriologica* 52 (3): 323–336.

Pitelka, F. A., P. Q. Tomich, and G. W. Treichel. 1955. "Ecological relations of jaegers and owls as lemming predators near Barrow, Alaska." *Ecological Monographs* 25:85–117.

Platt, J. R. 1964. "Strong inference." *Science* 146:347–353.

Poikonen, T., E. Koskela, T. Mappes, and S. C. Mills. 2008. "Infanticide in the evolution of reproductive synchrony effects on reproductive success." *Evolution* 62 (3): 612–621. doi: 10.1111/j.1558-5646.2007.00293.x.

Pollock, K. H., J. D. Nichols, C. Brownie, and J. E. Hines. 1990. "Statistical inference for capture-recapture experiments." *Wildlife Monographs* 107:1–97.

Popper, K. R. 1963. *Conjectures and Refutations: The Growth of Scientific Knowledge*. London: Routledge and Kegan Paul.

Porter, J. H., and R. D. Dueser. 1990. "Selecting a body-mass criterion for measuring dispersal." *Journal of Mammalogy* 71 (3): 470–473.

Previtali, M. Andrea, E. M. Lehmer, J. M. C. Pearce-Duvet, J. D. Jones, C. A. Clay, B. A. Wood, P. W. Ely, S. M. Laverty, and M. D. Dearing. 2010. "Roles of human disturbance, precipitation, and a pathogen on the survival and reproductive probabilities of deer mice." *Ecology* 91 (2): 582–592. doi: 10.1890/08-2308.1.

Pulliam, H. R. 1988. "Sources, sinks, and population regulation." *American Naturalist* 132:652–661.

Pusenius, J., J. Viitala, T. Marienberg, and S. Ritvanen. 1998. "Matrilineal kin clusters and their effect on reproductive success in the field vole *Microtus agrestis*." *Behavioral Ecology* 9 (1): 85–92.

Ranta, E., V. Kaitala, J. Lindstrom, and E. Helle. 1997. "The Moran effect and synchrony in population dynamics." *Oikos* 78 (1): 136–142.

Ranta, E., V. Kaitala, J. Lindstrom, and H. Linden. 1995. "Synchrony in population dynamics." *Proceedings of the Royal Society of London, Series B* 262:113–118.

Ranta, E., P. Lundberg, and V. Kaitala. 2006. *Ecology of Populations*. Cambridge: Cambridge University Press.

Ratcliffe, W. C., P. Hawthorne, M. Travisano, and R. F. Denison. 2009. "When stress predicts a shrinking gene pool, trading early reproduction for longevity can increase fitness, even with lower fecundity " *PLoS One* 4 (6): e6055. doi: 10.1371/journal.pone.0006055.

Redfield, J. A., M. J. Taitt, and C. J. Krebs. 1978. "Experimental alteration of sex ratios in population of *Microtus townsendii*, a field vole." *Canadian Journal of Zoology* 56:17–27.

Redpath, S. M., F. Mougeot, F. M. Leckie, D. A. Elston, and P. J. Hudson. 2006. "Testing the role of parasites in driving the cyclic population dynamics of a gamebird." *Ecology Letters* 9 (4): 410–418.

Reid, D. G., and C. J. Krebs. 1996. "Limitations to collared lemming population growth in winter." *Canadian Journal of Zoology* 74 (7): 1284–1291. doi: 10.1139/z96-143.

Reid, D. G., C. J. Krebs, and A. J. Kenney. 1995. "Limitation of collared lemming population growth at low densities by predation mortality." *Oikos* 73 (3): 387–398.

———. 1997. "Patterns of predation on noncyclic lemmings." *Ecological Monographs* 67 (1): 89–108.

Reid, D. G., F. Bilodeau, C. J. Krebs, G. Gauthier, A. J. Kenney, B. S. Gilbert, M. C.-Y. Leung, D. Duchesne, and E. J. Hofer. 2011. "Lemming winter habitat choice: A snow-fencing experiment." *Oecologia* 168(4): 935–946.

Rhoades, D. F., and R. G. Cates. 1976. "Towards a general theory of plant antiherbivore chemistry." *Recent Advances in Phytochemistry* 19:168–213.

Ricker, W. E. 1954. "Effects of compensatory mortality upon population abundance." *Journal of Wildlife Management* 18:45–51.

Ripple, W. J., and R. L. Beschta. 2006. "Linking a cougar decline, trophic cascade, and catastrophic regime shift in Zion National Park." *Biological Conservation* 133 (4): 397–408.

Rodgers, A. R., and M. C. Lewis. 1985. "Diet selection in Arctic lemmings (*Lemmus sibericus* and *Dicrostonyx groenlandicus*): Food preferences." *Canadian Journal of Zoology* 63:1161–1173.

Roff, D. A., and D. J. Fairbairn. 2001. "The genetic basis of dispersal and migration, and its consequences for the evolution of correlated traits." In *Dispersal*, edited by J. Clobert, E. Danchin, A. A. Dhondt, and J. D. Nichols, 191–202. Oxford: Oxford University Press.

Rose, R. K., and M. S. Gaines. 1978. "The reproductive cycle of *Microtus ochrogaster* in eastern Kansas." *Ecological Monographs* 48 (1): 21–42.

Rosenzweig, M. L., and R. H. Macarthur. 1963. "Graphical representation and stability conditions of predator-prey interactions." *American Naturalist* 97:209–223.

Rousset, F. 2001. "Genetic approaches to the estimation of dispersal rates." In *Dispersal*, edited by J. Clobert, E. Danchin, A. A. Dhondt and J. D. Nichols, 18–28. Oxford: Oxford University Press.

Royama, T. 1992. *Analytical Population Dynamics*. London: Chapman and Hall.

Rule, S., B. W. Brook, S. G. Haberle, C. S. M. Turney, A. P. Kershaw, and C. N. Johnson. 2012. The aftermath of megafaunal extinction: Ecosystem transformation in Pleistocene Australia. *Science* 335:1483–1486.

Runge, J. P., M. C. Runge, and J. D. Nichols. 2006. "The role of local populations within a landscape context: Defining and classifying sources and sinks." *American Naturalist* 167 (6): 925–938. doi: 10.1086/503531.

Saitoh, T. 1981. "Control of female maturation in high density populations of the red-backed vole, *Clethrionomys rufocanus bedfordiae*." *Journal of Animal Ecology* 50 (1): 79–87.

———. 1987. "A time series and geographical analysis of population dynamics of the red-backed vole in Hokkaido, Japan." *Oecologia* 73:382–388.

Saitoh, T., N. C. Stenseth, and O. N. Bjørnstad. 1997. "Density dependence in fluctuating grey-sided vole populations." *Journal of Animal Ecology* 66:14–24.

Saitoh, T., N. C. Stenseth, and O. N. Bjørnstad. 1998. "The population dynamics of the vole *Clethrionomys rufocanus* in Hokkaido." *Researches on Population Ecology* 40 (1): 61–76.

Schaffer, W. M., and R. H. Tamarin. 1973. "Changing reproductive rates and population cycles in lemmings and voles." *Evolution* 27 (1): 111–124.

Schultz, A. M. 1964. "The nutrient-recovery hypothesis for arctic microtine cycles. II. Ecosystem variables in relation to the arctic microtine cycles." In *Grazing in Terrestrial and Marine Environments*, edited by D. J. Crisp, 57–68. Oxford: Blackwells.

———. 1969. "A study of an ecosystem: The arctic tundra." In *The Ecosytem Concept in Natural Resource Management*, edited by G. Van Dyne, 77–93. New York: Academic Press.

Selås, V., E. Framstad, and T. K. Spidsø. 2002. "Effects of seed masting of bilberry, oak and spruce on sympatric populations of bank vole (*Clethrionomys glareolus*) and wood mouse (*Apodemus sylvaticus*) in southern Norway." *Journal of Zoology (London)* 258 (4): 459–468.

Selonen, V., and I. K. Hanski. 2010. "Condition-dependent, phenotype-dependent and genetic-dependent factors in the natal dispersal of a solitary rodent." *Journal of Animal Ecology* 79 (5): 1093–1100. doi: 10.1111/j.1365–2656.2010.01714.x.

Selye, H. 1936. "Syndrome produced by diverse nocuous agents." *Nature* 138:32.

———. 1955. "Stress and disease." *Science* 122 (3171): 625–631.

Sergio, F., and F. Hiraldo. 2008. "Intraguild predation in raptor assemblages: A review." *Ibis* 150 (Suppl. 1): 132–145.

Sheriff, M. J., C. J. Krebs, and R. Boonstra. 2009. "The sensitive hare: Sublethal effects of predator stress on reproduction in snowshoe hares." *Journal of Animal Ecology* 78:1249–1258.

Sheriff, M. J., C. J. Krebs, and R. Boonstra. 2010a. "Assessing stress in animal populations: Do feces and plasma glucocorticoids tell the same story?" *General and Comparative Endocrinology* 166 (3):614–619. doi: 10.1016/j.ygcen.2009.12.017.

———. 2010b. "The ghosts of predators past: Population cycles and the role of maternal programming under fluctuating predation risk." *Ecology* 91 (10): 2983–2994.

Sheriff, M. J., B. Dantzer, B. Delehanty, R. Palme, and R. Boonstra. 2011. "Measuring stress in wildlife: Techniques for quantifying glucocorticoids." *Oecologia* 166:869–887. doi: 10.1007/s00442–011–1943-y.

Sheriff, M. J., C. J. Krebs, and R. Boonstra. 2011. "From process to pattern: How fluctuating predation risk impacts the stress axis of snowshoe hares during the 10-year cycle." *Oecologia* 166:593–605. doi: 10.1007/s00442–011–1907–2.

Sherratt, J. A. 2001. "Periodic travelling waves in cyclic predator–prey systems." *Ecology Letters* 4 (1): 30–37. doi: 10.1046/j.1461–0248.2001.00193.x.

Sibly, R. M., J. Hone, and T. H. Clutton-Brock. 2003. *Wildlife Population Growth Rates*. Cambridge: Cambridge University Press.

Simmons, R., P. Barnard, B. MacWhirter, and G. L. Hansen. 1986. "The influence of microtines on polygyny, productivity, age, and provisioning of breeding Northern Harriers: A 5-year study." *Canadian Journal of Zoology* 64 (11): 2447–2456.

Sinclair, A. R. E. 1996. "Mammal populations: fluctuation, regulation, life history theory and their implications for conservation." In *Frontiers of Population*

Ecology, edited by R. B. Floyd, A. W. Sheppard and P. J. De Barro, 127–154. Melbourne: CSIRO Publishing.

———. 1998. "Natural regulation of ecosystems in protected areas as ecological baselines." *Wildlife Society Bulletin* 26 (3): 399–409.

Sinclair, A. R. E., and C. J. Krebs. 2002. "Complex numerical responses to top-down and bottom-up processes in vertebrate populations." *Philosophical Transactions of the Royal Society of London, Series B* 357:1221–1231.

Sinclair, A. R. E., and J. N. M. Smith. 1984. "Do plant secondary compounds determine feeding preferences of snowshoe hares?" *Oecologia* 61:403–410.

Sinervo, B., E. Svensson, and T. Comendant. 2000. "Density cycles and an offspring quantity and quality game driven by natural selection." *Nature* 406 (31 August 2000): 985–988.

Singleton, G. R., P. R. Brown, R. P. Pech, J. Jacob, G. J. Mutze, and C. J. Krebs. 2005. "One hundred years of eruptions of house mice in Australia: A natural biological curio." *Biological Journal of the Linnean Society* 84:617–627.

Singleton, G. R., A. J. Kenney, C. R. Tann, Sudarmaji, and N. Q. Hung. 2003. "Myth, dogma and rodent management: Good stories ruined by data?" In *Rats, Mice and People: Rodent Biology and Management*, edited by G. R. Singleton, L. A. Hinds, C. J. Krebs and D. M. Spratt, 554–560. Canberra: Australian Centre for International Agricultural Research.

Singleton, G. R., Sudarmaji, J. Jacob, and C. J. Krebs. 2005. "Integrated management to reduce rodent damage to lowland rice crops in Indonesia." *Agriculture, Ecosystems & Environment* 107 (1): 75–82.

Smith, J. E., and G. O. Batzli. 2006. "Dispersal and mortality of prairie voles (*Microtus ochrogaster*) in fragmented landscapes: A field experiment." *Oikos* 112 (1): 209–217. doi: 10.1111/j.0030-1299.2006.13431.x.

Smith, M. J., A. White, X. Lambin, J. A. Sherratt, and M. I. Begon. 2006. "Delayed density-dependent season length alone can lead to rodent population cycles." *American Naturalist* 167 (5): 695–704.

Smith, M. J., A. White, J. A. Sherratt, S. Telfer, M. Begon, and X. Lambin. 2008. "Disease effects on reproduction can cause population cycles in seasonal environments." *Journal of Animal Ecology* 77 (2): 378–389.

Solberg, E. J., P. Jordhoy, O. Strand, R. Aanes, Loison A., B.-E. Saether, and J. D. C. Linnell. 2001. "Effects of density-dependence and climate on the dynamics of a Svalbard reindeer population." *Ecography* 24 (4): 441–451.

Solomon, M. E. 1949. "The natural control of animal populations." *Journal of Animal Ecology* 18 (1): 1–35.

Speakman, J., and E. Król. 2005. "Limits to sustained energy intake IX: A review of hypotheses." *Journal of Comparative Physiology B: Biochemical, Systemic, and Environmental Physiology* 175 (6): 375–394. doi: 10.1007/s00360-005-0013-3.

Stacy, J. E., P. E. Jorde, H. Steen, R. A. Ims, A. Purvis, and L. S. Jakobsen. 1997. "Lack of concordance between mtDNA gene flow and population density fluctuations in the bank vole." *Molecular Ecology* 6:751–759.

Steen, H., J. C. Holst, T. Solhoy, M. Bjerga, E. Klaussen, I. Prestegard, R. C. Sundt, and O. Johannesen. 1997. "Mortality of lemmings, *Lemmus lemmus*, at peak density in a mountainous area of Norway." *Journal of Zoology, London* 243:831–835.

Steen, H., R. A. Ims, and G. A. Sonerud. 1996. "Spatial and temporal patterns of small-rodent population dynamics at a regional scale." *Ecology* 77:2365–2372.

Stenseth, N. C. 1988. "The social fence hypothesis: A critique." *Oikos* 52 (2): 169–177.

———. 1999. "Population cycles in voles and lemmings: Density dependence and phase dependence in a stochastic world." *Oikos* 87:427–461.

Stenseth, N. C., O. N. Bjørnstad, and T. Saitoh. 1996. "A gradient from stable to cyclic populations of *Clethrionomys rufocanus* in Hokkaido, Japan." *Proceedings of the Royal Society of London, Series B* 263:1117–1126.

Stenseth, N. C., and W. Z. Lidicker, Jr. 1992. "Where do we stand methodologically about experimental design and methods of analysis in the study of dispersal?" In *Animal Dispersal: Small Mammals as a Model*, edited by N. C. Stenseth and W. Z. Lidicker Jr., 295–318. London: Chapman & Hall.

Stenseth, N. C., O. N. Bjørnstad, and T. Saitoh. 1998. "Seasonal forcing on the dynamics of *Clethrionomys rufocanus*: Modeling geographic gradients in population dynamics." *Researches on Population Dynamics* 40, in press.

Stenseth, N. C., H. Leirs, S. Mercelis, and P. Mwanjabe. 2001. "Comparing strategies for controlling an African pest rodent: An empirically based theoretical study." *Journal of Applied Ecology* 38 (5): 1020–1031.

Stenseth, Nils C., D. Ehrich, E. Knispel Rueness, O. C. Lingjaerde, K.-S. Chan, S. Boutin, M. O'Donoghue, D. A. Robinson, H. Viljugrein, and K. S. Jakobsen. 2004. "The effect of climatic forcing on population synchrony and genetic structuring of the Canadian lynx." *Proceedings of the National Academy of Sciences of the USA* 101 (16): 6056–6061.

Stromayer, K. A. K., and R. J. Warren. 1997. "Are overabundant deer herds in the eastern United States creating alternate stable states in forest plant communities?" *Wildlife Society Bulletin* 25 (2): 227–234.

Sullivan, T. P., C. Nowotny, R. A. Lautenschlager, and R. G. Wagner. 1998. "Silvicultural use of herbicide in sub-boreal spruce forest: Implications for small mammal population dynamics." *Journal of Wildlife Management* 62 (4): 1196–1206.

Summerhayes, V. S. 1941. "The effect of voles (*Microtus agrestis*) on vegetation." *Journal of Ecology* 29 (1): 14–48.

Sutherland, D. R., and G. R. Singleton. 2006. "Self-regulation within outbreak populations of feral house mice: A test of alternative models." *Journal of Animal Ecology* 75 (2): 584–594.

Taitt, M. J., J. H. W. Gipps, C. J. Krebs, and Z. Dundjerski. 1981. "The effect of extra food and cover on declining populations of *Microtus townsendii*." *Canadian Journal of Zoology* 59 (8): 1593–1599.

Taitt, M. J., and C. J. Krebs. 1981. "The effect of supplementary food on small rodent populations. 2. Voles (*Microtus townsendii*)." *Journal of Animal Ecology* 50 (1): 125–137.

———. 1983. "Predation, cover, and food manipulations during a spring decline of *Microtus townsendii*." *Journal of Animal Ecology* 52:837–848.

Tamarin, R. H., L. M. Reich, and C. A. Moyer. 1984. "Meadow vole cycles within fences." *Canadian Journal of Zoology* 62:1796–1804.

Tamarin, R. H. 1977a. "Demography of the beach vole (*Microtus breweri*) and the meadow vole (*Microtus pennsylvanicus*) in southeastern Massachusetts." *Ecology* 58:1310–1321.

———. 1977b. "Reproduction in the island beach vole, *Microtus breweri,* and the mainland meadow vole, *Microtus pennsylvanicus,* in southeastern Massachusetts." *Journal of Mammalogy* 58 (4): 536–548.

———. 1978a. "A defense of single-factor models of population regulation." *Pymatuning Laboratory of Ecology, Special Publication* 5:159–166.

———. 1978b. "Dispersal, population regulation, and K-selection in field mice." *American Naturalist* 112 (985): 545–555.

———. 1984. "Body mass as a criterion of dispersal in voles: A critique." *Journal of Mammalogy* 65:691–692.

Tamarin, R. H., R. S. Ostfeld, S. R. Pugh, and G. Bujalska. 1990. *Social Systems and Population Cycles in Voles*. Basel: Birkhäuser Verlag.

Telfer, S., M. Bennett, K. Bown, R. Cavanagh, L. Crespin, S. Hazel, T. Jones, and M. Begon. 2002. "The effects of cowpox virus on survival in natural rodent populations: Increases and decreases." *Journal of Animal Ecology* 71 (4): 558–568.

Telfer, S., M. Bennett, K. Bown, D. Carslake, R. Cavanagh, S. Hazel, T. Jones, and M. Begon. 2005. "Infection with cowpox virus decreases female maturation rates in wild populations of woodland rodents." *Oikos* 109 (2): 317–322.

Thompson, C. M., and E. M. Gese. 2007. "Food webs and intraguild predation: community interactions of a native mesocarnivore." *Ecology* 88 (2): 334–346.

Thompson, I. D., and W. J. Curran. 1995. "Habitat suitability for marten of second-growth balsam fir forests in Newfoundland." *Canadian Journal of Zoology* 73:2059–2064.

Thompson, W. R. 1929. "On natural control." *Parasitology* 21:269–281.

Tkadlec, E. 2000. "The effects of seasonality on variation in the length of breeding season in arvicoline rodents." *Folia Zoologica* 49 (4): 269-286.

Tkadlec, E., and P. Krejcova. 2001. "Age-specific effects of parity on litter size in the common vole (*Microtus arvalis*)." *Journal of Mammalogy* 82 (2): 545–550.

Tkadlec, E., and N. C. Stenseth. 2001. "A new geographical gradient in vole population dynamics." *Proceedings of the Royal Society of London, Series B* 268:1547–1552. doi: 10.1098/rspb.2001.1694.

Travina, I. V. 2002. "Long-term dynamics of lemming numbers on Wrangel Island." *Arctic Birds* 4:30–34.

Turchin, P. 2003. *Complex Population Dynamics: A Theoretical/Empirical Synthesis*. Princeton, NJ: Princeton University Press.

Turchin, P., and G. O. Batzli. 2001. "Availability of food and the population dynamics of arvicoline rodents." *Ecology* 82 (6): 1521–1534.

Turchin, P., and I. Hanski. 1997. "An empirically based model for latitudinal gradient in vole population dynamics." *American Naturalist* 149 (5): 842–874.

Turchin, P., L. Oksanen, P. Ekerholm, T. Oksanen, and H. Henttonen. 2000. "Are lemmings prey or predators?" *Nature* 405:562–565.

Verrill, S. 2003. "Confidence bounds for normal and lognormal distribution coefficients of variation." *United States Department of Agriculture, Forest Service, Forest Products Laboratory* Research Paper FPL–RP–609:1–13.

Volterra, V. 1926. "Fluctuations in the abundance of a species considered mathematically." *Nature* 118:558–560.

Watson, A. 1964. "Aggression and population regulation in red grouse." *Nature* 202:506–507.

———. 1967. "Population control by territorial behaviour in red grouse." *Nature* 215:1274–1275.

Watson, A., and D. Jenkins. 1968. "Experiments on population control by territorial behaviour in red grouse." *Journal of Animal Ecology* 37:595–614.

Watson, A., and R. Moss. 1970. "Dominance, spacing behavior and aggression in relation to population limitation in vertebrates." In *Animal Populations in Relation to Their Food Resources*, edited by A. Watson, 167–220. Oxford: Blackwell.

Weiner, J. 1995. "On the practice of ecology." *Journal of Ecology* 83:153–158.

White, G. C., D. R. Anderson, K. P. Burnham, and D. L. Otis. 1982. "Capture-recapture and removal methods for sampling closed populations." *Los Alamos National Laboratory* LA-8787-NERP, 235 pp.

White, G. C. . 2008. "Closed population estimation models and their extensions in program MARK." *Environmental and Ecological Statistics* 15 (1): 89–99.

White, T. C. R. 1993. *The Inadequate Environment: Nitrogen and the Abundance of Animals*. New York: Springer-Verlag.

Wingfield, J. C., and R.M. Sapolsky. 2003. "Reproduction and resistance to stress: When and how." *Journal of Neuroendocrinology* 15:711–724.

Wolff, J. O. 1993. "What is the role of adults in mammalian juvenile dispersal?" *Oikos* 68 (1): 173–176.

———. 1997. "Population regulation in mammals: An evolutionary perspective." *Journal of Animal Ecology* 66 (1): 1–13.

———. 2003a. "Density dependence and the socioecology of space use in rodents." In *Rats, Mice and People: Rodent Biology and Management*, edited by G. R. Singleton, L. A. Hinds, C. J. Krebs and D. M. Spratt, 124–130. Canberra: Australian Centre for International Agricultural Research.

———. 2003b. "Laboratory studies with rodents: facts or artifacts?" *BioScience* 53 (4): 421–427.

———. 2007. "Social biology of rodents." *Integrative Zoology* 2 (4):193–204.

Wolff, J. O., W. D. Edge, and G. Wang. 2002. "Effects of adult sex ratios on recruitment of juvenile gray-tailed voles, *Microtus canicaudus*." *Journal of Mammalogy* 83 (4): 947–956.

Wolff, J. O., and J. A. Peterson. 1998. "An offspring-defense hypothesis for territoriality in female mammals." *Ethology, Ecology and Evolution* 10:227–239.

Wolff, J. O., E. M. Schauber, and W. D. Edge. 1996. "Can dispersal barriers really be used to depict emigrating small mammals?" *Canadian Journal of Zoology* 74:1826–1830.

Wolff, J. O., and P. W. Sherman. 2007. *Rodent Societies: An Ecological and Evolutionary Perspective*. Chicago: University of Chicago Press.

Ydenberg, R. C. 1987. "Nomadic predators and geographical synchrony in microtine population cycles." *Oikos* 50:270–272.

Ylönen, H., and J. S. Brown. 2007. "Fear and the foraging, breeding, and sociality of rodents." In *Rodent Societies: An Ecological and Evolutionary Perspective*, edited by J. O. Wolff and P. W. Sherman, 328–341. Chicago: University of Chicago Press.

Ylönen, H., E. Koskela, and T. Mappes. 1997. "Infanticide in the bank vole (*Clethrionomys glareolus*): Occurrence and the effect of familiarity on female infanticide." *Annales Zoologici Fennici* 31:259–266.

Ylönen, H., T. Mappes, and J. Viitala. 1990. "Different demography of friends and strangers: An experiment on the impact of kinship and familiarity in *Clethrionomys glareolus*." *Oecologia* 83:333–337.

Yudina, V. F., and T. A. Maksimova. 2005. "Dynamics of yielding capacity of small cranberry in southern Karelia." *Russian Journal of Ecology* 36:239–242.

Zimov, S. A., V. I. Chuprynin, A. P. Oreshko, F. S. Chapin, III, J. F. Reynolds, and M. C. Chapin. 1995. "Steppe-tundra transition: A herbivore-driven biome shift at the end of the Pleistocene." *American Naturalist* 146 (5): 765–794.

Index